Introduction to Probability with R

CHAPMAN & HALL/CRC
Texts in Statistical Science Series

Series Editors
Bradley P. Carlin, *University of Minnesota, USA*
Julian J. Faraway, *University of Bath, UK*
Martin Tanner, *Northwestern University, USA*
Jim Zidek, *University of British Columbia, Canada*

Texts in Statistical Science

Introduction to Probability with R

Kenneth Baclawski

Northeastern University
Boston, Massachusetts, U.S.A.

CRC Press

Taylor & Francis Group
Boca Raton London New York

CRC Press is an imprint of the
Taylor & Francis Group, an **informa** business

A CHAPMAN & HALL BOOK

CRC Press
Taylor & Francis Group
6000 Broken Sound Parkway NW, Suite 300
Boca Raton, FL 33487-2742

First issued in paperback 2022

Version Date: 20110725

ISBN 13: 978-1-03-247780-0 (pbk)
ISBN 13: 978-1-4200-6521-3 (hbk)

DOI: 10.1201/9781420065220

Library of Congress Cataloging-in-Publication Data

Baclawski, Kenneth.
    Introduction to probability with R / Kenneth Baclawski.
        p. cm. -- (Chapman & Hall/CRC texts in statistical science series ; 75)
    Includes bibliographical references and index.
    ISBN-13: 978-1-4200-6521-3 (hardcover : alk. paper)
    ISBN-10: 1-4200-6521-1 (hardcover : alk. paper)
    1. Probabilities. 2. R (Computer program language)--Mathematical models. I. Title.

    QA273.B2535 2008
    519.20285'5133--dc22   2007041288

Contents

Foreword

Probability theory is one of the pivotal branches of the mathematical sciences and has numerous applications in a variety of disciplines, such as economics, engineering, computer science, and biology. In this book, based on the lecture notes of a popular course taught at MIT, Professor Kenneth Baclawski introduces the fundamentals of probability theory by tackling the major themes. The objective of this book is to make probability theory widely accessible to a broad range of modern readers, and therefore, the book has focused on providing intuitive explanations of concepts in addition to rigorous mathematical formulations and proofs.

The text consists of ten chapters and two appendices, encompassing a broad spectrum of probability and statistics topics ranging from set theory to statistics and the normal distribution to Poisson process to Markov chains. The author has covered each topic with an ample depth and with an appreciation of the problems faced by the modern world. The book contains a rich collection of exercises and problems, and readers can test the extent of knowledge they have grasped.

As a biostatistician, a bioinformatician, and a genetic epidemiologist, I have been fascinated by the grand challenges of unraveling the mystery of life using mathematical theories, in particular, probability theory. We are in the midst of a paradigm shift in the use of contemporary quantitative approaches and state-of-the-art information technology in solving life science puzzles. The emergence of the unusually multidisciplinary discipline of bioinformatics, which focuses on curating, storing, extracting, analyzing, and interpreting gargantuan volumes of biological sequence and functional assay data, has played a key role in driving that shift. In order to comprehend the multifaceted biological processes and myriads of biochemical reactions, it has become indispensable to perform large-scale quantitative modeling and simulation. For example, the Human Genome Project and the International HapMap Project have resulted in an exponential growth of databases of DNA sequences and single nucleotide polymorphisms (SNPs) for bioinformaticians to analyze. Understanding probability theory by reading this book, and applying the theory to address bewildering biological questions, could be central to utilizing such sequence and SNP repositories.

In 2003, MIT established a Computational and Systems Biology Initiative, and Harvard Medical School launched a new Department of Systems Biology, which are aiming at addressing the emerging demands of how each organism works as a system. Systems biology is a new discipline that strives for a complete understanding of the complex interactions of genes, proteins, and other molecules at the subcellular, cellular, organ, or organism levels. Because of the high-content nature of system biology, it draws from the foundations of probability and statistics, physics, computer science, biochemistry, and molecular and cell biology. We can gain great insights and acumen through reading this book.

It is worth noting that an excellent introduction to the open source software R is given in the book. R is an increasingly popular programming language in probability and statistics upon which the Bioconductor Project is built, and is now extensively used for the analysis of genomic and gene expression microarray data. The book comes with real R codes that are used to solve probability and statistics questions. Learning R is indeed a fun part accompanying the reading of the content of the book, and frequent exercises help the reader to make steadfast progresses confidently in writing scripts using R.

This book showcases interesting, classic puzzles throughout the text, and readers can also get a glimpse of the lives and achievements of important pioneers in mathematics. Therefore, it is highly suitable for students undertaking a first course in probability and statistics. This book is also a very useful reference book for advanced undergraduate and beginning graduate courses in economics, engineering, and computer and information science. Further, it will be of great value to researchers, and practitioners in medicine, who have some basic knowledge of probability, because the reader's intuition can be shaped and strengthened by reading the book to see how reasoning with probability is critical for making judicious decisions.

Tianhua Niu
Director of Bioinformatics
Division of Preventive Medicine
Brigham and Women's Hospital

Assistant Professor of Medicine
Department of Medicine
Harvard Medical School

Assistant Professor of Epidemiology
Department of Epidemiology
Harvard School of Public Health

Preface

This book began over 30 years ago when Gian-Carlo Rota and I teamed up to create a course on probability at MIT. The course proved to be very popular, and I wrote up my notes, so they could be used as the textbook for the course when I left MIT. Gian-Carlo continued to teach the course for many years, and it continued to be very popular. We remained close friends, and it had always been our intention to eventually publish a textbook based on my notes, but our careers diverged, and we never quite managed to start the project. From time to time, I would write new material in the hope that someday it might be of use, when we did start. With Gian-Carlo's tragic and untimely death in 1999, it seemed that the prospects for the project were permanently ended. However, I eventually realized that I should start the project on my own. It would not have the benefit of Gian-Carlo's input, but at least I could hope to capture some of his ideas for a wider audience than just MIT.

Nevertheless, I should make it clear that my notes and this textbook are mine alone. All errors, omissions and misrepresentations are my responsibility. There are also a great many topics that Gian-Carlo never considered, especially those relating to the use of computers, because the necessary hardware and software infrastructure was still in its infancy 30 years ago. Needless to say, Gian-Carlo had no part in developing any of these in the present textbook.

Probability theory is now over 350 years old, and the first significant theorem of probability theory, the Law of Large Numbers, was discovered almost 300 years ago, but progress in probability theory was not steady over this whole period. Indeed, there have been large gaps in which the field was nearly forgotten, followed by periods during which great strides have been made. The last few decades are one of these periods, and I endeavored to capture some of the new ideas and to convey some of the current excitement. The development of computers has had an especially strong impact. Many techniques that were developed in the past primarily to simplify complicated computations are now largely irrelevant because computers have made it so easy to do the computations in their original form. More significantly, the computer has made it feasible to construct and to analyze far more elaborate and more accurate models of phenomena.

Perhaps one of the most significant recent developments was the creation of the R language. Of course, one can certainly perform probabilistic computations using any programming language, and there are many statistical programming languages that have been developed, some of which have been available for more than 30 years. R is not new in any of these senses. What is special about R is the convergence of the community around R as the base for building probabilistic and statistical applications. No previous language has engendered such a broad consensus. Building R into the textbook was a natural fit, and I incorporated a great many programs into the text when it helped convey the subject matter. Students are encouraged to install R, and to try the programs that are presented in the text. The programs are so small, averaging only a few lines of code, that it seems like an exaggeration to call them programs at all. Yet the programs are as sophisticated and powerful as programs several pages long in other languages. Because of the brevity of the programs and the simplicity of the syntax, students should be able to become proficient in R even if they have never programmed before. The animations are a special feature that not only convey a great deal of information about important topics in probability, but also encourage experimentation, since the programs are so easily understood and therefore so easily modified. All of the programs in the book as well as many others are available on the textbook website: `http://stochas.org`.

The textbook is organized around a number of themes, and the subject matter is richly woven around these themes. One of the most important themes is that probability is a way of looking at the world. There are many ways of approaching natural problems. Some, like the geometrical point of view, have been with us for thousands of years. Probability, by contrast, is relatively recent, having first emerged as a distinct mathematical theory only about 350 years ago. The probabilistic point of view is another way of focusing on problems that has been highly successful. Much of modern science and engineering is dependent on probability theory. One of the main purposes of this textbook is to help students to learn to think probabilistically. Unfortunately, the only way to achieve this is to learn the theorems of probability. As one masters the theorems, the probabilistic point of view begins to emerge, even as the specific theorems fade away: like the grin on the Cheshire Cat.

Another theme is the notion of a stochastic process. Probability can be regarded as the study of random variables. However, random variables never occur in isolation. They invariably arise within a context. The contexts are the stochastic processes. Surprisingly, there are only a small number of major stochastic processes, from which all of the major probability distributions arise. Thoroughly understanding these processes and the relationships among them is an essential part of learning to think probabilistically. Processes can be combined and linked together to form other processes using mechanisms such as compounding, randomization and recursion. By analogy with computer programs, we refer to these process combinations as stochastic programs.

Another important aspect of *thinking* probabilistically is *reasoning* probabilistically. This is another important theme. While processes permeate the entire book, probabilistic reasoning emerges only gradually. There are a few allusions to the logic of probability in early chapters, but we only feel the full force of reasoning in Chapters 5 and 6, which deal with statistics and conditional probability.

The other themes are even more localized. Transforms, such as the Laplace, Fourier and z-transform, are powerful tools in engineering and science. Transforms are given a unified treatment in Chapter 8, where they emerge as a byproduct of randomization. Entropy and information furnish the most compelling reason why such a relatively small number of fundamental stochastic processes can explain such a large variety of natural phenomena. This topic is covered in Chapter 9. Finally, Markov chains are a versatile tool for modeling natural phenomena. We develop this topic from a variety of points of view and link it with the previous themes in Chapter 10.

Each chapter ends with exercises and answers to selected exercises. The section just before the exercises is optional, except for Chapters 1 and 2. Optional sections may be included or omitted independently, according to the time available and the interests of the teacher and students. There are also two appendices containing some proofs and advanced topics that are also optional.

The web domain for this book is `stochas`, short for "stochastic." The website can be accessed using either `stochas.org` or `stochas.com`. This site provides a variety of materials to assist teachers, students, scientists, statisticians and mathematicians, including:

- All of the R programs shown in the textbook, the R programs that were used to generate diagrams and graphs (such as the timeline below), and many others

- PowerPoint slides for presentations and lectures

- Errata

- Links to related sites and projects

Each chapter begins with a short biographical note about one of the contributors to probability theory up to the beginning of the twentieth century. They were chosen primarily for their relevance to the subject matter of the chapters, and there was limited space, so many individuals who played an important role in the development of probability theory had to be omitted. For those who like historical timelines, here is when they lived:

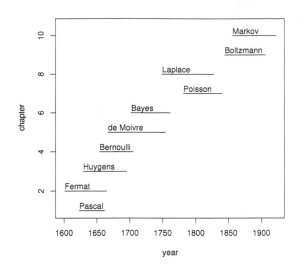

While Gian-Carlo was the most important influence on my notes, many other individuals influenced and supported this book. I had the privilege of knowing Mark Kac who, along with Gian-Carlo, taught me the essence of probability theory. The use of R dates back to when I was visiting the Harvard Medical School. I wish to thank JoAnn Manson, Tianhua Niu, and other members at the Division of Preventive Medicine, Brigham and Women's Hospital and Simin Liu at the Department of Epidemiology, UCLA, for their help and encouragement. My work on scale-invariant distributions was sponsored by a grant from Amdahl Corporation. I wish to acknowledge the contribution of Dieter Gawlick and Jim Gray in this area. The material on the Metropolis-Hastings algorithm originated from when I worked with Nick Metropolis, Stan Ulam and Gian-Carlo Rota at Los Alamos. I also wish to acknowledge the support of Vistology, Inc., especially Mieczyslaw M. Kokar and Christopher J. Matthews, for some of my work on Bayesian networks and influence diagrams. Special thanks go to my friend, colleague and co-author Tianhua Niu of the Harvard Medical School and the Harvard School of Public Health who generously took time out of his very busy schedule to give a careful and thorough review of my book. Finally, I wish to thank my family for their love, support and encouragement to complete this book.

Ken Baclawski
Boston, Massachusetts

CHAPTER 1

Sets, Events and Probability

Blaise Pascal was a French physicist and mathematician. His early work was in the natural and applied sciences where he made important contributions to the study of fluids, to the concepts of pressure and vacuum, and to many other fields. In 1654 he began a correspondence with another mathematician, Pierre de Fermat, to create the mathematical theory of probability. The motivation for creating this new theory was to solve a problem of his friend, the Chevalier de Méré. The problem was that of two players who were playing a game of chance and who wished to divide the stakes fairly when the game was interrupted prior to its normal completion and could not be continued. The division of the stakes should be based on the chance that each player has of winning the game from the point where the game was interrupted. While Pascal and Fermat created this new field, they did not develop it very far. Nevertheless, their correspondence had a significant influence on their contemporaries who then developed the field. Pascal was sickly and frail, but had a well-developed sense of humor. Some of his most popular quotations include, "The heart has its own reason which reason does not know," "It is not certain that everything is uncertain," and one which is especially apt as we begin our study of probability, "The excitement that a gambler feels when making a bet is equal to the amount he might win times the probability of winning it."

Suppose that we toss a coin any number of times and that we list the information of whether we got heads or tails on the successive tosses:

H	T	H	H	H	T	H	T	T	T	T	T	T	H	H	$\cdots$
1	2	3	4	5	6	7	8	9	10	11	12	13	14	15	$\cdots$

The act of tossing this coin over and over again as we have done is an example of an *experiment,* and the sequence of H's and T's that we have listed is called its *experimental outcome.* We now ask what it means, in the context of our experiment, to say that we got a head on the fourth toss. We call this an **event**. While it is intuitively obvious what an event represents, we want to find a precise meaning for this concept. One way to do this which is both

1

obvious and subtle is to *identify* an event with the set of all ways that the event can occur. For example, the event "the fourth toss is a head" is the same as the set of all sequences of H's and T's whose fourth entry is H. At first it appears that we have said little, but in fact we have made a conceptual advance. We have made the intuitive notion of an event into a concrete notion: *events are a certain kind of set*.

However, a warning is appropriate here. An event is not the same concept as that of an experimental outcome. An outcome consists of the total information about the experiment after it has been performed. Thus while an event may be easy to describe, the set to which it corresponds consists of a great many possible experimental outcomes, each being quite complex. In order to distinguish the concept of an event from the concept of an experimental outcome we will employ an artificial term for the latter. We will call it a **sample point**. Now a sample point will seldom look like an actual point in the geometric sense. We use the word "point" to suggest the "indivisibility" of one given experimental outcome, in contrast to an event which is made up of a great many possible outcomes. The term "sample" is suggestive of the random nature of our experiment, where one particular sample point is only one of many possible outcomes.

We will begin with a review of the theory of sets, with which we assume some familiarity. We will then extend the concept of a set by allowing elements to occur more than just once. We call such an entity a **multiset**. By one more conceptual step, the notion of a probability measure emerges as an abstraction derived from the multiset concept. Along the way we will repeatedly return to our coin-tossing experiment. We do this not only because it is an important process for modeling phenomena but also because we wish to emphasize that probability deals with very special kinds of sets.

1.1 The Algebra of Sets

In probability, we always work within a *context,* which we define by specifying the set of all possible experimental outcomes or equivalently all possible sample points. We call this set the **sample space**, typically denoted Ω. The term "sample space" does not help one to visualize Ω any more than "sample point" is suggestive of an experimental outcome. But this is the term that has become standard. Think of Ω as the "context" or "universe of discourse." It does not, however, in itself define our experiment. Quite a bit more will be required to do this. One such requirement is that we must specify which subsets of Ω are to be the events of our experiment. In general, not every subset will be an event. The choice of subsets which are to be the events will depend on the phenomena to be studied.

We will specify the events of our experiment by specifying certain very simple events which we will call the "elementary events," which we then combine to form more complicated events. The ways we combine events to form other events are called the *Boolean* or *logical operations*. Let A and B denote two events. The most important Boolean operations are the following:

Operation	Set	Definition
Union	$A \cup B$	elements either in A or in B (or both)
Intersection	$A \cap B$	elements both in A and in B
Complement	$\overline{A}$	elements not in A

Each of these has a natural interpretation in terms of events.

Operation	Event	Interpretation
OR	$A \cup B$	"either A or B (or both) occur"
AND	$A \cap B$	"both A and B occur"
NOT	$\overline{A}$	"A does not occur"

Several other Boolean operations are defined in the exercises.

When two events A and B have the property that if A occurs then B does also (but not necessarily vice versa), we will say that A is a **subevent** of B and will write $A \subseteq B$. In set-theoretic language one would say that A is a *subset* of B or that B *contains* A. The three Boolean operations and the subevent relation satisfy a number of laws such as commutativity, associativity, distributivity and so on, which we will not discuss in detail, although some are considered in the exercises. For example **de Morgan's laws** are the following:

$$\overline{(A \cap B)} = \overline{A} \cup \overline{B}$$

$$\overline{(A \cup B)} = \overline{A} \cap \overline{B}$$

In terms of events, the first of these says that if it is not true that both A and B occur, then either A does not occur or B does not occur (or both), and conversely. One has a similar statement for the second law. Generally speaking, drawing a Venn diagram suffices to prove anything about these operations. For example, here is the Venn diagram proof of de Morgan's first law. First draw the two events, A and B:

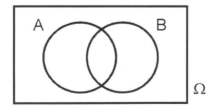

If we shade in the event $A \cap B$:

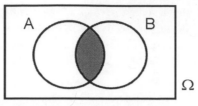

then the event $\overline{(A\cap B)}$ consists of the shaded portion of the following diagram:

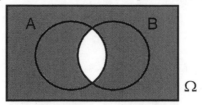

Next, shade in the events $\overline{A}$ and $\overline{B}$, respectively:

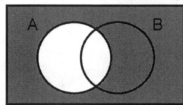
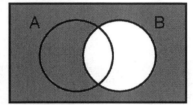

Combining these gives us the event $\overline{A}\cup\overline{B}$:

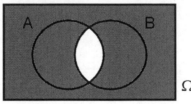

If we compare this with the event $\overline{(A\cap B)}$ we see that $\overline{(A\cap B)} = \overline{A}\cup\overline{B}$.

For more complicated expressions, involving many sets, and for which the Venn diagram would be extremely complex, it is very useful to know that there is a way we can simplify such an expression into an essentially unique expression. The idea is that every Boolean expression is a union of the smallest subevents obtainable by intersecting events occurring in the expression. To be more precise, suppose that we have an expression involving the events $A_1, A_2, \ldots A_n$, and unions, intersections and complements in any order nested as deeply as required. The simplified expression we obtain can be described in two steps.

Step 1. Write $A^{+1} = A$ and $A^{-1} = \overline{A}$. The events:

$$A_1^{i_1}\cap A_2^{i_2}\cap\cdots\cap A_n^{i_n},$$

as $i_1, i_2, \ldots, i_n$ take on all possible choices of ± 1, are the smallest events obtainable from the events $A_1, A_2, \ldots, A_n$ using Boolean operations. We call these events the *atoms* defined by $A_1, A_2, \ldots, A_n$. Notice that in general there can be 2^n of them, but that in particular cases some of the atoms may be empty so there could be fewer than 2^n in all. For example, the events A_1, A_2, A_3 break up Ω into (at most) eight atoms:

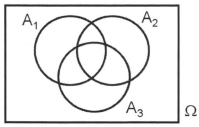

Step 2. Any expression involving $A_1, A_2, \ldots, A_n$, which uses Boolean operations, can be written as a union of certain of the atoms.

There are many procedures that can be used to determine which of the atoms are to be used. We leave it as an exercise to describe such a procedure. The resulting expression will be called the "atomic decomposition." By using Venn diagrams and atomic decompositions, any problem involving a finite number of events can be analyzed in a straightforward way. Unfortunately, many of our problems will involve infinitely many events, and for these we will later need some new ideas.

1.2 The Bernoulli Sample Space

We now return to the example in the beginning of this chapter: tossing a coin. A sample point for this experiment is an infinite sequence of ones and zeroes or equivalently of H's and T's. The toss itself is called a *Bernoulli trial*, a term that is used whenever a random action can have two possible outcomes. The word *trial* is used generically for an action that may have any number of outcomes. Even if we are only concerned with a finite sequence of Bernoulli trials, which is seemingly more realistic, it is nevertheless easier for computational reasons to imagine that we *could* go on tossing the coin forever. Moreover, we will find that certain seemingly rather ordinary events can only be expressed in such a context.

The set of all possible sequences of ones and zeroes is called the **Bernoulli sample space** Ω and the experimental process of which it forms the basis is called the **Bernoulli process**. For the moment we will be a little vague about what a "process" means, but we will make it precise later. The events of the Bernoulli sample space consist of certain subsets of Ω. To describe which subsets these are, we first describe some very simple events called **elementary events**. They are the events "the first toss comes up heads," "the second toss comes up heads," etc. We will write H_n for the event "the nth toss comes up heads." The complement of the event H_n will be denoted $T_n = \overline{H_n}$; it is the event "the nth toss comes up tails." One must be careful here. It is obvious that the complementary event to "the nth toss is heads" is "the nth toss is tails." However, it is less obvious that as *sets* $\overline{H_n} = T_n$, since both H_n and T_n are infinite sets and it is not easy to imagine what they "look like." As a rule, it is much easier to think in terms of the events themselves rather than in terms of their representations as sets.

We can now describe in general what it means for a subset of Ω to be an event of the Bernoulli sample space: an event is any subset of Ω obtainable from the elementary events by using the operations of complement as well as of unions and intersections of infinite sequences of events. The fact that we allow infinite unions and intersections will take some getting used to. What we are saying is that we allow any statements about the Bernoulli process which may, in principle, be expressed in terms of tosses of heads and tails (elementary events) using the words "and" (intersection), "or" (union), "not" (complement) and "ever" (infinite sequences).

To illustrate this, we consider the following example of a Bernoulli event: "A sequence of two successive H's occurs before a sequence of two T's ever occurs." We will call a specified finite sequence of H's and T's a **run**. So the event in question is "The run HH occurs before the run TT ever occurs." Here is a sample point for which this event occurs:

H   T   H   H   H   T   H   T   T   T   T   T   T   H   H $\cdots$

and here is one where it does not:

H   T   H   T   T   T   T   T   T   H   H   H   T   H   H $\cdots$

Write A for this event. The essence of the event A is that we will continue flipping that coin until either an HH or a TT occurs. When one of them happens we may then quit, or we may not; but it is nevertheless computationally easier to conceive of the experiment as having continued forever. Now it ought to be conceptually clear that it is possible to express A in terms of elementary Bernoulli events, but at first it may seem mysterious how to do it. The idea is to break apart A into simpler events which can each be expressed relatively easily in terms of elementary events The whole art of probability is to make a judicious choice of a manner of breaking up the event being considered. In this case, we break up the event A according to when the first run of HH occurs. Let A_n be the event "The run HH occurs first at the nth toss and the run TT has not yet occurred." The event A is the (infinite) union of all the A_n's, and in turn each A_n can be expressed in terms of the elementary events as follows:

$$A_2 = H_1 \cap H_2 \qquad\qquad \text{HH}\ldots$$
$$A_3 = H_1 \cap H_2 \cap H_3 \qquad\quad \text{THH}\ldots$$
$$A_4 = H_1 \cap \overline{H_2} \cap H_3 \cap H_4 \qquad \text{HTHH}\ldots$$
$$A_5 = \overline{H_1} \cap H_2 \cap \overline{H_3} \cap H_4 \cap H_5 \quad \text{THTHH}\ldots$$
etc.

Note that not only is A the union of the A_n's but also none of the A_n's overlap with any other. In other words, no sample point of A has been counted twice. This property will be very important for probabilistic computations.

As an exercise, one might try to calculate the expression, in terms of elementary events, of the event "A run of k heads occurs before a run of n tails occurs." Later we will develop tools for computing the probability of such an event quite easily, and this exercise will quickly convince one of the power of these tools.

1.3 The Algebra of Multisets

We now go back to our study of set theory. Our objective is to extend the concept of a set by allowing elements of sets to be repeated. This more general concept is called a **multiset**. To give an example, suppose that a committee of 10 members has an election to determine its chairperson. Of the votes that are cast, 7 are for candidate A, 2 for B and 1 for C. The set of votes is most easily expressed as a multiset consisting of 10 elements: 7 of type A, 2 of type B and 1 of type C. In set-builder notation, we write this $\{a, a, a, a, a, a, a, b, b, c\}$. We can write this more economically as $\{a^7, b^2, c^1\}$, the exponents denoting the number of copies of the element that are in the multiset. Notice that a set is a special kind of multiset.

As with sets, we can combine several multisets to form new multisets. In some ways, these operations are more natural than the analogous ones for sets. The operations are *addition* and *multiplication*. In the exercises, we describe one more operation: *subtraction*. Given two multisets M and N, their *sum* $M + N$ is obtained by combining all the elements of M and N, counting multiplicities. For example, if a occurs three times in M and twice in N, then it occurs five times in $M + N$. The *product* MN of M and N is obtained by multiplying the multiplicities of elements occurring in both M and N. For example, if a occurs three times in M and twice in N, then it occurs six times in MN. Here are some more examples:

$$\{a, b\} + \{b, c\} = \{a, b, b, c\}$$
$$\{a, b\} \times \{b, c\} = \{b\}$$
$$\{a, a, a, b, b\} + \{a, b, b, b, c\} = \{a, a, a, a, b, b, b, b, b, c\}$$
$$\{a, a, a, b, b\} \times \{a, b, b, b, c\} = \{a, a, a, b, b, b, b, b, b\}$$

or using exponent notation:

$$\{a^1, b^1\} + \{b^1, c^1\} = \{a^1, b^2, c^1\}$$
$$\{a^1, b^1\} \times \{b^1, c^1\} = \{b^1\}$$
$$\{a^3, b^2\} + \{a^1, b^3, c^1\} = \{a^4, b^5, c^1\}$$
$$\{a^3, b^2\} \times \{a^1, b^3, c^1\} = \{a^3, b^6\}$$

When A and B are two sets, it now makes sense to speak of their sum $A + B$ and their product AB. What do these mean in terms of sets? The product is easy to describe: it coincides precisely with the intersection $A \cap B$. For this reason, it is quite common to write AB for the intersection of two events. On the other hand, the sum of two sets is not so easy to describe. In general $A + B$ will not be a set even when both A and B are. The reason is that those elements occurring both in A and in B will necessarily occur *twice* in $A + B$. However, if A and B are *disjoint,* that is, when $A \cap B$ is empty, then $A + B$ is a set and coincides with $A \cup B$. As this situation is quite important in probability, we will often write $A + B$ to denote the union of A and B when A is disjoint from B, and we will then refer to $A + B$ as the **disjoint union** of A and B.

1.4 The Concept of Probability

Consider once again the election multiset that we introduced in Section 1.3: $\{a^7, b^2, c^1\}$. What percentage of the votes did each of the candidates receive? An easy calculation reveals that A received 70% of the total, B received 20% and C received only 10%. The process of converting "raw counts" into percentages loses some of the information of the original multiset, since the percentages do not reveal how many votes were cast. However, the percentages do contain all the information relevant to an election.

By taking percentages, we are replacing the complete information of the number of votes cast for each candidate by the information of what *proportion* of votes were cast for each candidate *relative* to the total number of votes cast. The concept of probability is an abstraction derived from this situation. Namely, a probability measure on a set tells one the proportion or size of an element or a subset *relative* to the size of the set as a whole. We may intuitively think of a probability as an assignment of a nonnegative real number to every element of the set in such a way that the sum of all such numbers is 1. The multiset $\{a^7, b^2, c^1\}$ gives rise to a probability measure which will be denoted in the following manner. For every *subset S* of $\{a, b, c\}$, we write $P(S)$ for the *proportion* of the multiset $\{a^7, b^2, c^1\}$, which has elements from S. For example, $P(\{a\})$ is 0.7 because 70% of the elements of $\{a^7, b^2, c^1\}$ are a's. Similarly, $P(\{a, b\})$ is 0.9, $P(\{b, c\})$ is 0.3, $P(\{a, b, c\})$ is 1.0 and so on. We call P a *probability measure.* It is important to realize that P is defined not on elements but on *subsets.* We do this because we observed that events are subsets of the sample space, and we wish to express the concept of a probability *directly* in terms of events. As we have seen, it is easier to think directly in terms of events rather than in terms of sets of outcomes. For this reason, we henceforth decree that a **probability measure** P on a sample space Ω is a function which assigns a real number $P(A)$ to every event A of Ω such that

Probabilities are nonnegative: $P(A) \geq 0$

The total probability is 1: $P(\Omega) = 1$

Probability is additive:

If $A_1, A_2, \ldots$ is a (finite or infinite) sequence of *disjoint* events, then

$$P(A_1 + A_2 + \ldots) = P(A_1) + P(A_2) + \ldots$$

The combination of a probability measure and a sample space is often abbreviated as *probability space*.

At first it may not be easy to see that these three conditions capture the concept of "proportion" we described above. The first two conditions, however, are easy to understand: we do not allow outcomes to occur a negative number of times, and the measure of Ω itself is 1 because it is the totality of all possible outcomes. It is the third condition that is the most difficult to justify. This condition is called **countable additivity**. When the sequence of events consists of just two events A and B, the condition should be intuitively obvious. Let C be the union $A \cup B$. Since A and B are assumed to be disjoint, C is the same $A + B$. Probabilistically, this says that A and B are mutually exclusive alternatives: C occurs if and only if exactly one of A or B occur. Clearly, if this is so, then the probability of C is "distributed" between A and B, i.e., $P(C) = P(A) + P(B)$. The extension of this rule to an infinite sequence of events is somewhat counterintuitive, but one can get used to it when one sees concrete examples.

1.5 Properties of Probability Measures

We now show three important facts about probability measures. These facts relate the concept of probability to the Boolean concepts of subevent, union, intersection and complement.

Subevents. If A is a subevent of B, then $P(A) \leq P(B)$.

Proof. Although this should be intuitively clear, we will *prove* it from the three conditions for P to be a probability. If we write $B \backslash A$ for $B \cap \overline{A}$, as shown in the following Venn diagram:

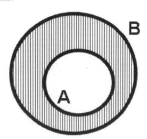

Then $A \subseteq B$ means that B is the disjoint union of A and $B \backslash A$. This should be clear form the Venn diagram, or just think of what it says: every element of B is either in A or it is not, and these alternatives are mutually exclusive. Therefore, countable additivity implies that

$$P(B) = P(A + (B \backslash A)) = P(A) + P(B \backslash A)$$

By the first axiom of probability, $P(B \backslash A) \geq 0$. Therefore,

$$P(B) = P(A) + P(B \backslash A) \geq P(A)$$

As a consequence, we find that since every event A is a subevent of Ω,

$$0 \leq P(A) \leq P(\Omega) = 1$$

This corresponds to our intuitive feeling that probability is a measure of likelihood, ranging from extremely unlikely (probability zero) to extremely likely (probability one).

Union and Intersection. If A and B are two events, then

$$P(A \cup B) = P(A) + P(B) - P(A \cap B)$$

Proof. To prove this we first write $A \cup B$ as a disjoint union of atoms. From the Venn diagram it is clear that

$$A \cup B = (A \cap B) + (A \backslash B) + (B \backslash A)$$

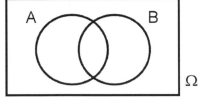

Similarly, we can write A and B as (disjoint) unions of atoms:

$$A = (A \cap B) + (A \backslash B)$$
$$B = (A \cap B) + (B \backslash A)$$

By countable additivity,

$$P(A \cup B) = P(A \cap B) + P(A \backslash B) + P(B \backslash A)$$
$$P(A) = P(A \cap B) + P(A \backslash B)$$
$$P(B) = P(A \cap B) + P(B \backslash A)$$

Now solve for $P(A \backslash B)$ and $P(B \backslash A)$ in the last two expressions and substitute these into the first. This gives our formula. The usefulness of this formula is that it applies even when A and B are not disjoint.

Here is a concrete example. Suppose that we have two coins. Let A be the event that the first shows heads, and let $a = P(A)$ be the probability of this. Similarly, let B be the event that the second shows heads, and let b be $P(B)$. What is the probability that when we toss both of them at least one shows heads? Clearly we want $P(A \cup B)$. By the above formula, we find that $P(A \cup B) = P(A) + P(B) - P(A \cap B) = a + b - P(A \cap B)$. However, we do not yet know how to compute $P(A \cap B)$ in terms of $P(A)$ and $P(B)$. We will return to this problem in Section 1.6.

Complement. If A is an event, then $P(\overline{A}) = 1 - P(A)$.

Proof. To see this simply note that Ω is the disjoint union of A and $\overline{A}$. By the second and third conditions for probability, we have

$$1 = P(\Omega) = P(A + \overline{A}) = P(A) + P(\overline{A})$$

Thus, we see that the probability for an event *not* to occur is "complementary" to the probability for its occurrence. For example, if the probability of getting heads when we toss a coin is p, then the probability of getting tails is $q = 1 - p$.

1.6 Independent Events

The notion of independence is an intuitive one derived from experience: two events are independent if they have no effect on one another. More precisely, if we have two independent events A and B, then knowing A has occurred does not change the probability for B to occur and vice versa. When we have the notion of conditional probability in Chapter 6, we can make this statement completely rigorous. Nevertheless, even with the terminology we have so far, the concept of independence is easy to express. We say two events A and B are *independent* when

$$P(A \cap B) = P(A)\, P(B)$$

If we use multiset notation, writing AB for $A \cap B$, then this rule is very suggestive: $P(AB) = P(A)\, P(B)$. It is important to realize that only independent events satisfy this rule just as only disjoint events satisfy additivity: $P(A + B) = P(A) + P(B)$.

Consider the case of coin tossing. The individual tosses of the coin are independent: the coin is the same coin after each toss and has no memory of having been tossed before. As a result, the probability of getting two heads in two tosses is the square of the probability of getting one head on one toss. As an application, consider the two-coin toss problem in Section 1.5. Since we are tossing two different coins, it seems reasonable to expect A and B to be independent. Therefore, the probability that both coins show heads is $P(A \cap B) = P(A)\, P(B) = ab$, and the probability that one of the coins shows heads is

$$P(A \cup B) = P(A) + P(B) - P(A \cap B) = a + b - ab$$

For example, if the probability of getting a head on each toss is $1/2$, then the probability of getting a head on one of two tosses is $1/2 + 1/2 - 1/4 = 3/4$.

A set of three events, A, B and C, is *independent* when:

1. Any pair of the three are independent, and

2. $P(A \cap B \cap C) = P(A)\, P(B)\, P(C)$.

It is possible for three events to satisfy (1) but not (2). This is an important point that is easily missed. Consider again the two-coin toss problem above. Let C be the event that the two coins show different faces (one heads the other tails). Then A, B and C are pairwise independent; for example, knowing that the first coin shows heads tells one nothing about whether the other will be the same or different. However, the three events are not independent: the occurrence of any two of them precludes the third from occurring.

Similarly, given any number of events (even an infinite number), we say that they are *independent* when

$$P(A_1 \cap A_2 \cap \cdots \cap A_n) = P(A_1)\, P(A_2) \cdots P(A_n)$$

for *any* finite subcollection $A_1, A_2, \ldots, A_n$ of the events.

1.7 The Bernoulli Process

This is the process of tossing a biased coin. In a given toss, we suppose that the probability is p for heads and q for tails, where $p + q = 1$. Generally speaking, we will also be implicitly assuming that both p and q are nonzero, but other than this we shall make no restriction on what value p could have. We call p the *bias* of the coin. A *fair coin* is a special kind of biased coin; namely, one with bias $p = 1/2$.

We want to assign a probability to every elementary event and show how this allows us to compute the probability of every other event. This is done in two steps.

Step 1. $P(H_n) = p$ and $P(T_n) = P(\overline{H_n}) = q = 1 - p$.

This assignment is made irrespective of n. In effect, we assume that we are using the same coin (or at least have the same bias) during each toss.

We have now defined the probability of the elementary events. But we still cannot determine the probability of an arbitrary event because none of the three conditions determines what, for example, $P(H_1 \cap H_2)$ can be, although they limit the possible choice. This leads us to our second assumption:

Step 2.

$$P\big(H_1^{i_1} \cap H_2^{i_2} \cap \cdots \cap H_n^{i_n}\big) = P\big(H_1^{i_1}\big)\, P\big(H_2^{i_2}\big) \cdots P\big(H_n^{i_n}\big)$$

where $i_1, i_2, \ldots, i_n$ take on all possible choices of ± 1.

Here we have drawn from the physical nature of the phenomenon of tossing a coin. The question of whether tosses of a real coin are independent is another question entirely: the question of the validity of the model as a means of describing the actual physical experiment. We will not consider this question until Chapter 5.

For any other event A of the Bernoulli process, the probability is calculated by expanding A in terms of the elementary events (possibly using countable unions) and by using the three conditions for probability. It is not always obvious what the best way to do this might be. There is an "art" to this part of the subject, and developing one's skill in this art is the whole idea of learning probability.

Let us return to the event: "a run of HH occurs before a run of TT." Recall that this event can be expanded as a disjoint union:

$$A = (H_1 \cap H_2) + (T_1 \cap H_2 \cap H_3) + (H_1 \cap T_2 \cap H_3 \cap H_4) + \cdots$$

By countable additivity, we compute:

$$
\begin{aligned}
P(A) &= P(H_1 \cap H_2) + P(T_1 \cap H_2 \cap H_3) + P(H_1 \cap T_2 \cap H_3 \cap H_4) + \cdots \\
&= P(H_1)\,P(H_2) + P(T_1)\,P(H_2)\,P(H_3) \\
&\qquad + P(H_1)\,P(T_2)\,P(H_3)\,P(H_4) + \cdots \\
&= p^2 + qp^2 + pqp^2 + qpqp^2 + \cdots \\
&= \left(p^2 + pqp^2 + \cdots\right) + \left(qp^2 + qpqp^2 + \cdots\right) \\
&= p^2\left(1 + pq + (pq)^2 + \cdots\right) + qp^2\left(1 + pq + (pq)^2 + \cdots\right) \\
&= p^2\left(\frac{1}{1-pq}\right) + qp^2\left(\frac{1}{1-pq}\right) \\
&= \frac{p^2 + qp^2}{1-pq}
\end{aligned}
$$

We are assuming here that we know how to sum a geometric series:

$$
1 + r + r^2 + r^3 + \cdots = \frac{1}{1-r}, \quad \text{if } |r| < 1
$$

We can check our computation of $P(A)$ by a simple expedient: suppose that the coin is *fair*, i.e., $p = q = 1/2$. In this case, it is easy to see that $P(A)$ must be 1/2, because either HH will occur before TT or TT will occur before HH, and since the coin is fair either one of these is equally likely. And indeed setting $p = q = 1/2$ in our formula above shows that this is the case.

A probability measure is a function P on events such that

Nonnegative. For every event A, $P(A) \geq 0$

Normalized. $P(\Omega) = 1$

Countably Additive. If $A_1, A_2, \ldots$ is a (finite or infinite) sequence of *disjoint* events, then $P(A_1 + A_2 + \ldots) = P(A_1) + P(A_2) + \ldots$

If A and B are disjoint events	$P(A+B) = P(A) + P(B)$
If A and B are independent	$P(AB) = P(A)\,P(B)$
If A is a subevent of B	$P(A) \leq P(B)$
If A and B are any two events	$P(A \cup B) = P(A) + P(B) - P(A \cap B)$
If A is any event	$P(\overline{A}) = 1 - P(A)$

1.8 The R Language

The R language is an open-source programming language and software environment for statistical computing and graphics. It was originally created in 1995 by Ross Ihaka and Robert Gentleman (hence the name R) at the University of Auckland, New Zealand. By 1997 the original pair of developers had grown to a much larger team consisting of a core team that can modify the core functionality and a much larger group of individuals who contribute packages for specialized purposes. For more information about how to download the core software and the contributed packages, see www.r-project.org. Over 1000 contributed packages are now available. One can freely obtain both the source code and precompiled binaries for most operating systems.

R is widely used for statistical software development and data analysis, and has become a de-facto standard among statisticians for statistical analysis and for the development of statistical software. Another of R's strengths is its graphical facilities, which produce publication-quality graphs. R uses a command line interface, and this will be how it is shown in this book. However, several graphical user interfaces are available.

Examples of R programs appear throughout this book. We will begin with a very brief introduction to the R language, but the emphasis will be on "learning by doing." Techniques and features of R will be developed as they are appropriate for solving tasks. All of the graphs presented in the book were plotted using R, and the programs for creating those graphs are explained.

Getting Started

The first step is to install R on your computer. This is easily done either by downloading the precompiled software appropriate to your operating system from www.r-project.org or, for the more sophisticated programmer, by downloading the source code and compiling it yourself. One starts R as one would start any other program on your computer. If you start R in command line mode, it will write some introductory text to the screen, and then present you with a prompt character, like this:

```
>
```

If you start R with a graphical user interface, there should be a window within the interface where the prompt appears much as for the command line mode. At this point you can type an R command, such as

```
> 1+1
```

You then press the enter key, and the R environment will execute your command. R will either write something to your screen as it will in this case, or it will display a graph in a window on your screen as we will see later. In this case, it will write

```
[1] 2
```

The reason for the "[1]" will be explained in a moment. When we show an

R session in this book, the lines without a prompt character are the ones written by R. Incidentally, if you do not type a complete command, R will respond with a different kind of prompt character (a plus sign) to indicate that it cannot execute your request because it is not yet complete. This will continue until your command is complete.

When you are finished with your R session, you can quit from R by typing this:

```
> q()
```

R will ask you whether you want to save your current values so that the next time you run R it will start where you left it. Then R will stop running.

Vectors

In this book, and in probability in general, we will continually be dealing with sequences of numbers $x_1, x_2, \ldots, x_n$. Consequently, it is no surprise that this is also the most fundamental concept in R. Such a sequence is called a *vector*. The simplest vector is the sequence $1, 2, \ldots, n$. For example, if you want n to be 100, then type this command:

```
> 1:100
```

This will create a vector with 100 elements, from 1 to 100. R will respond by listing elements of the vector you just created:

```
[1]   1   2   3   4   5   6   7   8   9  10  11  12  13  14  15  16  17  18
[19]  19  20  21  22  23  24  25  26  27  28  29  30  31  32  33  34  35  36
[37]  37  38  39  40  41  42  43  44  45  46  47  48  49  50  51  52  53  54
[55]  55  56  57  58  59  60  61  62  63  64  65  66  67  68  69  70  71  72
[73]  73  74  75  76  77  78  79  80  81  82  83  84  85  86  87  88  89  90
[91]  91  92  93  94  95  96  97  98  99 100
```

The brackets denote the subscripts of the elements of your vector. Vectors can be operating upon using the usual arithmetic operators such as addition, subtraction, multiplication, division and exponentiation. For example, if you want the sequence $1, 4, \ldots, 100$ of squares up to 100, just type this:

```
> (1:10)^2
```

R will

reply with:

```
[1]   1   4   9  16  25  36  49  64  81 100
```

R has a large variety of functions, such as **sum**, **max** and **min** which compute the sum, maximum and minimum of all elements of the vector, respectively. Most R functions can be applied to vectors, as well as to other types of objects such as matrices and grouped data tables. Some of the functions, like **sum**, **max** and **min**, produce just one number as their result. However, most functions apply their operations to every element of the vector, and produce another vector of the same size. To obtain information about an R function, just type a question mark followed by the function name, or equivalently, use the help function. For example, **log** is the natural logarithm. To find out all of its features type the command:

```
> help(log)
```

Applying the `log` function to the set of squares up to 100 produces this:

```
> log((1:10)^2)
 [1] 0.000000 1.386294 2.197225 2.772589 3.218876 3.583519
 [7] 3.891820 4.158883 4.394449 4.605170
```

Note that by common convention, if no base is specified for log, then it is the natural logarithm. Any other base must be explicitly specified. For example, $\log_{10}$ is the base-10 logarithm. In R this can be specified either as `log10` or by specifying the base as a parameter as in

```
> log(10^(2:5), base=10)
[1] 2 3 4 5
```

Incidentally, ordinary (nonvector) numbers were not forgotten! They are just vectors of length 1. This is why R writes the [1] in front of 2 when you ask for 1+1.

When we need a vector that is not a simple progression, one can explicitly specify its elements using the c function. For example,

```
> c(20,22,18,-20,10)
[1]   20   22   18  -20   10
```

This is similar to the set-builder notation in mathematics, except that one is building a sequence, not a set. It also differs in another respect. If a parameter is a vector, the c function will extract its numbers and add them to the others as in

```
> c(20,22,c(15,1,c(6,3),4),-1)
[1] 20 22 15  1  6  3  4 -1
```

The end result is a vector consisting only of numbers, not numbers and vectors. The name for this function is an abbreviation of the word "concatenation" because it concatenates all of the numbers from all of its parameters.

Another important R function is selecting an element of a vector. This is done by specifying the one you want in brackets. For example:

```
> ((1:10)^2)[5]
[1] 9
```

is the fifth square. This function is called "indexing" and it is a versatile feature of R that allows one to select a set of elements from a vector by a large variety of criteria. We will see many examples of indexing later.

The Bernoulli Process

We started our study of probability with the process of tossing a coin. The function that tosses a coin in R is called `rbinom`. This function implements a more general process than the Bernoulli process, and we are only using a special case here. To toss a fair coin 10 times, we use this command:

```
> rbinom(10,1,1/2)
 [1] 1 0 1 1 1 0 1 0 0 0 0 0 0 1 1
```

This function has three parameters. The first parameter is the number of tosses. The second parameter selects the special case of this function that implements the Bernoulli process. The general process will be explained in Section 4.6. The third parameter is the bias of the coin. Each time you call this function you will get another sequence of 0's and 1's. The `rbinom` function is one of the R functions for generating random numbers. We will see many more later in the book.

Boolean expressions

As in all programming languages, one can compare numbers using equal, not equal, less than, less than or equal, and so on. Such comparisons can then be combined using the Boolean operations "and," "or," and "not." The symbols used for these operations are the same as in most programming languages today:

Operation	R Notation
and	&&
or	||
not	!
equal	==
not equal	!=
less than	<
less than or equal	<=
greater than	>
greater than or equal	>=

However, R differs from most languages in allowing one to compare vectors as well as individual numbers. The result of such a comparison is another vector. For example, comparing a Bernoulli process with 1 can be done like this:

```
> rbinom(10,1,1/2)==1
 [1]  TRUE FALSE  TRUE  TRUE FALSE FALSE FALSE  TRUE FALSE FALSE
```

and results in a vector of Boolean values. Instead of 0's and 1's, we now have a Boolean vector of `FALSE`'s and `TRUE`'s. This vector can now be used with the bracket operator to select the places where the heads have occurred:

```
> (1:10)[rbinom(10,1,1/2)==1]
[1] 1 3 4 8
```

Names and Functions

When an expression starts getting big or complicated, it is useful to be able to specify names. For example,

```
> squares <- (1:10)^2
```

assigns the name `squares` to the vector of the first 10 squares. This is roughly the same as assigning names in mathematics by using a phrase beginning with "Let S be ..." In computer languages, these names are called *variables*, although they are not very similar to variables in mathematics. In R one can assign names in either direction, such as

```
> (1:10)^2 -> squares
```

If an expression involves some parameters, one must declare the parameters; otherwise R expects that the names must already have meanings. For example, here is how one would compute the sequence of heads for a randomly generated Bernoulli sample point:

```
> (1:n)[rbinom(n,1,p)==1]
```

However, if we typed the command exactly the same as shown above, then R would complain that we have not said what n and p are supposed to be. The correct way to say this is:

```
> heads <- function(n,p) (1:n)[rbinom(n,1,p)==1]
```

We can now invoke this function any number of times by just giving the two parameters, as in

```
> heads(10,3/4)
[1]  1  3  6  8  9 10
```

It is a common practice in computer languages for a parametrized expression to be called a *function*. However, it has little in common with the mathematical notion of a function. For example, if one runs the **heads** expression again, one would usually get a different answer even if the parameters were the same. As a result, an R function with no parameters is not the same as an ordinary name. The former could give different answers each time it is invoked, while the latter will always return the same value until that value is reassigned to another one.

Technical Details

As one might imagine, a large language like R will have a lot of technical details and special features that most people should never bother with. In this section, we show a few of these. Unless you have a fondness for looking "under the hood," you probably will not need to read this section.

When R performs arithmetic vector operations such as addition and multiplication, it operates on vectors that have the same length. The operation is performed on the corresponding elements, and the result has the same length as the vectors that were operated on. But if this is true, then how can one explain how R can perform an operation like this:

```
> (1:10) + 5
 [1]  6  7  8  9 10 11 12 13 14 15
```

The answer is a mechanism called "recycling." If one of the vectors in an operation is too short, then it starts over from the beginning. Normally, you will never need to be concerned about the fact that R is recycling. However, sometimes R will give an answer that looks anomalous. For example,

```
> (1:5) + (0:9)
 [1]   1   3   5   7   9   6   8  10  12  14
```

does not make sense until one realizes that the first vector is being recycled. This kind of anomaly is especially difficult to track down when mismatched vectors occur in the middle of some much longer computation whose intermediate results are not being displayed. Many commonly used R functions will check that the longer vector is a multiple of the size of the recycled one, and this will catch many of the problems that can arise due to mismatched vectors.

1.9 Exercises

Exercise 1.1 As a event $A\backslash B$ stands for "A occurs but B does not." Show that the operations of union, intersection and complement can all be expressed using only this operation.

Exercise 1.2 The *symmetric difference* of A and B, written $A\Delta B$, is defined by
$$A\Delta B = (A\backslash B)\cup(B\backslash A)$$
As an event, $A\Delta B$ means "either A occurs or B occurs but not both." For this reason, this operation is also called the "exclusive or." Use a Venn diagram to illustrate this operation.

Exercise 1.3 The *set of elements where A implies B*, denoted A/B, is the set
$$A/B = \overline{A}\cup B$$
As an event, A/B stands for "if A occurs then B does also." Use a Venn diagram to illustrate this operation.

Exercise 1.4 Use Venn diagrams to prove the following:

1. $(A/B)\cap(B/C)\subseteq A/C$, i.e., if A implies B and B implies C, then A implies C.
2. $(A/B)\cap(A/C) = A/(B\cap C)$, i.e., A implies B and A implies C if and only if A implies B and C.
3. $(A/B)\cap(B/A) = \overline{(A\Delta B)}$.

Exercise 1.5 Show that for any four sets A, B, C and D, the following is true:
$$(A\cup B)\backslash(C\cup D)\subseteq(A\backslash C)\cup(B\backslash D)$$

Exercise 1.6 Prove that any Boolean (logical) expression in $A_1, A_2, \ldots A_n$ is a union of atoms. In what sense is this union unique?

Exercise 1.7 Express $P(A\cup B\cup C)$ in terms of $P(A)$, $P(B)$, $P(C)$, $P(A\cap B)$, $P(A\cap C)$, $P(B\cap C)$, and $P(A\cap B\cap C)$.

Exercise 1.8 Let D_1 be the event "exactly one of the events A, B and C occurs." Express $P(D_1)$ in terms of $P(A)$, $P(B)$, $P(C)$, $P(A\cap B)$, $P(A\cap C)$, $P(B\cap C)$, and $P(A\cap B\cap C)$.

Exercise 1.9 Give an explicit expression for the event "a run of three heads occurs before a run of two tails" in terms of elementary Bernoulli events. Suggest how this might be extended to "a run of k heads occurs before a run of n tails."

Exercise 1.10 Let B be a multiset. We stay that A is a *submultiset* of B if every element of A occurs at least as many times in B as it does in A. For example, $\{a, a, b\}$ is a submultiset of $\{a, a, b, b, c\}$ but not of $\{a, b, b\}$. When A is a submultiset of B, it makes sense to speak of the *difference* of A and B, written $B - A$; namely, define $B - A$ to be the unique multiset such that $A + (B - A) = B$. For example, $\{a, a, b, b, c\} - \{a, a, b\} = \{b, c\}$. Suppose that A and B are sets. When does $B - A$ coincide with $B \backslash A$? Prove that $A \cup B = A + B - AB$.

Exercise 1.11 In a certain town, there are exactly 1000 families that have exactly 3 children. Records show that 11.9% have 3 boys, 36.9% have 2 boys, 38.1% have 2 girls and 13.1% have 3 girls. Use a multiset to describe this situation. Give an interpretation in terms of probability. What is the probability that in one of the above families all the children have the same sex?

Exercise 1.12 In a factory there are 100 workers. Of the total, 65 are male, 77 are married and 50 are both married and male. How many workers are female? What fraction of the female workers is married? Ask the same questions for male workers.

Exercise 1.13 The countable additivity condition for a probability measure can be stated in several other ways. Prove that countable additivity implies each of (a) and (b) below and that either of these imply countable additivity.

(a) If $A_1 \subseteq A_2 \subseteq \cdots$ is an ascending sequence of events and if $A = A_1 \cup A_2 \cup \cdots$, then $P(A) = \lim_{n \to \infty} P(A_n)$.

(b) If $A_1 \supseteq A_2 \supseteq \cdots$ is a descending sequence of events and if $A = A_1 \cap A_2 \cap \cdots$, then $P(A) = \lim_{n \to \infty} P(A_n)$.

Exercise 1.14 If A and B form a pair of independent events, show that the pair $(A, \overline{B})$, the pair $(\overline{A}, B)$, and the pair $(\overline{A}, \overline{B})$ are each a pair of independent events.

Exercise 1.15 In Exercise 1.12, are the properties of being male and of being female independent? Is the property of being male independent of being married?

Exercise 1.16 The probability of throwing a "6" with a single die is $1/6$. If three dice are thrown independently, what is the probability that exactly one shows a "6"? Use Exercise 1.8.

Exercise 1.17 A student applies for two national scholarships. The probability that he is awarded the first is $1/2$, while the probability for the second is only $1/4$. But the probability that he gets both is $1/6$. Are the events that he gets the two scholarships independent of one another? Discuss what this means.

Exercise 1.18 A baseball player has a 0.280 batting average. What is the probability that he is successful the next three times at bat? See Exercise 1.16. To do this exercise one must assume that the player's times at bat are independent. Is this a reasonable assumption?

Exercise 1.19 Three pigeons have been trained to peck at one of two buttons in response to a visual stimulus and do so correctly with probability p. Three pigeons are given the same stimulus. What is the probability that the majority (i.e., at least two) peck at the correct stimulus? Suppose that one of the pigeons sustains an injury and subsequently pecks at one or the other button with equal probability. Which is more likely to be the correct response, the button pecked by one of the normal pigeons or the button pecked by a majority of the three pigeons?

Exercise 1.20 (Prendergast) Two technicians are discussing the relative merits of two rockets. One rocket has two engines, the other four. The engines used are all identical. To ensure success the engines are somewhat redundant: the rocket will achieve its mission even if half the engines fail. The first technician argues that the four-engine rocket ought to be the better one.

The second technician then says, "Although I cannot reveal the failure probability of an engine because it is classified top secret, I can assure you that either rocket is as likely to succeed as the other."

The first technician replies, "Thank you. What you just told me allows me to compute the failure probability both for an engine and for a rocket."

Can you do this computation also?

Exercise 1.21 Rewrite the following vague conversation using the language of probability theory. You may assume that it is possible to distinguish between "good weather" and "bad weather" unambiguously.

"The Weather Bureau is not always right, but I would say that they are right more often than not," said Alice thoughtfully.

"Ah, but what comfort is it during miserable weather to know that the forecast was right? If it's wrong that is not going to affect the likelihood of good weather," retorted Bob.

"You may be right, but that does not contradict what I said, even though the forecast is pessimistic only about twice a week," answered Alice persuasively.

Show that these statements, when translated into the language of probability, allow one to determine that the Weather Bureau will predict good weather before a miserable day no more than about once a month.

Exercise 1.22 How many times must one roll a die in order to have a 99% chance of rolling a "6"? If you rolled a die this many times and it never showed a "6," what would you think?

Exercise 1.23 A dice game commonly played in gambling casinos is Craps. In this game two dice are repeatedly rolled by a player, called the "shooter," until either a win or a loss occurs. It is theoretically possible for the play to continue for any given number of rolls before the shooter either wins or loses. Computing the probability of a win requires the full use of condition (3). Because of the complexity of this game, we will consider here a simplified version. The probability of rolling a "4" using two (fair) dice is 1/12, and the probability of rolling a "7" is 1/6. What is the probability of rolling a "4" before rolling a "7"? This probability appears as part of the calculation in Exercise 3.18.

Exercise 1.24 Let A be the event in the Bernoulli process that the coin forever alternates between heads and tails. This event consists of just two possible sample points.

$$A = \{HTHTH\cdots, THTHT\cdots\}$$

Using Exercise 1.13, prove that $P(A) = 0$. Does this mean that A is impossible? More generally, prove that if an event B in the Bernoulli process consists of only a finite number of sample points, then $P(B) = 0$.

Use R to answer the remaining exercises.

Exercise 1.25 Compute the sum of the integers from 1 to 100. Who was famous for having done this very rapidly as a youth? Next, compute the sum of the squares of the numbers $1, 2, \ldots, 100$.

Exercise 1.26 Show how to compute the numbers from -6 to 6 in increments of 0.001.

Exercise 1.27 Run the Bernoulli process with bias 0.2 for 1000 tosses, and compute the number of heads. Do this a few times. About how many heads do you get?

Exercise 1.28 Run the Bernoulli process with bias 0.8 for 1000 tosses, and select just the last toss.

1.10 Answers to Selected Exercises

Answer to 1.1.

1. $\overline{A} = \Omega \backslash A$

2. $A \cap B = A \backslash (\Omega \backslash B)$

3. $A \cup B = \overline{(\overline{A} \cap \overline{B})} = \Omega \backslash ((\Omega \backslash A) \backslash B)$

Answer to 1.2.   The Venn diagram for $A \triangle B$ is

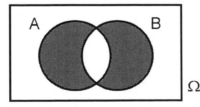

Answer to 1.3.   The Venn diagram for A/B is

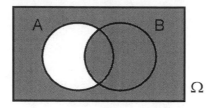

Answer to 1.5.

$$(A \cup B) \setminus (C \cup D) = (A \cup B) \cap \overline{((C \cup D))}$$
$$= (A \cup B) \cap (\overline{C} \cap \overline{D})$$
$$= (A \cup B) \cap \overline{C} \cap \overline{D}$$
$$= (A \cap \overline{C} \cap \overline{D}) \cup (B \cap \overline{C} \cap \overline{D})$$
$$\subseteq (A \cap \overline{C}) \cup (B \cap \overline{D})$$
$$= (A \setminus C) \cup (B \setminus D)$$

Here is another proof. This one uses atomic decomposition:

$$(A \cup B) \setminus (C \cup D) = (A \cap \overline{B} \cap \overline{C} \cap \overline{D}) \cup (A \cap B \cap \overline{C} \cap \overline{D}) \cup (\overline{A} \cap B \cap \overline{C} \cap \overline{D})$$
$$(A \setminus C) = A \cap \overline{C} = (A \cap B \cap \overline{C} \cap D) \cup (A \cap \overline{B} \cap \overline{C} \cap D)$$
$$\cup (A \cap B \cap \overline{C} \cap \overline{D}) \cup (A \cap \overline{B} \cap \overline{C} \cap \overline{D})$$
$$(B \setminus D) = B \cap \overline{D} = (A \cap B \cap C \cap \overline{D}) \cup (\overline{A} \cap B \cap C \cap \overline{D})$$
$$\cup (A \cap B \cap \overline{C} \cap \overline{D}) \cup (\overline{A} \cap B \cap \overline{C} \cap \overline{D})$$

Answer to 1.6. There are many ways to see this intuitively. Here is a formal proof that every Boolean expression using the events $A_1, A_2, \ldots, A_n$ is a union of atoms.

Step 0. Every set A_m is a union of atoms as follows:

$$A_m = \cup \left(A_1^{i_1} \cap \ldots \cap A_m^{i_m} \cap \ldots \cap A_n^{i_n} \right)$$

where the $i_1, i_2, \ldots, i_n$ take on all possible values of ± 1, except for i_m which is always $+1$.

Step 1. If E_1 and E_2 can be expressed as unions of atoms, then their union $E_1 \cup E_2$ can also.

This is obvious.

Step 2. If E is a union of atoms, then $\overline{E}$ is also.

This is true because $\overline{E}$ is the union of all atoms not in the expression for E.

Step 3. If E_1 and E_2 can be expressed as unions of atoms, then their intersection $E_1 \cap E_2$ can also.

To prove this, apply de Morgan's law to get:

$$E_1 \cap E_2 = \overline{(\overline{E_1} \cup \overline{E_2})}$$

Then apply Steps 1 and 2 to this expression.

Using steps 0 to 3, we see that every Boolean expression in $A_1, A_2, \ldots, A_n$ can be expressed as a union of atoms, by induction. However, the expression so obtained need not be unique, since some of the atoms could be empty, and such atoms can always be either included or excluded.

Answer to 1.7.

$$P(A \cup B \cup C) = P(A) + P(B) + P(C)$$
$$- P(A \cap B) - P(A \cap C) - P(B \cap C) + P(A \cap B \cap C)$$

This can be proved by splitting everything into atoms.

Answer to 1.8.　As in Exercise 1.7, this can be computed by atomic decomposition. The formula is:

$$P(D_1) = P(A) + P(B) + P(C)$$
$$- 2P(A \cap B) - 2P(A \cap C) - 2P(B \cap C) + 3P(A \cap B \cap C)$$

The Venn diagram for D_1 is

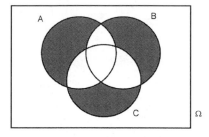

Answer to 1.9.　Let A be the event "A run of HHH occurs before a run TT," and let A_i be the event "A run HHH occurs at the ith toss and no run of HHH or TT has yet occurred." Then the first six A_i's are:

$$A_1 = \emptyset$$
$$A_2 = \emptyset$$
$$A_3 = H_1 \cap H_2 \cap H_3$$
$$A_4 = T_1 \cap H_2 \cap H_3 \cap H_4$$
$$A_5 = H_1 \cap T_2 \cap H_3 \cap H_4 \cap H_5$$
$$A_6 = (T_1 \cap H_2 \cap T_3 \cap H_4 \cap H_5 \cap H_6) \cup (H_1 \cap H_2 \cap T_3 \cap H_4 \cap H_5 \cap H_6)$$

In general, let B_i be the event "The ith toss was a tail and no runs of HHH or TT have occurred up to the ith toss." The we have the following inductive formula for B_i for $i>3$:

$$B_i = (B_{i-3}\cap H_{i-2}\cap H_{i-1}\cap T_i)\cup(B_{i-2}\cap H_{i-1}\cap T_i)$$

Furthermore, for $i>3$, we have:

$$A_i = B_{i-3}\cap H_{i-2}\cap H_{i-1}\cap H_i$$

The event A is then the disjoint union of the A_i's. A similar approach can be used for the event "A run of k heads occurs before a run of n tails," but *two* auxiliary sequences of events are needed.

Answer to 1.13.   To prove part (a), let $A_1\subseteq A_2\subseteq\cdots$ be an ascending sequence of events. Here is the Venn diagram:

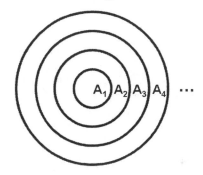

If we define a sequence of events $B_1 = A_1, B_2 = A_2\backslash A_1,\ldots, B_n = A_n\backslash A_{n-1}$, then each B_n is a "ring" in the above Venn diagram. Here are some examples:

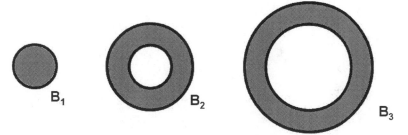

Therefore, $A_n = B_1\cup B_2\cup\cdots B_n$ is a *disjoint* union of events. By countable additivity, $P(A_n) = P(B_1) + P(B_2) + \cdots + P(B_n)$. Similarly, A is the disjoint union $B_1\cup B_2\cup B_3\cup\cdots$ of *all* the B_n's. Applying countable additivity again, $P(A_n) = P(B_1) + P(B_2) + P(B_3) + \cdots$ By definition, such an infinite series converges if and only if its sequence of partial sums converges:

$$P(A) = \lim_{n\to\infty} P(B_1) + P(B_2) + \cdots + P(B_n). = \lim_{n\to\infty} P(A_n)$$

It is easy to see how to reverse the above argument.

Part (b) follows from part (a) by a trick. Let $A_1\supseteq A_2\supseteq\cdots$ be a descending sequence of events, and let $A = A_1\cap A_2\cap\cdots$ Then the sequence of complements is ascending: $\overline{A_1}\subseteq\overline{A_2}\subseteq\cdots$ and $\overline{A} = \overline{A_1}\cup\overline{A_2}\cup\cdots$ by de Morgan's law. Now

$P(A) = \lim\limits_{n\to\infty} P(A_n)$ if and only if $1 - P(A) = \lim\limits_{n\to\infty} (1 - P(A_n))$ if and only if $P(\overline{A}) = \lim\limits_{n\to\infty} P(\overline{A_n})$, and this limit is covered by part (a).

Answer to 1.14. A and B are independent means that $P(A\cap B) = P(A)\,P(B)$. Now $P(A) = P(A\cap B) + P(A\cap\overline{B})$ by additivity so

$$
\begin{aligned}
P(A\cap\overline{B}) &= P(A) - P(A\cap B) \\
&= P(A) - P(A)\,P(B) \quad \text{by independence} \\
&= P(A)\,(1 - P(B)) \\
&= P(A)\,P(\overline{B})
\end{aligned}
$$

Interchanging the roles of A and B gives us $P(\overline{A}\cap B) = P(\overline{A})\,P(B)$. Replacing A by $\overline{A}$, and using the independence of $\overline{A}$ and B, we conclude, as above, that $P(\overline{A}\cap\overline{B}) = P(\overline{A})\,P(\overline{B})$.

Answer to 1.16. Let A, B, C be the events that the first, second, third die shows "6." We were given that $P(A) = P(B) = P(C) = 1/6$. By independence we then have that $P(A\cap B) = P(A\cap C) = P(B\cap C) = \frac{1}{36}$, and $P(A\cap B\cap C) = \frac{1}{216}$. By Exercise 1.8,

$$
P(D_1) = \frac{1}{6} - 2\frac{1}{36} + 3\frac{1}{216} = \frac{25}{72}.
$$

There are similar formulas for $P(D_i)$, where D_i is the event "Exactly i of the events $A_1, A_2, \ldots, A_n$ occur."

Answer to 1.21. Write G for the event that the weather is good tomorrow. Similarly, assume that the Weather Bureau either predicts good weather or predicts miserable weather, and that the event that the Bureau is correct is C. There are four quantities of interest in this problem:

$a = P(C\cap G)$ The prediction is correct and the weather is good

$b = P(C\cap\overline{G})$ The prediction is correct and the weather is miserable

$c = P(\overline{C}\cap G)$ The prediction is wrong and the weather is good

$d = P(\overline{C}\cap\overline{G})$ The prediction is wrong and the weather is miserable

These are the probabilities of the atoms. Because P is a probability measure, these quantities must satisfy these conditions:

$$
a, b, c, d \geq 0 \quad \text{and} \quad a + b + c + d = 1
$$

The statement that the Weather Bureau is not always right can be expressed probabilistically as $P(C) < 1$ or $a + b < 1$. That the Bureau is right more often than not means that $a + b > \frac{1}{2}$.

The statement that the forecast being wrong does not affect the likelihood of good weather is ambiguous, but one interpretation is that while the forecast is dependent on the state of the weather (one would hope!), the *correctness* of the forecast is independent of the weather. In probabilistic terms, this says that C and G are independent. By definition, this is the requirement that $P(C \cap G) = P(C)\,P(G)$ or in terms of atoms that $a = (a + b)(a + c)$.

Finally, the event that the forecast is pessimistic is $(C \cap \overline{G}) \cup (\overline{C} \cap G)$, and the assertion that this is true only about twice a week is that

$$P\big((C \cap \overline{G}) \cup (\overline{C} \cap G)\big) \approx \frac{2}{7}$$

or that $b + c \approx \frac{2}{7}$. Putting all of the constraints together, we have numbers a, b, c and d such that

$$
\begin{array}{ll}
a, b, c, d \geq 0 & a + b + c + d = 1 \\
a + b > \frac{1}{2} & a + b < 1 \\
a = (a + b)(a + c) & b + c \approx \frac{2}{7}
\end{array}
$$

These conditions do not determine a, b, c and d; but they do place strong constraints on what values they can take. Moreover, it is not obvious that there is a solution at all.

Nevertheless, there are, in fact, many solutions in this case. Here is an especially optimistic one:

$$a = \frac{5}{7}, \quad b = 0, \quad c = \frac{2}{7} \quad \text{and} \quad d = 0$$

In this solution, the weather is always good, and the Weather Bureau predicts it correctly $\frac{5}{7}$ of the time. It is easy to see that all of the conditions are satisfied, including $b + c = c = \frac{2}{7}$.

More generally, if the condition $b + c \approx \frac{2}{7}$ is taken to be an equality, then the value of b must be in the interval $\left[0, \frac{2}{7}\right]$, and c is determined by b. From the independence condition one can then solve for a. It is a quadratic polynomial, so there are, in principle, two solutions for a, but only one satisfies the other constraints. Once a is known, then the fact that the probabilities add to 1 allows one to compute d. One can also differentiate the formula for a as a function of b to find the minimum value of a. This happens when $b = \frac{1}{7}$, and the minimum value of a is $\dfrac{5 + \sqrt{21}}{14}$. Consequently, the maximum value of d is $\dfrac{5 - \sqrt{21}}{14} \approx 0.02982$. So there is less than a 3% chance that the Weather Bureau will predict good weather before a miserable day, which is less than once a month, on average.

Answer to 1.22. Look at the complementary event, "Roll the die n times and never roll a 6." We want this to be at most 0.01. This has probability $(5/6)^n$, assuming the die is a fair die. Set this to 0.01, and solve for n. This is done by taking logarithms. The solution is

```
> log(.01)/log(5/6)
[1] 25.25851
```

So one must roll the die 26 times. If one rolled the die 26 times and never showed a "6," then one would suspect that the die is faulty or loaded in some way.

Answer to 1.23. Let A be the event "A 4 appears before a 7 appears." Let A_i be the event "The first time that either a 4 or a 7 appears is the ith roll and at the ith roll a 4 appears." The probability of rolling *neither* a 4 nor a 7 is $1 - \frac{1}{12} - \frac{1}{6} = \frac{3}{4}$, so $P(A_i) = \left(\frac{3}{4}\right)^{i-1}\frac{1}{12}$. Now A is the disjoint union of the A_i's, so

$$P(A) = \sum_{i=1}^{\infty}\left(\frac{3}{4}\right)^{i-1}\frac{1}{12} = \left(\frac{1}{1-\frac{3}{4}}\right)\frac{1}{12} = 4\frac{1}{12} = \frac{1}{3}$$

Answer to 1.24. Let $A_1 = H_1 \cap T_2$, $A_2 = H_1 \cap T_2 \cap H_3 \cap T_4$, and so on. Then by Exercise 1.13,

$$P(\{HTHTHT\ldots\}) = P(H_1 \cap T_2 \cap H_3 \cap T_4 \cap \cdots) = \lim_{n\to\infty} P(A_n)$$

Clearly, $P(A_n) = (pq)^n$. Since $0 \le pq \le \frac{1}{4}$, the probability $(pq)^n$ approaches 0 as $n\to\infty$, and we can conclude that $P(\{HTHTHT\ldots\}) = 0$. Similarly, $P(\{THTHTH\ldots\}) = 0$. Thus if $A = \{HTHTHT\ldots, THTHTH\ldots\}$, then $P(A) = 0$. Note that this does *not* mean that A is impossible. It only means that it is extremely unlikely. The same argument can be applied to any finite set of Bernoulli sample points.

Answer to 1.25.

```
> sum(1:100)
[1] 5050
```

Carl Friedrich Gauss was famous for performing this feat very quickly as a child.

```
> sum((1:100)^2)
[1] 338350
```

Answer to 1.26. Use either

```
> (-6000:6000)/1000
```

or

```
> seq(from=-6,to=6,by=0.001)
```

CHAPTER 2

Finite Processes

Pierre de Fermat was a French lawyer and government official at the Parlement of Toulouse, France, and a mathematician who is given credit for early developments that led to modern calculus. For example, the integral of x^n was first evaluated by Fermat. Fermat and Descartes were the two leading mathematicians of the first half of the seventeenth century. Fermat independently discovered the fundamental principles of analytic geometry, although the principles are now commonly referred to as "Cartesian geometry" after Descartes. Fermat was secretive and a recluse. His only contact with the wider mathematical community was exchanges of letters with Pascal and Mersenne. It was his brief exchange of letters with Pascal that founded probability theory. Fortunately, Pascal and especially Mersenne, operated a correspondence network with other European thinkers, and Fermat's ideas were widely distributed. Although Fermat claimed to have proved all his arithmetic theorems, few records of his proofs have survived. Many mathematicians, including Gauss, doubted several of his claims, especially given the difficulty of some of the problems and the limited mathematical tools available to Fermat. His famous Last Theorem was first discovered by his son in the margin on his father's copy of an edition of *Arithmeticus* by Diophantus, "To divide a cube into two other cubes, a fourth power or in general any power whatever into two powers of the same denomination above the second is impossible, and I have assuredly found an admirable proof of this, but the margin is too narrow to contain it." He had not bothered to inform even Mersenne of it.

Because the theory of probability originated in the analysis of games of chance, the oldest probability models have only a finite number of sample points. In this chapter, we introduce the techniques for computing probabilities of events in a finite process. We will later see that these techniques are useful in many other processes as well.

2.1 The Basic Models

The basic situation for a finite process is the following. Imagine that we have a *population* consisting of *individuals*. They may be people, cards or whatever. Now choose one individual from the population. How many ways can this be done? The answer, of course, is the number of individuals in the population. The individual we choose is called a **sample** (*of size* 1) from the population. More generally, suppose that we choose not just one but a whole sequence of k individuals from the population. The sampling procedure we envision here is successive: we choose one individual at a time. How many ways can we do this? The answer will depend on whether the same individual can be chosen more than once, or equivalently on whether a chosen individual is returned to the population before another is chosen. The two kinds of sampling are called **sampling with replacement** and **sampling without replacement**. The sequence of chosen individuals is called a *sample* of size k.

To illustrate this, we consider dice and cards. A single roll of a die samples one of its 6 faces. If we roll the die k times, or equivalently if we roll k dice, we are choosing a sample of size k *with* replacement from the population of 6 faces, Similarly, a choice of 1 card from the standard 52-card deck is a sample from a population of 52 individuals. However, if we deal out k cards, we are choosing a sample of size k *without* replacement from the population of cards. Note that in a sample, the order matters, so that a *sample* of cards is not the same as a *hand* of cards, for which the order does not matter.

The description of a finite process given above is an abstraction called the **sampling model**. It occurs concretely in a large variety of situations. For example, the population could be an alphabet A, whose individuals are the letters. A random word of length k is just a sample of size k from the alphabet. Because the same letter can occur more than once, it is a sample with replacement. If we do not allow a letter to be used more than once in a word, then we have a sample without replacement. Words are an important feature of modern programming languages, except that words are normally called "strings," and letters are called "characters." This manifestation of the finite process is called the **distribution model** because one is distributing the letters to the positions in the word. If we number the positions in the word by the integers $1, 2, \ldots, k$, then a word is a mathematical function from the set $\{1, 2, \ldots, k\}$ to the alphabet A. More generally, given any two finite sets B and C, a sample with replacement is a function that maps B to C. A sample without replacement is a one-to-one function. This interpretation of finite sampling is called the *mathematical model*. In yet another model, the elements of the set C are regarded as being containers. When an element b of B is mapped to an element c of C, one regards the element b as being *placed* in the container c. The function is called a *placement* of the individuals in the containers. In this model, the individuals in B are called *objects* or *balls*, the containers are often called *urns*, and the model is called the **occupancy model**, because the objects *occupy* the containers. For each placement, the number of objects in one container is called the **occupancy number** of the container. In the distribution model, the occupancy number is the number of repetitions of a letter in a word. All of these models and many others are logically equivalent, so there is no harm in mixing terminology taken from the various models.

2.2 Counting Rules

Finite sampling, in its many variations and equivalent forms, occurs frequently in probability computations. A roll of dice, a hand of cards, even a configuration of particles in chemistry or physics are all forms of finite sampling. In this section, we will concentrate on the most basic rules for counting collections of samples. The most fundamental rule of counting is one so obvious that it does not seem necessary to dwell on it:

First Rule of Counting. If a placement is formed by making a succession of choices such that there are

n_1 possibilities for the first choice

n_2 possibilities for the second choice

and so on

Then the total number of placements that can be made by making a set of choices is the product

$$n_1 \times n_2 \times n_3 \times \cdots$$

Note that by a "succession of choices" we mean that *after* the first choice is made, there are n_2 choices for the second choice, and similarly for subsequent choices.

As a simple example of this rule, consider how many ways 3 dice can be rolled. A roll of 3 dice requires 3 "choices," 1 for each die. Since each die has 6 possibilities, there are $6^3 = 216$ rolls of 3 dice. Notice that it does not matter whether we view the 3 dice as being rolled one at a time or all at once. We will call such choices *independent:* the rolls of the other dice do not affect the set of possibilities available to any given die.

As another example, how many ways can we deal 5 cards from a deck of 52 cards? Note that we are considering *deals*, not *hands*, so we consider as different 2 hands having the same cards but dealt in different orders. A deal of 5 cards consists of 5 choices. There are 52 choices for the first card, 51 choices for the second card, etc. The total number of deals is then $52 \cdot 51 \cdot 50 \cdot 49 \cdot 48 = 311,865,200$. Unlike the situation for rolls of 3 dice, the cards dealt in the above example are not independent choices. The earlier cards we dealt do affect the set of possibilities for later cards. However, the earlier cards do not affect the *number* of possibilities for a later deal of a card. Hence the first rule still applies.

Before we consider more complex counting problems we restate the above two examples in the general language of finite sampling.

Samples with replacement The total number of functions from k elements to a set of n elements, or equivalently the total number of k-letter words made from an alphabet of n letters is n^k: each element or letter can be chosen in n ways independently of the others.

Samples without replacement The total number of one-to-one functions from k elements to a set of n elements, or equivalently the total number of k-letter words with no repeated letters, made from an alphabet of n letters is

$$n(n-1)\cdots(n-k+1)$$

There are k factors in this product, one for each choice. The factors decrease by one after each choice because the chosen element or letter can no longer be used for subsequent choices.

An important special case of the second formula is the one for which $k = n$. Such a function has a special name: it is a **permutation**. Permutations occur so frequently in computations that we have a special notation for the total number of them.

Definition. The total number of permutations of an n-element set, or equivalently the number of n-letter words using all the letters from an n-letter alphabet is called *n-factorial* and is written

$$n! = n(n-1)\cdots 3\cdot 2\cdot 1$$

For example, if we deal all 52 cards from a standard deck, there are 52! ways for this to happen.

Now that we have the concept of a factorial, we can give a simpler formula for the number of samples without replacement. As we computed above, the number of k element samples without replacement from an n-element set is: $n(n-1)\cdots(n-k+1)$. Now it is easy to see that if we multiply this formula by $(n-k)! = (n-k)(n-k-1)\cdots(2)(1)$, we will get the product of all numbers from n down to 1. In other words, we get $n!$ Therefore,

$$n(n-1)\cdots(n-k+1) = \frac{n!}{(n-k)!}$$

We can then summarize the computations in this table:

Sample of k from n with replacement	n^k
Sample of k from n without replacement	$\dfrac{n!}{(n-k)!}$

2.3 Computing Factorials

The computation of factorials is common not only in probability but also in many other scientific and engineering disciplines. As a result, most programming languages support this function. For example, the number of shuffles of a standard 52-card deck of playing cards is 52! = `factorial(52)` = `8.065818e+67`. Because the factorial function grows so rapidly, one cannot compute the factorial on very large numbers within the limits of the usual floating point representations of numbers. To deal with this problem, R has a function `lfactorial` that is the natural logarithm of the factorial function. For example, R will fail to compute 1000000!, but it has no difficulty computing this:

```
> lfactorial(1000000)
[1] 12815518
```

Converting this to base 10,

```
> lfactorial(1000000)/log(10)
[1] 5565709
```

tells us that 1000000! is approximately $10^{5565709}$. The factorial function can be defined on noninteger and even complex numbers.

2.4 The Second Rule of Counting

Anyone who has played cards knows that one normally does not care about the order in which one is dealt a hand of cards. That is, a *hand* of cards is not the same as a *deal* of cards. A poker hand is defined to be a *subset* of five cards from a standard deck, and the order in which the cards are obtained is immaterial. We cannot count the number of poker hands using the first rule of counting because they violate the fundamental premise of that rule: the placements must be obtained by successive choices. However, every poker hand can be dealt in precisely 5! = 120 ways. Therefore, the number of poker hands is

$$\frac{52 \cdot 51 \cdot 50 \cdot 49 \cdot 48}{5!} = \frac{311,875,200}{120} = 2,598,960$$

This illustrates the second rule of counting, also called the "shepherd principle": if it is too difficult to count sheep, count their legs and then divide by four.

Second Rule of Counting. If we wish to count a set of placements obtained by making choices but for which the order of choice does not matter, count as if it did matter and then divide by the number of ordered placements per unordered placement.

Let us illustrate this in a more complicated situation. How many bridge situations are possible? By definition a bridge game consists of 4 people being dealt 13 cards each. However, the order in which each of the 4 hands is dealt does not matter. So we first count as if the order did matter: this gives 52! possible deals. But each hand can be dealt in 13! ways, and there are 4 hands for a total of $(13!)^4$ ways to deal a given bridge game. Therefore, there are

$$\frac{52!}{(13!)^4} \approx 5.36447 \times 10^{28}$$

possible bridge situations.

One must be careful when applying the second rule to make certain that the number of ordered placements is the same for any unordered placement. In Section 2.7 we will give an example of this kind of difficulty.

Before we end this section, we cannot resist one more generalization suggested by the bridge game example. Using the language of functions, a bridge game maps all 52 cards of the deck to the set of 4 players so that exactly 13 cards are mapped to each player. In the occupancy model, a bridge game is a placement of 52 cards in 4 containers so that each container has an occupancy number equal to 13. More generally, we would like to count the number of ways that k objects can be placed in a set of n containers so that θ_1 are placed in the first container, θ_2 objects are in the second container, etc. Since each of the k objects is placed in some container, we must have $\theta_1 + \theta_2 + \cdots \theta_n = k$. Now by the second rule of counting, the number of such placements is:

$$\frac{k!}{\theta_1! \theta_2! \cdots \theta_n!}$$

This expression, called the **multinomial coefficient**, is written

$$\binom{k}{\theta_1, \theta_2, \ldots, \theta_n} = \frac{k!}{\theta_1! \theta_2! \cdots \theta_n!}$$

For example,

$$\binom{7}{2, 2, 3} = \frac{7!}{2!2!3!} = \frac{5040}{2 \times 2 \times 6} = 210$$

This multinomial coefficient is pronounced "seven choose two, two, three." Because the factorial can be defined for positive real numbers, the multinomial coefficient is also defined for any positive real numbers.

An important special case of the multinomial coefficient is the case $n = 2$, when it is called the **binomial coefficient**. This number should be a familiar concept from the binomial expansion in algebra and calculus. Because a placement into two containers is completely determined by the placement in either one of the containers, we can also interpret the binomial coefficient $\binom{k}{\theta_1, \theta_2}$ as the number of θ_1-element subsets of a k-element set. The binomial coefficient is often abbreviated to $\binom{k}{\theta_1} = \binom{k}{\theta_1, \theta_2}$ and is then pronounced simply "k choose θ_1." Using this notation, we can quickly compute the number of poker hands because a poker hand consists of five cards chosen from 52 cards:

$$\binom{52}{5} = \binom{52}{5, 47} = \frac{52!}{5!47!}$$

Furthermore, we can use binomial coefficients and the first rule of counting to find another formula for the multinomial coefficient. A placement of k objects into n containers with occupancy numbers $\theta_1, \theta_2, \ldots, \theta_n$ can be made using $n - 1$ choices: first choose θ_1 objects from the k objects and put them in the first container, next choose θ_2 objects from the remaining $k - \theta_1$ objects and put them in the second container,..., continue in this way until we choose θ_{n-1} objects from the remaining $k - \theta_1 - \cdots - \theta_{n-2}$ objects and put them in the next to last container, and finally put the last $\theta_n = k - \theta_1 - \cdots - \theta_{n-1}$ objects in the last container. No choice is necessary for the last step. Therefore:

Placement of k individuals into n containers with occupancy numbers $\theta_1, \theta_2, \ldots, \theta_n$	$\binom{k}{\theta_1, \theta_2, \ldots, \theta_n}$
Subsets of size θ of a set of k individuals	$\binom{k}{\theta} = \binom{k}{\theta, k-\theta}$
Relationship between multinomial and binomial coefficients $$\binom{k}{\theta_1, \theta_2, \ldots, \theta_n} = \binom{k}{\theta_1}\binom{k-\theta_1}{\theta_2} \cdots \binom{k-\theta_1 - \cdots - \theta_{n-2}}{\theta_{n-1}}$$	

Multinomial and Binomial Coefficients

2.5 Computing Probabilities

Consider a finite sample space Ω. The events of Ω are the subsets if Ω. We would like to see what a probability measure P on Ω means. Remember that P is defined on events of Ω. An event $A \subseteq \Omega$ may be written as a finite set

$$A = \{\omega_1, \omega_2, \ldots, \omega_n\}$$

and, by additivity, any probability measure P must satisfy

$$P(A) = P(\omega_1) + P(\omega_2) + \cdots P(\omega_n)$$

We call the events $\{\omega\}$, having just one outcome $\omega \in \Omega$, the atoms of Ω. It is important to distinguish between an outcome and an atom: the atom $\{\omega\}$ of an outcome ω is the event "ω occurs." The distinction is roughly the same as that between a noun and a simple sentence containing it, e.g., between "tree" and "This is a tree."

What we have just shown above is that *every probability measure P on a finite sample space Ω is determined by its values on atoms*. The value on an arbitrary event $A \subseteq \Omega$ is then computed by the formula:

$$P(A) = \sum_{\omega \in A} P(\{\omega\})$$

The values of P on the atoms may be assigned arbitrarily so long as:

1. For every atom $\{\omega\}$, $0 \le P(\{\omega\}) \le 1$, and

2. $\displaystyle\sum_{\omega \in \Omega} P(\{\omega\}) = 1.$

Whenever (1) and (2) hold, P defines a consistent probability measure on Ω.

The simplest example of a probability measure on a finite sample space Ω is the one we will call the **equally likely probability**; it is the unique probability measure P on Ω for which every atom is assigned the same probability as any other. Hence, for every atom $\{\omega\}$, $P(\{\omega\}) = 1/|\Omega|$. For more general events $A\subseteq\Omega$ this probability has a simple formula:

$$P(A) = \frac{|A|}{|\Omega|} = \frac{\text{number of outcomes in } A}{\text{total number of outcomes in } \Omega}$$

The equally likely probability is quite common in gambling situations as well as in sampling theory in statistics, although in both cases great pains are taken to see to it that it really is the correct model. For this reason, this probability has something of an air of artificiality about it, even though it or probability measures close to it, do occur often in nature. Unfortunately, nature seems to have a perverse way of hiding the proper definition of Ω from casual inspection.

The phrases "completely at random" or simply "at random" are used to indicate that a given problem is assuming the equally likely probability. The latter phrase is misleading because every probability measure defines a concept of randomness for the sample space in question. Even certainty for one outcome is a special case of a probability measure. In Chapter 9 we will justify the use of the description "completely random" for the equally likely probability measure. For now we consider some concrete examples of this probability measure.

Rolling Dice. What is the probability that in a roll of 3 dice no 2 show the same face? We already computed $|\Omega|$ in the Section 2.2: $|\Omega| = 216$. The event A in question is "The faces of the 3 dice are all different." We think of an outcome in A as a placement of 3 objects into 6 containers so that no 2 objects are in the same container. There are $\dfrac{6!}{3!} = 6\cdot5\cdot4 = 120$ placements with this property. Hence

$$P(A) = \frac{6!}{3!6^3} = \frac{720}{1296} = 0.555\ldots$$

Birthday Coincidences. If n students show up at random in a classroom, what is the probability that at least two of them have the same birthday? In order to solve this problem we will make some simplifications. We will assume that there are only 365 days in every year; that is, we ignore leap years. Next, we will assume that every day of the year is equally likely to be a birthday. Lastly, we assume that the students' birthdays are independent dates. These are innocuous assumptions.

Let B be the event in question, and let $A = \overline{B}$ be the complementary event "no two students have the same birthday." Now just as we computed in the dice rolling problem above, $P(A) = \dfrac{365!}{(365 - n)!365^n}$, and hence $P(B) = 1 - \dfrac{365!}{(365 - n)!365^n}$. We would like to write this in R as

```
> pb <- function(n)
+     1 - factorial(365)/(factorial(365-n)*365^n)
```

But this will result in numbers that will cause overflows to occur. So we take the logarithm of each factor and then exponentiate the final result using the exp function:

```
> pb <- function(n)
+      1-exp(lfactorial(365)-lfactorial(365-n)-n*log(365))
> pb(c(20,22,25,30))
[1] 0.4114384 0.4756953 0.5686997 0.7063162
```

So in a class of 30 students the odds are 7 to 3 in favor of at least 2 having a common birthday. To produce a graph of this function, we use the plot function. This function has two required parameters and many optional parameters. The two parameters are vectors with the x and y coordinates of the points that are to be plotted on the graph. The two vectors must have the same number of elements. We will introduce some of the optional parameters later. Here is a graph of the probabilities:

```
> plot(1:100, pb(1:100))
```

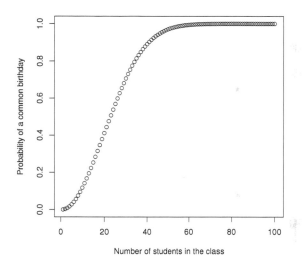

Number of students in the class

Random Committees. In the U.S. Senate, a committee of 50 senators is chosen at random. What is the probability that Massachusetts is represented? What is the probability that every state is represented?

In any real committee the members are not chosen at random. What this question is implicitly asking is whether random choice is "reasonable" with respect to two criteria of reasonableness. Note that the phrase "at random" is ambiguous. A more precise statement would be that every 50-senator committee is as probable as any other.

We first count $|\Omega|$, the number of possible committees. Since 50 senators are being chosen from 100 senators, there are

$$|\Omega| = \binom{100}{50}$$

committees. Let A be the event "Massachusetts is *not* represented." The committees in A consist of 50 senators chosen from the 98 non-Massachusetts senators. Therefore,

$$|A| = \binom{98}{50}$$

The probability of A is easily evaluated by using the R function `choose` that computes the binomial coefficient:

```
> choose(98,50)/choose(100,50)
[1] 0.2474747
```

So the answer to our original question is that Massachusetts is represented with probability $1 - 0.247 = 0.753$ or 3 to 1 odds. This seems reasonable.

Now consider the event $A = $ "Every state is represented." Each committee in A is formed by making 50 *independent* choices from the 50 state delegations. Hence $|A| = 2^{50}$ and the probability is

```
> 2^50/choose(100,50)
[1] 1.115953e-14
```

This probability is so small that the event A is essentially impossible. By this criterion, random choice is not a reasonable way to choose a committee. Some binomial coefficients are so large that `choose` will cause the computation to fail. In such a case, one should use the `lchoose` function, which computes the logarithm of the binomial coefficient.

2.6 Exercises

Exercise 2.1 Flip a coin three times. How many ways can this be done? List them. Convert each to the corresponding word from a two-letter alphabet. Are any of these words one-to-one functions?

Exercise 2.2 Roll a die twice. List in a column the ways that this can be done. In the next column list the corresponding words from the six-letter alphabet $\{A, B, C, D, E, F\}$. Mark the ones that have no repeated letters.

Exercise 2.3 You are interviewing families in a certain district. In order to ascertain the opinion held by a given family, you sample two persons from the family. Recognizing that the order matters in which the two persons from one family are interviewed, how many ways can one sample two persons from a six-person family? List the ways and compare with the lists in Exercise 2.2 above. If the two persons are interviewed simultaneously, so that the order no longer matters, how many ways can one sample two persons from a six-person family?

Exercise 2.4 A small college has a soccer team that plays eight games during its season. In how many ways can the team end its season with five wins, two losses and one tie? Use a multinomial coefficient.

Exercise 2.5 Ten students are traveling home from college in Los Angeles to their homes in New York City. Among them they have two cars, each of which will hold six passengers. How many ways can they distribute themselves in the two cars?

Exercise 2.6 Compute the order of magnitude of 1000!. That is, compute the integer n for which 1000! is approximately equal to 10^n.

Exercise 2.7 How many ways can a 100-member senate be selected from a country having 300,000,000 inhabitants?

Exercise 2.8 Have the students in your probability class call out their birthdays until someone realizes there is a match. Record how many birthdays were called out. We will return to this problem in Exercise 3.25.

Exercise 2.9 Give a formula for the probability that in a class of n students at least two have adjacent or identical birthdays. Ignore leap years. Calculate this probability using R for $n = 10, 15, 20$ and 25.

Exercise 2.10 Compute the probabilities for two dice to show n points, $2 \leq n \leq 12$. Do the same for three dice.

Exercise 2.11 It is said that the Earl of Yarborough used to bet 1000 to 1 against being dealt a hand of 13 cards containing no card higher than 9 in the whist or bridge order. Did he have a good bet? In bridge the cards are ranked in each suit from 2 (the lowest) through 10, followed by the Jack, Queen, King and Ace in this order.

Exercise 2.12 May the best team win! Let us suppose that the two teams that meet in the World Series are closely matched: the better team wins a given game with a probability of 0.55. What is the probability that the better team will win the World Series? Do this as follows. Treat the games as tosses of a biased coin. Express the event "the better team wins" in terms of elementary Bernoulli events, and then compute the probability. How long must a series of games be in order to be reasonably certain (with 95% probability) that the best team will win?

Exercise 2.13 Although Robin Hood is an excellent archer, getting a "bullseye" nine times out of ten, he is facing stiff opposition in the tournament. To win he finds that he must get at least four bullseyes with his next five arrows. However, if he gets five bullseyes, he runs the risk of exposing his identity to the sheriff. Assume that, if he wishes to miss the bullseye, he can do so with probability 1. What is the probability that Robin wins the tournament?

Exercise 2.14 A smuggler is hoping to avoid detection by mixing some illegal drug tablets in a bottle containing some innocuous vitamin pills. Only 5% of the tablets are illegal in a jar containing 400 tablets. The customs official tests 5 of the tablets. What is the probability that she catches the smuggler? Is this a reasonable way to make a living?

Exercise 2.15 Every evening a man either visits his mother, who lives downtown, or visits his girlfriend, who lives uptown (but not both). In order to be completely fair, he goes to the bus stop every evening at a random time and takes either the uptown or the downtown bus, whichever comes first. As it happens, each of the 2 kinds of buses stops at the bus stop every 15 minutes with perfect regularity (according to a fixed schedule). Yet he visits his mother only around twice each month. Why?

Exercise 2.16 In a small college, the members of a certain Board are chosen randomly each month from the entire student body. Two seniors who have never served on the Board complain that they have been deliberately excluded from the Board because of their radical attitudes. Do they have a case? There are 1000 students in the college and the Board consists of 50 students chosen eight times every year.

Exercise 2.17 The smuggler of Exercise 2.14 passes through customs with no difficulty even though they test 15 tablets. But upon reaching home, he discovers to his dismay that he accidentally put too many illegal drug tablets in with the vitamin pills, for he finds that 48 of the remaining 385 tablets are illegal. Does he have reason to be suspicious? The question he should ask is the following: given that he packed exactly 48 illegal pills, what is the probability that none of the 15 tested were illegal?

Exercise 2.18 Compute the probability that a fair coin tossed 200 times comes up heads exactly half of the time. Similarly, what is the probability that in 600 rolls of a fair die, each face shows up exactly 100 times?

Exercise 2.19 The following is the full description of the game of Craps. On the first roll of a pair of dice, 7 and 11 win, while 2, 3 and 12 lose. If none of these occur, the number of dots showing is called the "point," and the game continues. On every subsequent roll, the point wins, 7 loses and all other rolls cause the game to continue. You are the shooter; what is your probability of winning?

Exercise 2.20 Compute the probability of each of the following kinds of poker hand, assuming that every five-card poker hand is equally likely.

	kind of hand	definition
(a)	"nothing"	none of (b)-(j)
(b)	one pair	two cards of the same rank
(c)	two pair	two cards of one rank and two of another
(d)	three-of-a-kind	three cards of the same rank
(e)	straight	ranks in ascending order (ace may be low card or high card but not both at once)
(f)	full house	three of one rank and two of another
(g)	flush	all cards of the same suit
(h)	straight flush	both (e) and (g)
(i)	four-of-a-kind	four cards of the same rank
(j)	royal flush	(h) with ace high

Poker Hands

Note that the kinds of hand in this table are pairwise disjoint. For example, a straight hand does not include the straight flush as a special case.

Exercise 2.21 It is an old Chinese custom to play a dice game in which six dice are rolled and prizes are awarded according to the pattern of the faces shown, ranging from "all faces the same" (highest prize) to "all faces different." List the possible patterns obtainable and compute the probabilities. Do you notice any surprises?

Exercise 2.22 Some environmentalists want to estimate the number of whitefish in a small lake. They do this as follows. First 50 whitefish are caught, tagged and returned to the lake. Some time later another 50 are caught and they find 3 tagged ones. For each n compute the probability that this could happen if there are n whitefish in the lake. For which n is this probability the highest? Is this a reasonable estimate for n?

Exercise 2.23 A group of astrologers has, in the past few years, cast some 20,000 horoscopes. Consider only the positions (houses) of the sun, the moon, Mercury, Venus, Earth, Mars, Jupiter and Saturn. There are 12 houses in the Zodiac. Assuming complete randomness, what is the probability that at least 2 of the horoscopes were the same?

Exercise 2.24 The three-person duel is a difficult situation to analyze in full generality. We consider just a simple special case. Three individuals, X, Y and Z, hate other so much they decide to have a duel only one of which can survive. They stand at the corners of an equilateral triangle. The probability of a hit by each of the three participants is 0.5, 0.75 and 1, respectively. For this reason they decide that they will each shoot at whomever they wish, taking turns cyclically starting with X and continuing with Y, then Z, then X again and so on. All hits are assumed to be fatal. What strategy should each employ, and what are their probabilities of survival?

Exercise 2.25 The theory of probability began with the correspondence between Pascal and Fermat which came about as a result of problems put to Pascal by the Chevalier de Méré, a noted gambler of the time. Here is one of those problems. Two gamblers are playing a game which is interrupted. How should the stake be divided? The winner of the game was to be the one who first won 4 deals out of 7. One gambler has so far won 1 deal and the other 2 deals. They agree to divide the stake according to the probability each had of winning the game, and this probability is to be computed by assuming that each player has an equal probability of winning a given game. Express the event that the first gambler wins in terms of elementary Bernoulli events. Then compute the probability.

Exercise 2.26 In what was apparently Isaac Newton's only excursion into probability, he answered a question put to him by Samuel Pepys. The problem was to determine which is more likely, showing at least one six in 6 throws of a die, at least two sixes in 12 throws or at least three sixes in 18 throws. Compute these probabilities and consider the general question of the probability of showing at least n sixes in $6n$ throws of a die.

2.7 Answers to Selected Exercises

Answer to 2.10. Two dice show n points with probability $\dfrac{6-|n-7|}{36}$. For three dice, the counting may be simplified by noticing that we can think of the first two dice as being a set to which the third die is added. Thus, the number of ways to get n with three dice is a sum of six consecutive numbers in this list:

$$\overbrace{0,0,0,0,0,1}^{n=3},2,\underbrace{3,4,5,6,5,4}_{n=10},3,2,\overbrace{1,0,0,0,0,0}^{n=18}$$

Thus for $n = 3$, there is 1 way; for $n = 10$ there are 27 ways; etc. To get probabilities divide by 216. This procedure is an example of a convolution. We will see it again in more generality in Section 6.7.

Answer to 2.14.

```
> 1-(1-0.05)^5
[1] 0.2262191
```

The smuggler has a 1 in 4 chance of getting caught. It is not a very good way for the smuggler to make a living. On the other hand, it is a reasonable way for the customs official to make a living. The reasons are discussed in more detail in Section 10.2 on the ruin problem.

Answer to 2.15. The paradox here is thinking that a completely random *time* (uniform) implies that the man will take a completely random *bus* (equally likely). If the following is the bus schedule, one can see that the man will take a random bus, but it will not be completely random:

Downtown	Uptown
$4:00PM$	$4:14PM$
$4:15PM$	$4:29PM$
$4:30PM$	$4:44PM$
$\vdots$	$\vdots$

This paradox will appear again in various other disguises. It is important that you understand it thoroughly. Such paradoxes are a common feature when one uses "fuzzy thinking" about probabilities.

Answer to 2.19. One wins or loses immediately with probabilities 2/9 and 1/9, respectively. The game goes on with probability 2/3. However, the subsequent probabilities depend on the "point." Suppose that the probability of rolling the point in a single roll is p. Then the probability of getting and winning with that point is the infinite series

$$p\left(\sum_{i=1}^{\infty}\frac{1}{\left(1-p-\frac{1}{6}\right)^{i-1}}\right)p = \frac{p^2}{p+\frac{1}{6}}$$

where the first p is for "getting" the point and the last p is for "winning" it. The other factor is the probability that one rolls neither the point nor 7 a sequence of times. The total probability of winning is 2/9 plus the sum of $\dfrac{p^2}{p+\frac{1}{6}}$ over all possible points. The probability of winning is thus computed by first setting p to the vector of probabilities for the points, and then adding up the formula we have just derived:

```
> p <- c(3,4,5,5,4,3)/36
> 2/9 + sum(p^2/(p+1/6))
[1] 0.4929293
```

Answer to 2.20. Cards have 13 ranks, (Ace, 2, 3, ..., King) and 4 suits (Clubs, Diamonds, Hearts, Spades). We count poker hands by choosing ranks and then suits or vice versa. Thus, one pair can be achieved in

$$\binom{13}{1}\binom{4}{2}\binom{12}{3}\binom{4}{1}\binom{4}{1}\binom{4}{1}$$

ways. The $\binom{13}{1}$ chooses the rank for the pair, the $\binom{4}{2}$ chooses the suits of the pair, $\binom{12}{3}$ chooses the remaining ranks, and $\binom{4}{1}^3$ chooses their suits. The computations for two pair, three-of-a-kind, four-of-a-kind and full house are similar. For cards in ascending order, we may start with 1 of 10 ranks, and the suits may be chosen arbitrarily for a total of $10\binom{4}{1}^5$ ways. The suits will all be the same in $10\binom{4}{1}$ ways, and one can get a royal flush in just $\binom{4}{1}$ ways. These cases form a descending (nested) sequence so each must be subtracted from the more general case. To get all cards of the same suit, we first choose the suit and then their ranks for a total of $\binom{4}{1}\binom{13}{5}$ ways. Again one must subtract the overlaps. Finally, to count the hands that have "none of the above," one counts the total number of hands $\binom{13}{5}\binom{4}{1}^5$, and then subtracts all the other cases. We now compute these probabilities using R. One could do this with an ordinary vector, but then the elements would just be labeled with indexes starting at 1, rather than with the names of the kinds of hands. The *data frame* is a vector or matrix that allows one to specify any kind of label for each element. Except for the labels, a data frame is just like any other vector or matrix. In fact, ordinary vectors and matrices are implemented in R by means of data frames. Here are the probabilities in R expressed as a data frame:

```
> poker <- data.frame(
+    none = choose(13,5)*4^5 - 10*4^5 - 4*choose(13,5) + 10*4,
+    pair = 13*choose(4,2)*choose(12,3)*4^3,
+    two.pair = choose(13,2)*choose(4,2)^2*11*4,
+    triple = 13*choose(4,3)*choose(12,2)*4^2,
+    straight = 10*4^5 - 10*4,
+    full.house = 13*choose(4,3)*12*choose(4,2),
+    flush = 4*choose(13,5) - 10*4,
+    four = 13*choose(4,4)*12*4,
+    straight.flush = 10*4 - 4,
+    royal.flush = 4)
```

```
> sum(poker) - choose(52,5)
> poker/choose(52,5)
      none      pair   two.pair      triple     straight
1 0.5011774 0.422569 0.04753902 0.02112845 0.003924647
    full.house          flush            four straight.flush
1  0.001440576 0.001965402 0.0002400960    1.385169e-05
    royal.flush
1 1.539077e-06
```

Answer to 2.21. The patterns are: 6, 51, 42, 411, 33, 321, 3111, 222, 2211, 21111, 111111. For pattern 6 (all the same) there are $\binom{6}{6}\binom{6}{1} = 6$ choices; for the pattern 51 there are $\binom{6}{5}\binom{6}{1}\binom{1}{1}\binom{5}{1}$ choices; etc. As with cards, if two of the numbers in the pattern are the same, e.g., 2211, one cannot distinguish between them. So there are $\binom{6}{2}\binom{4}{2}\binom{6}{2}\binom{2}{1}\binom{1}{1}\binom{4}{2}$ ways to get the pattern 2211. Read this as follows: "Choose two dice $\binom{6}{2}$, choose two more $\binom{4}{2}$, now choose what will show on these two pairs $\binom{6}{2}$, choose another die $\binom{2}{1}$, choose the last die $\binom{1}{1}$, finally choose what will show on these last two dice $\binom{4}{2}$." Here it is in R:

```
> dice <- data.frame(
+   "6" = choose(6,6)*choose(6,1),
+   "51" = choose(6,5)*choose(6,1)*choose(1,1)*choose(5,1),
+   "42" = choose(6,4)*choose(6,1)*choose(2,2)*choose(5,1),
+   "411" = choose(6,4)*choose(6,1)*choose(2,1)*choose(1,1)
+         *choose(5,2),
+   "33" = choose(6,3)*choose(3,3)*choose(6,2),
+   "321" = choose(6,3)*choose(6,1)*choose(3,2)*choose(5,1)
+         *choose(1,1)*choose(4,1),
+   "3111" = choose(6,3)*choose(6,1)*choose(3,2)*choose(5,1)
+          *choose(1,1)*choose(4,1),
+   "222" = choose(6,2)*choose(4,2)*choose(2,2)*choose(6,3),
+   "2211" = choose(6,2)*choose(4,2)*choose(6,2)*choose(2,1)
+          *choose(1,1)*choose(4,2),
+   "21111" = choose(6,2)*choose(6,1)*choose(4,1)*choose(3,1)
+           *choose(2,1)*choose(1,1)*choose(5,4),
+   "111111" = choose(6,1)*choose(5,1)*choose(4,1)*choose(3,1)
+            *choose(2,1)*choose(1,1)*choose(6,6)
+ )
> sum(dice) - 6^6
> dice/6^6
            X6          X51          X42         X411          X33
1 0.0001286008 0.003858025 0.009645062 0.03858025 0.006430041
        X321      X3111       X222      X2211      X21111     X111111
1 0.154321 0.154321 0.03858025 0.3472222 0.2314815 0.0154321
```

The only surprise is that one pair (21111) is less probable than two pair (2211).

In fact, two pair will occur about 35% of the rolls, while one pair only occurs about 23% of the time, so this should be quite noticeable.

Answer to 2.22. After tagging, there are 50 tagged fish and $(n - 50)$ untagged ones. Catching 50 fish can be done in $\binom{n}{50}$ ways, while one can catch 3 tagged and 47 untagged fish in $\binom{50}{3}\binom{n-50}{47}$ ways. So if this event is written A_n, then we can write the probability of this event in R as follows:

```
> fish <- function(n) choose(50,3)*choose(n-50,47)/choose(n,50)
```

So far all of our graphs have been simple point plots. This time, instead of drawing points, we want to draw a curve. In R, this is called a "line graph," and it is specified by using the `type="l"` parameter. Here is a line graph of our `fish` function from 50 to 2000:

```
> n <- 50:2000
> plot(n, fish(n), type="l")
```

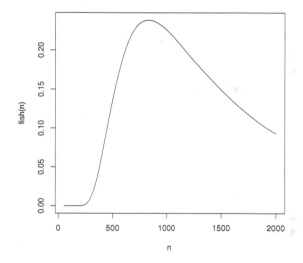

The maximum is easily computed to be `max(fish(n))`. Finding where this maximum occurs is easily done by selecting the components of the vector `n` where the maximum occurs. This will, in general, select all of the values of the vector `n` where the maximum occurs, but in this case it occurs just once:

```
> m <- max(fish(n))
> m
[1] 0.2382917
> n[fish(n)==m]
[1] 833
```

Notice that while $n = 833$ is the value of n that maximizes $P(A_n)$, it is only slightly better than neighboring values. For example, fish(830) = 0.238287. Indeed, A_n is not a very likely event for any n in the interval from 450 to 1900.

It is never higher than 24%. We will discuss this problem in the context of statistical hypothesis testing. The value $\hat{n} = 833$ for which $P(A_n)$ is maximized is called the **maximum likelihood estimator** of n, a term introduced by R.A. Fisher.

Answer to 2.23. This is another "birthday coincidences" problem, but there are $N = 12^8 = 429,981,696$ possible horoscopes, which is much higher than the 366 possible birthdays. Nevertheless, the chance of a coincidence is

$$1 - \frac{N!}{(N-M)!N^M},$$

where M=20,000. Here is the computation:

```
> M <- 20000
> N <- 12^8
> q <- factorial(N)/(factorial(N-M)*N^M)
> 1-q
```

However, this will fail because the factorials are much too large. So we need to "take logarithms":

```
> M <- 20000
> N <- 12^8
> lq <- lfactorial(N)
+      - lfactorial(N-M) + M*log(N)
> 1-exp(lq)
[1] 0.3719400
```

Surprisingly, there is about a 1 in 3 chance that 2 of the horoscopes are the same.

Answer to 2.25. Use a Bernoulli process in which the tosses are the deals. Let heads be a win for the first gambler. The elementary events that constitute a win for the first gambler are:

$$H_1 \cap H_2 \cap H_3$$

$$T_1 \cap H_2 \cap H_3 \cap H_4$$

$$H_1 \cap T_2 \cap H_3 \cap H_4$$

$$H_1 \cap H_2 \cap T_3 \cap H_4$$

The probability that the first gambler wins is easily computed to be 5/16.

CHAPTER 3

Discrete Random Variables

Christiaan Huygens was a Dutch mathematician, astronomer and physicist. He was the first to propose that Saturn has rings, and he discovered Titan, the first moon of Saturn to be identified. At about this time, Huygens independently solved a number of the earliest problems in probability theory, and wrote *De ludo aleae*, the first book on the subject, which he published in 1657. Among his contributions to probability theory, Huygens introduced the distinction between sampling with replacement and sampling without replacement. His book remained the only book on probability theory for 50 years, until it was superseded by other books, especially Jacob Bernoulli's *Ars Conjectandi*, which incorporated Huygens' work. While Huygens continued to work on probability theory for the rest of his life, his later contributions were not significant. On the other hand, his work on clocks and optics was very important. In his *Treatise on Light*, published in 1690, he proposed that light was emitted in all directions as a series of waves in a medium called the *luminiferous ether*. This theory competed with, and eventually overturned, Newton's corpuscular theory of light. The wave theory of light based on luminiferous ether remained the generally accepted theory for 200 years, until the late nineteenth and early twentieth centuries when it was replaced by quantum mechanics.

While Pascal and Fermat were the founders of probability theory, it was only known to the relatively small group of mathematicians and scientists who were part of their correspondence network. Huygens introduced the subject to the scientific public. Huygens is often cited by historians as being associated with what has been called the "Scientific Revolution." He certainly was not the only one who was responsible for this historical event. However, in one respect he was both unique and pivotal. His book established a new way of looking at and answering questions about nature, in which uncertainty is represented in an explicit and quantifiable manner.

The basis for this new way of answering questions is the notion of a random variable. For example, suppose we toss a coin. How long will it take to get the first head? How can one answer such a question? Sometimes the first head will appear on the first toss, sometimes on the next and so on. Clearly, we cannot answer such a question with a single number. The originality of the probabilistic point of view is that it answers such a question with a series of possible answers, each with its own probability.

The intuitive idea of a random variable is that it is the strengthening of the notion of a variable. Recall from calculus and algebra that a *variable* is a symbol together with a set over which the symbol ranges. For example, in calculus one often says "let x be a variable ranging over the real numbers" or more succinctly "let x be a real variable." This range is the foundation on which the variable built, so it is called the *support* of the variable. Now a **random variable** (sometimes abbreviated to RV) is a variable *together with* the probability that it takes each of its possible values. These values can, in principle, be from any set. For example, the random variables in the Bernoulli process take values in the set $\{H, T\}$. An event can also be regarded as a random variable that takes values in the set $\{TRUE, FALSE\}$. We first consider random variables whose values range over a finite or countably infinite set. Such a random variable is called a **discrete random variable**. The most important kind of discrete random variable is one that is supported on the integers or a subset of the integers. Any discrete random variable can be converted into one that uses integers by simply enumerating the possible values. For example, the value "TRUE" is usually assigned the integer value 1, and "FALSE" is usually assigned the integer value 0, and this convention is common in modern programming languages. For this reason, we will focus on integer random variables.

An integer random variable is a random variable supported on the integers, together with the probability p_n that it takes each value n. Implicit in this is that $\sum_n p_n = 1$, which means that the random variable always takes some value or other. Some of the p_n's can be zero, which means that these integers do not occur as values of the random variable. For example, if $p_n = 0$ whenever $n \leq 0$, then the random variable is said to be *positive*, i.e., it takes only positive integer values.

We are now ready for the precise mathematical definition. Do not be surprised if at first this notion does not appear to match what we have just been led to expect. It has taken an enormous amount of time and effort to make this notion rigorous so it will require some effort and many examples to make this concept clear.

An **integer random variable** is a function X defined on a sample space Ω, and supported on the integers. In other words, for every sample point $\omega \in \Omega$, $X(\omega)$ is an integer. The **probability distribution** of X is the sequence of numbers p_n such that p_n is the probability of the event "X equals n." The event "X equals n" is usually written $(X = n)$. As a subset of Ω, this event is $(X = n) = \{\omega \in \Omega \mid X(\omega) = n\}$. We shall generally avoid writing out this set explicitly each time. One should develop an intuitive feeling for the event $(X = n)$.

Of course, we have implicitly assumed that the subsets $(X = n)$ really are events of the sample space Ω. This is a technical point that will never be of direct concern to us. Suffice it to say that a fully rigorous definition of an integer random variable is: a function X on a sample space Ω such that its values are all integers and such that all the subsets $(X = n)$ are events of Ω.

The probability distribution of an integer random variable X always satisfies these two properties:

Probabilities are nonnegative: $p_n \geq 0$ for all n, and

The total probability is 1: $\sum_{n} p_n = 1$

The former property expresses the fact that the p_n's are probabilities, while the latter says that X always takes some value. We use the word "distribution" because the probability is like a mass of weight 1 being broken up and *distributed* among the various integer values. For this reason, the probability distribution of an integer random variable X is also called a **probability mass function**.

The intuitive idea of a random variable relates to the precise definition of a random variable in the following way. Whenever we have some measurement with probabilities, look for a sample space and a function on it. The random variable then *really* comes from observing some phenomenon on this sample space. The fact that we only had a probability distribution at first arose from the fact that we had forgotten about the phenomenon from which the measurement came. Of course, all this means little until we have seen examples.

3.1 The Bernoulli Process: Tossing a Coin

Our first collection of examples will be taken from the process of tossing a coin. For each example, we ask a question. The question is answered by finding the probability distribution of a random variable. We then show how to generate the random variable using R.

Waiting for a toss to be heads.   Recall that the Bernoulli sample space Ω is the set of all infinite sequences of zeros and ones corresponding to the tosses of a biased coin with probability p of coming up heads (or one) and q of coming up tails (or zero). Let W_1 be the waiting time for the first head. In other words, we ask the question: how long do we have to wait to get the first head? The answer is a probability distribution p_n, where p_n is the probability that we must wait until the nth toss to get the first head. In terms of random variables:

$$p_n = P(W_1 = n)$$

How can we compute this? Well, the event $(W_1 = n)$ is the event: "at the nth toss we get a head and the preceding $n - 1$ tosses are all tails." In terms of elementary events:

$$(W_1 = n) = T_1 \cap T_2 \cap \cdots \cap T_{n-1} \cap H_n$$

Therefore, $p_n = P(W_1 = n) = q^{n-1}p$.

Just for once, let us satisfy ourselves that $\sum_n p_n = 1$:

$$\sum_n p_n = \sum_{n=1}^{\infty} q^{n-1}p = p\sum_{n=1}^{\infty} q^{n-1} = p\frac{1}{1-q} = p\frac{1}{p} = 1$$

So it checks. Of course it is not really necessary that we do this. The very definition of a probability distribution requires that the sum of the p_n's is equal to 1.

Notice that in understanding W_1 as a random variable, we worked completely probabilistically. We never spoke of W_1 as a function. What is W_1 as a function? For each $\omega \in \Omega$, $W_1(\omega)$ is the first position of the sequence ω such that at that position ω has an H. For example, $W_1(TTTHHTHH...)$ is 4. However, looking at W_1 as a function is quite unnatural. One should try to think of W_1 purely probabilistically. Indeed, one might say that probability theory gives one a whole new way of looking at sets and functions.

We already know how to a generate Bernoulli sample point using R. Let us compute the random variable W_1. The idea is to use the indexing capability of R. Here is a program that generates a Bernoulli sample point where the bias is $p = 1/10$, and then computes W_1:

```
> support <- 1:1000
> sample <- rbinom(1000, 1, 0.1)
> support[sample==1][1]
[1] 17
```

The expression `sample==1` produces a vector of `TRUE`'s and `FALSE`'s corresponding to the 1's and 0's of the sample (i.e., the H's and T's of the sample point). Then the indexing operation `support[sample==1]` selects the places where the H's occurred. Finally, we select the first one to get W_1, the time when the first head occurred. Next write this as a function so that we can run it more than once:

```
> w1 <- function() (1:1000)[rbinom(1000, 1, 0.1)==1][1]
> w1()
[1] 5
> w1()
[1] 1
> w1()
[1] 12
```

We can then run this program repeatedly by using the `replicate` function:

```
> replicate(10,w1())
 [1]  6  1  2  2  2  4  6 15  3 18
```

Finally, we tabulate the values up to some maximum value, and graph them:

```
> values <- tabulate(replicate(1000,w1()), 40)
> plot(1:40, values)
```

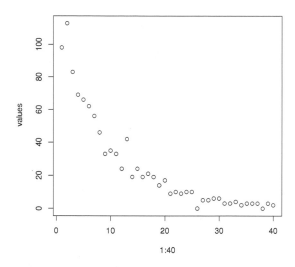

The distribution for W_1 is an important one. It is called the **geometric distribution**. One can generate a random number according to this distribution directly by using the `rgeom` function.

```
> plot(1:40, tabulate(rgeom(1000,0.1),40))
```

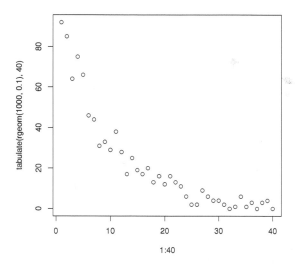

R supports a great variety of probability distributions. For each of these distributions, R has four functions. We have already seen the first kind of function: the ones for generating random numbers according to the distribution. The other three are concerned with the probability distribution function.

Suppose that X is a random variable whose distribution is named `xyz`. The R functions for this distribution are:

`dxyz(n, ...)` computes the probability $P(X = n)$. The other parameters define the specific distribution. These parameters are the same for all four functions of one distribution. For example, the geometric distribution has one parameter: the bias p. The reason for using the letter `d` for this function will be clarified when we consider general probability distributions in Chapter 4.

`pxyz(m, ...)` is the cumulative probability $\sum_{n \le m} P(X = n)$. Again, the choice of the letter `p` to designate this function will be clarified when we consider general probability distributions in Chapter 4.

`qxyz(x, ...)` This is the inverse of the `pxyz` function. The letter `q` stands for **quantile**. It gives the the smallest value of m for which $\sum_{n \le m} P(X = n)$ is at least equal to `x`.

`rxyz(n, ...)` This is the random number generator for the distribution. The first parameter is the number of random values desired.

We can use `dgeom` to graph the geometric distribution for various choices of the bias. To draw all of these graphs on top of each other so we can compare them, use the `points` function. It has the same parameters as the `plot` function:

```
> support <- 1:40
> plot(support, dgeom(support, 0.5), type="l")
> points(support, dgeom(support, 0.3), type="l")
> points(support, dgeom(support, 0.1), type="l")
```

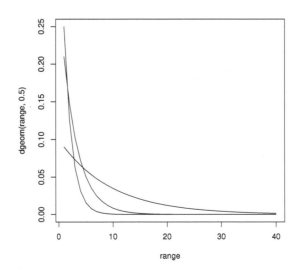

Note that when you type in the lines above, R will proceed to show the first graph immediately without waiting for the other ones. One can either type the commands in a separate window and then cut and paste them into the R window, or else one can put all of the commands within a pair of braces like this:

```
> {
+ support <- 1:40
+ plot(support, dgeom(support, 0.5), type="l")
+ points(support, dgeom(support, 0.3), type="l")
+ points(support, dgeom(support, 0.1), type="l")
+ }
```

The braces tell R that all of the commands are to be grouped together and treated as if they were a single command. Braces are also useful for defining complicated functions that require more than one command. We will see examples later.

Waiting for several tosses to be heads. Consider another example. Let W_k be the waiting time for the kth head. The event $(W_k = n)$ is the event: "a head occurs at the nth toss and exactly $k-1$ heads occur during the preceding $n-1$ tosses." The probability distribution is:

$$p_n = P(W_k = n) = \binom{n-1}{k-1} p^{k-1} q^{n-k} p = \binom{n-1}{k-1} p^k q^{n-k}$$

How does one see this? Well, the $k-1$ heads can occur in any $(k-1)$-subset of the first $n-1$ tosses. There are $\binom{n-1}{k-1}$ such subsets. For each such subset, the probability of getting heads in those positions and tails in the others is $p^{k-1} q^{n-k}$. Finally, the probability of getting a head on the nth toss is p.

Needless to say, it is not very easy to write an explicit expression for the events $(W_k = n)$ in terms of elementary events although that is implicit in our computation above. Notice too that $\sum_{n=k}^{\infty} \binom{n-1}{k-1} p^k q^{n-k} = 1$, a fact that is not very easy to prove directly.

The program we already wrote for finding W_1 actually found all of the W_k's:

```
> support <- 1:40
> sample <- rbinom(length(support), 1, 0.1)
> support[sample==1]
[1] 17 23 24 38
```

Note the use of the `length` function which computes the number of elements in a vector. As before, one can select a particular W_k, by using the indexing function. For example, here is the computation of W_3:

```
> support[sample==1][3]
[1] 24
```

The only problem with this program is that we are necessarily limited to a finite sequence of H's and T's, so it is possible that we did not compute the sequence far enough for a particular W_k to be known. If this is the case, the indexing operation will give this answer:

```
> support[sample==1][5]
[1] NA
```

One can interpret this answer as being "unknown."

The distribution of W_k is another important probability distribution, called the *negative binomial* distribution or the **Pascal distribution**. The function **rnbinom** can be used to generate numbers according to this distribution, except that the **rnbinom** function only counts the number of T's that occurred before k H's occurred. In other words, it does not include the H's. However, this is easily fixed by just adding k to each value produced by **rnbinom**. Here is the graph of W_3, similar to the earlier one for W_1:

```
> plot(1:100, tabulate(rnbinom(10000,3,0.1)+3,100))
```

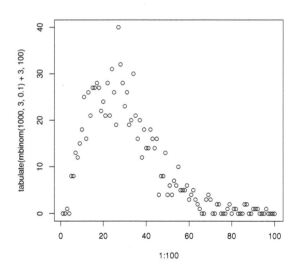

The negative binomial distribution has two parameters, the number of heads, k, and the bias p. We first graph the distribution for $k = 3$, but with several choices of the bias:

```
> support <- 0:40
> plot(support+3, dnbinom(support,3,0.5), type="l")
> points(support+3, dnbinom(support,3,0.3), type="l")
> points(support+3, dnbinom(support,3,0.1), type="l")
```

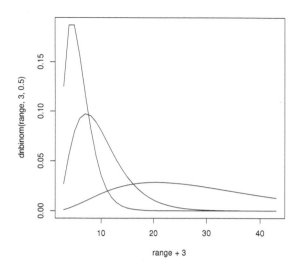

Next, we graph the distribution for a fixed bias $p = 1/2$, but with several choices of k:

```
> support <- 0:40
> plot(support+3, dnbinom(support,3,0.5), type="l")
> points(support+3, dnbinom(support,3,0.3), type="l")
> points(support+3, dnbinom(support,3,0.1), type="l")
```

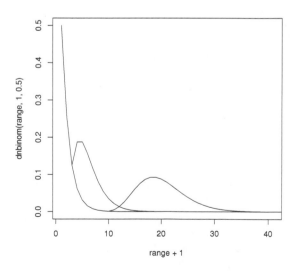

One toss of the coin. Consider the random variable

$$X_n = \begin{cases} 1 & \text{if the } n^{th} \text{ toss is H} \\ 0 & \text{if the } n^{th} \text{ toss is T} \end{cases}$$

or more succinctly, X_n is the nth toss expressed numerically. The distribution of X_n is

$$p_0 = P(X_n = 0) = q$$
$$p_1 = P(X_n = 1) = p$$

and all other p_n are zero.

The number of heads. Next, let S_n be the number of heads in the first n tosses. The distribution of S_n is:

$$p_k = P(S_n = k) = \binom{n}{k} p^k q^{n-k}$$

because the event $(S_n = k)$ means that k heads and $n - k$ tails occur in the first n tosses. There are $\binom{n}{k}$ ways that the k heads can appear and each pattern has the probability $p^k q^{n-k}$ of occurring. The fact that $\sum_k p_k = 1$ is just the binomial theorem:

$$\sum_{k=0}^{n} \binom{n}{k} p^k q^{n-k} = (p+q)^n = 1^n = 1$$

Indeed, this is a probabilistic *proof* of the binomial theorem. For this reason, the distribution of S_n is called the *binomial distribution*. It is one of the most frequently used distributions in probability theory.

Incidentally, the event $(S_n = k)$ is not the same as the event $(W_k = n)$. The distinction is that $(W_k = n)$ requires that there be k heads in the first n tosses *and* that the kth head occur at the nth toss. $(S_n = k)$ is only the event that k heads occur in the first n tosses. Since both names for the distributions and their formulas are very similar, one must be careful to distinguish them.

The binomial and negative binomial distributions are collections of distributions. One sometimes uses the term "family" when talking about such collections because the distributions in one family share common characteristics. The distributions within a family are distinguished from one another by "parameters." For example, the binomial family has two parameters: n and the bias p. The R functions for randomly generating numbers require that these parameters be specified so that R can determine the specific distribution that is required. Because of the dependence on parameters, a family of distributions is also called a *random* or *stochastic function*. One can think of a random function as a kind of machine or algorithm that takes the parameters as inputs and produces a probability distribution as its output. Here is a pictorial representation:

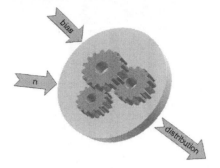

Be careful not to think of a random function as a "randomly chosen function" any more than a random variable is a "randomly chosen variable."

Another way to represent S_n is:

$$S_n = X_1 + X_2 + \cdots + X_n$$

This illustrates the fact that we may combine random variables using algebraic operations. After all, random variables are functions on Ω and as such may be added, subtracted, etc. Thus if X and Y are integer random variables on Ω, then, as a function, the random variable $X + Y$ takes the value $X(\omega) + Y(\omega)$ on the sample point $\omega \in \Omega$. For example $(W_k = n)$ is the event

$$(X_1 + X_2 + \cdots + X_{n-1} = k - 1) \cap (X_n = 1) = (S_{n-1} = k - 1) \cap (X_n = 1)$$

Unfortunately, the use of large quantities of symbolism tends to obscure the underlying simplicity of the question we asked. We shall try to avoid doing this if possible.

Once again, we consider how to compute this new random variable when we have a Bernoulli sample point. The relevant R function is `cumsum` which computes all sums of a vector. If the vector x has elements $x_1, x_2, \ldots x_n$, then `cumsum(x)` has elements x_1, $x_1 + x_2$, $x_1 + x_2 + x_3, \ldots, x_1 + x_2 + \ldots + x_n$. In other words, `cumsum` computes exactly the sequence

$$S_1, S_2, S_3, \ldots, S_n$$

However, generating the random variables S_n in R is especially easy because the `rbinom` function generates this random variable directly. The second parameter of this function is n, the number of X_i's that are added to obtain S_n. The random variable S_n can take values from 0 to n. So we can graph the entire probability distribution for S_n, unlike the case of the distributions W_k, which theoretically can take arbitrarily large values. Here is the graph of S_6 when the bias is $p = 1/2$:

```
> plot(0:6, dbinom(0:6, 6, 1/2))
```

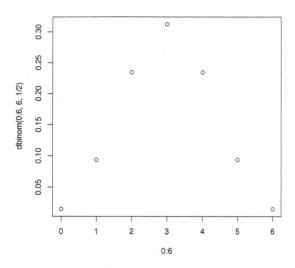

To see how the graph varies with respect to the parameters, one can "animate" the evolution by drawing a series of graphs. This requires that we create a series of graphs in a quick succession. We do this by defining a function and applying it to a sequence of parameters. There are several R functions that are available for applying a function multiple times. The most commonly used is `sapply`. This function has two parameters. The first is a vector of parameters, and the second is an R function that has exactly one parameter. The `sapply` function applies every element of the first parameter to the second parameter, and returns a vector with the values computed by the function. Since we sometimes do not care about this vector, it is convenient to assign it so it does not get written to the screen. To emphasize that we are ignoring the vector, we assign it the name `ignore`.

The function we want to apply in this case is the `plot` function. However, because the graphs may be drawn so quickly that one cannot see them, it may be necessary on your computer to add a pause between the graphs. This is done in R with the `Sys.sleep` function. It causes your program to stop running (i.e., pause or sleep) for a specified number of seconds. Our first animation shows how the distribution varies with the bias when n is fixed at 10:

```
> binom <- function(p) {
+   plot(0:10, dbinom(0:10, 10, p))
+   Sys.sleep(0.1)
+ }
> ignore <- sapply((0:100)/100, binom)
```

This animation has the disadvantage that the scale of the y-axis keeps changing. To fix the scale use the `ylim` parameter:

```
> binom <- function(p) {
```

```
+   plot(0:10, dbinom(0:10, 10, p), ylim=c(0,0.5))
+   Sys.sleep(0.1)
+ }
> ignore <- sapply((0:100)/100), binom)
```

Next, we look at what happens when n varies but the bias is fixed at $p = 1/2$:

```
> binom <- function(n) {
+   plot(0:n, dbinom(0:n, n, 1/2))
+   Sys.sleep(0.1)
+ }
> ignore <- sapply(1:100, binom)
```

The result is quite remarkable. The graph very quickly converges to a shape that changes very little except that it gradually becomes narrower. This is the graphical manifestation of one of the most important theorems in probability theory: the Central Limit Theorem. We will state this theorem in Chapter 5, and we will prove it in Chapter 8. To see the convergence even more dramatically, we need to control the scales of the graph as it evolves. The x-axis scale is specified by the xlim parameter that uses a 2-element vector to define where the x-axis of the graph should begin and end. The reason for the specific scaling we use here will be explained in Section 5.3:

```
> binom <- function(n) {
+    plot(0:n, dbinom(0:n, n, 1/2),
+       xlim=c((n/2)-2*sqrt(n),(n/2)+2*sqrt(n)))
+    Sys.sleep(0.05)
+  }
> ignore <- sapply(1:200, binom)
```

To see what happens when we use a different bias, here is the case $p = 1/4$:

```
> binom <- function(n) {
+    plot(0:n, dbinom(0:n, n, 1/4),
+       xlim=c((n/4)-2*sqrt(n),(n/4)+2*sqrt(n)))
+    Sys.sleep(0.05)
+  }
> ignore <- sapply(1:200, binom)
```

Gaps between heads. Now consider the random variable T_k, the length of the gap between the $(k-1)^{st}$ and k^{th} heads in the sequence of tosses. For example, here are the first few T_k's for one of the sample points:

$$\omega = \underbrace{TTTH}_{T_1=W_1}\underbrace{TTTTH}_{T_2}\underbrace{H}_{T_3}\underbrace{TTH}_{T_4}\cdots$$

The T_k's and W_k's are related to each other:

$$T_k = W_k - W_{k-1}$$
$$W_k = T_1 + T_2 + \cdots T_k$$

What is the distribution of T_k? When we later have the notion of conditional probability, we will have a very natural way to compute this. However, we can nevertheless easily compute the distribution of T_k because of the independence of the various tosses of the coin. In other words, when computing $P(T_k = n)$ we may imagine that we *start* just after the $(k-1)^{st}$ head has been obtained. Therefore, the distribution of T_k is

$$p_n = P(T_k = n) = P(W_1 = n) = q^{n-1}p$$

exactly the same distribution as that of W_1.

Notice that T_k for $k>1$ is *not* the same random variable as W_1, and yet their distributions are the same. How can this be? Actually we have already seen this phenomenon before but did not notice it because it was too trivial an example: the Bernoulli random variables $X_1, X_2, \ldots$ are different random variables, but they all have the same distributions. This phenomenon will occur frequently and is very important.

Definition. Two integer random variables X and Y are said to be **equidistributed** or **stochastically identical** when

$$P(X = n) = P(Y = n) \quad \text{for all integers } n.$$

Thus for example, W_1 and T_k are equidistributed random variables. Similarly, the X_n's are equidistributed random variables. Although X_1 and X_2 measure completely different phenomena, they have exactly the same probabilistic structure.

At first, it may seem like a challenge to compute the gaps T_k for a Bernoulli sample point using R, since it involves computing quantities *within* a vector rather than *between* two vectors. We have already encountered a computation of this kind when we needed to compute S_n. For this we can use the cumsum function. To compute the gaps, we use a "trick" to convert an operation within a vector to one between two vectors. Starting with the vector w of waiting times, we construct two new vectors: wl and wr. The names stand for "left" and "right." These vectors are constructed by using the vector builder function like this:

```
> wl <- c(0, w)
> wr <- c(w, 0)
```

The c function takes any number of vectors or numbers and converts them into a single vector. So if the elements of w are

$$w_1, w_2, \ldots, w_n$$

then the elements of wl will be

$$0, w_1, w_2, \ldots, w_n$$

and the elements of wr will be

$$w_1, w_2, \ldots, w_n, 0$$

We now compute the difference wr-wl. The elements will then be:

$$w_1, w_2 - w_1, w_3 - w_2, \ldots, w_n - w_{n-1}, -w_n$$

Except for the last element, these are exactly the gaps we wanted to compute.

3.2 The Bernoulli Process: Random Walk

Consider the random variables X'_n given by:

$$X'_n = \begin{cases} 1 & \text{if the } n^{th} \text{ toss is H} \\ -1 & \text{if the } n^{th} \text{ toss is T} \end{cases}$$

X_n and X'_n are closely related: $X'_n = 2X_n - 1$. However, if we form the random variable analogous to S_n, we measure a quite different phenomenon. If $S'_n = X'_1 + X'_2 + \cdots + X'_n$, then S'_n is the position of a *random walk* after n steps: a step to the right is $+1$, a step to the left is -1, so the sum of the first n steps is the position at that time.

What is the probability distribution of S'_n? This calculation is a good example of a "perturbation" (or change of variables) applied to a model. We want to compute $P(S'_n = x)$. Here we use x for an integer; think of it as a point on the x-axis. Let h be the number of heads and t the number of tails, both during the first n tosses. Then:

$$x = h - t \text{ and } n = h + t$$

Solving for h and t gives:

$$h = \frac{n+x}{2} \text{ and } t = \frac{n-x}{2}$$

Therefore:

$$P(S'_n = x) = P\left(S'_n = \frac{1}{2}(n+x)\right) = \binom{n}{\frac{1}{2}(n+x)} p^{\frac{1}{2}(n+x)} q^{\frac{1}{2}(n-x)}$$

provided that $\frac{1}{2}(n+x)$ is a nonnegative integer.

To animate a random walk, we first convert from the X_n's to the X'_n's. This is easily done by multiplying by 2 and subtracting 1. We then compute the S'_n's by using cumsum:

```
> n <- 100
> support <- 1:n
> walk <- cumsum(2*rbinom(n,1,0.5)-1)
>
```

Next, we define a function that draws the graph up to time x. For this we use the indexing function to perform the selection on the random walk, and then plot the graph with a suitable pause:

```
> graph <- function(k) {
+    plot(1:k, walk[support<=k], xlim=c(1,n),
+         ylim=c(-n/10,n/10), type="l")
+    Sys.sleep(0.1)
+ }
```

Finally, we perform the animation:

```
> ignore <- sapply(support, graph)
```
Try running it a few times to acquire some intuition about what a random walk "looks like."

3.3 Independence and Joint Distributions

Recall that two events A and B are independent when $P(A \cap B) = P(A) P(B)$. This definition is abstracted from experience; as for example when tossing a coin, the second time the coin is tossed, it does not remember what happened the first time. We extend this notion to random variables. Intuitively, two random variables X and Y are independent if the measurement of one does not influence the measurement of the other. In other words, the events expressible in terms of X are independent of those expressible in terms of Y. We now make this precise.

Definition. Two integer random variables X and Y are **independent** when

$$P((X = m) \cap (Y = n)) = P(X = m) P(Y = n)$$

for every pair of integers m, n.

We illustrate this with our standard example: the Bernoulli process. The random variables X_j and X_k are independent when $j \neq k$. This is obvious from the definition of the Bernoulli process. Less obvious is that T_j and T_k are independent when $j \neq k$. We check this for T_1 and T_2. By previous computations,

$$P(T_1 = m) = q^{m-1}p \qquad \text{and} \qquad P(T_2 = n) = q^{n-1}p$$

for every pair of positive integers m, n. Now compute $P((T_1 = m) \cap (T_2 = n))$. The event $(T_1 = m) \cap (T_2 = n)$ means that the first $m+n$ tosses have precisely the pattern:

$$\underbrace{TT...TH}_{m}\underbrace{TT...TH}_{n}$$

Therefore,

$$(T_1 = m) \cap (T_2 = n) = q^{m-1}pq^{n-1}p = P(T_1 = m) P(T_2 = n)$$

and we conclude that T_1 and T_2 are independent.

On the other hand, W_j and W_k are not independent random variables. This is intuitively obvious, but we will check it nevertheless in the case of W_1 and W_2. We previously computed:

$$P(W_1 = m) \quad = \quad (n-1)q^{m-1}p$$

$$P(W_2 = n) \quad = \quad (n-1)q^{n-2}p^2$$

Now $(W_1 = m) \cap (W_2 = n)$ is the same as the event $(T_1 = m) \cap (T_2 = n - m)$ both being the event that the first n tosses have this pattern:

$$\overbrace{\underbrace{TT...TH}_{m}\underbrace{TT...TH}_{n-m}}^{n}$$

Therefore,

$$P(W_1 = m) \cap (W_2 = n) = \begin{cases} q^{n-2}p^2 & \text{if } m < n \\ 0 & \text{if } m \geq n \end{cases}$$

which is very different from $P(W_1 = m)\, P(W_2 = n)$. Indeed, when $m \geq n \geq 2$, one probability is zero and the other is not. So W_1 and W_2 are *dependent random variables*. In other words, W_1 *influences* W_2.

When two random variables are not independent, is there a way to measure the dependence of one of them on the other? In more common parlance, how do we measure the "correlation" of two random variables? We measure this with the joint distribution of two random variables.

Definition. For two integer random variables X and Y, the **joint distribution** of X and Y is

$$c_{m,n} = P((X = m) \cap (Y = n))$$

The numbers $c_{m,n}$ cannot be computed in general from the individual distributions of X and Y. The joint distribution measures the total dependence of X and Y or, equivalently, the cause and effect of one random variable on the other. Joint distributions have the following properties:

1. $\displaystyle\sum_m \sum_n c_{m,n} = 1$, i.e., something must happen

2. $\displaystyle\sum_n c_{m,n} = P(X = m)$

3. $\displaystyle\sum_m c_{m,n} = P(Y = n)$

The distributions of X and Y considered *relative* to their joint distribution are called the **marginal distributions** or simply the **marginals**. Despite the fancy terminology, the marginals are simply the distributions of X and Y with which we are already familiar.

Just as we can have the joint distribution of two random variables, we can have the joint distribution of any finite collection of random variables. The formulas are so obvious that we will not bother to write them down explicitly.

We now compute some examples. If X and Y are independent random variables with distributions $p_m = P(X = m)$ and $r_n = P(Y = n)$, then their joint distribution is

$$c_{m,n} = P((X = m) \cap (Y = n)) = P(X = m)\, P(Y = n) = p_m r_n$$

In other words, the joint distribution of independent random variables is the product of the marginals.

Next, consider the random variables W_j and W_k for which $j < k$. Their joint distribution is

$$c_{m,n} = P((W_j = m) \cap (W_k = n))$$

For this to be nonzero we must have $m < n$. The event $(W_j = m) \cap (W_k = n)$

means that we have $j - 1$ heads in the first $m - 1$ tosses and $k - j - 1$ heads in the "gap" of length $n - m - 1$ between the jth and k^{th} heads.

$$\overbrace{\underbrace{THT...TT}_{m}\underbrace{TTH...T}_{n-m}}^{n}$$

In the first group of tosses above, there are m tosses ending with the jth head; and in the second group, there are $n - m$ tosses ending with the kth head. Writing all this out gives:

$$
\begin{aligned}
c_{m,n} &= P((W_1 = m) \cap (W_2 = n)) \\
&= \binom{m-1}{j-1} p^j q^{m-j} \binom{n-m-1}{k-j-1} p^{k-j} q^{n-m-(k-j)} \\
&= \binom{m-1}{j-1} \binom{n-m-1}{k-j-1} p^k q^{n-k}
\end{aligned}
$$

The total number of tosses involved is n: exactly k of them are heads and $n - k$ are tails. This furnishes a quick check that the exponents on p and on q are correct.

As a final example, we compute the joint distribution of the first k waiting times. For $n_1 < n_2 < \cdots < n_k$ the joint distribution is:

$$c_{n_1, n_2, \cdots, n_k} = P((W_1 = n_1) \cap (W_2 = n_2) \cap \cdots \cap (W_k = n_k))$$

This is actually quite easy to compute because there is only one "way" to get the event $(W_1 = n_1) \cap (W_2 = n_2) \cap \cdots \cap (W_k = n_k)$ up to the n_k^{th} toss. Therefore:

$$c_{n_1, n_2, \cdots, n_k} = pq^{n_1 - 1} pq^{n_1 - n_2 - 1} \ldots = p^k q^{n_k - k}$$

3.4 Expectations

A probability distribution can be regarded as being a kind of mass distribution. Think of a long, thin rod with weights evenly spaced along the rod, corresponding to the probabilities p_n of the distribution. When we later consider continuous random variables, the mass distribution will be continuously spread along the rod. If we now imagine that we try to balance the rod, there should be a position along the rod where the weights on either side of the position exactly counterbalance each other. This is the *center of mass*. For many purposes, it is as if all of the mass were located at that position. We can regard it as the probabilistic center of the distribution. There are many names for this position, but the one used most commonly in the probabilistic context is the *expectation*:

Definition. Suppose X is an integer random variable with distribution $p_n = P(X = n)$. The **expectation** (also called the *expected value*, the *mean* or the *average*) of X is

$$E(X) = \sum_n n p_n = \sum_n n P(X = n)$$

It can happen that this sum does not exist. We will not worry about this. Implicit in any statement about an expectation is the assumption that the expectation exists.

For example, if X_n is the nth trial in the Bernoulli process, the expectation of X_n is $E(X_n) = 1 \cdot p + 0 \cdot q = p$. The expected or average value of the nth toss is p. Needless to say, X_n will not ever take the value p (except in the trivial cases when p is 0 or 1). Intuitively, however, if we perform a large number of trials and then average the results, we will get roughly p.

By far the most important fact about the expectation is additivity. Throughout our investigation of probability theory, we will find this rule to be one of the most useful for analyzing stochastic phenomena and for answering probabilistic questions.

Addition Rule for Expectations. For any two integer random variables,

$$E(X + Y) = E(X) + E(Y)$$

The surprising thing about this fact is that it holds regardless of whether X and Y are independent or not.

Proof. Let $c_{m,n} = P((X = m) \cap (Y = n))$ be the joint distribution of X and Y. Now $X + Y$ is a new random variable. What is its distribution? Well, we must consider all possible ways that $X + Y$ can take on a given value:

$$
\begin{aligned}
P(X + Y = k) &= \sum_{\{m,n \mid m+n=k\}} P((X = m) \cap (Y = n)) \\
&= \sum_{m+n=k} c_{m,n}
\end{aligned}
$$

Therefore, the expectation of $X + Y$ is:

$$
\begin{aligned}
E(X + Y) &= \sum_k k P(X + Y = k) \\
&= \sum_k k \sum_{m+n=k} c_{m,n} \\
&= \sum_m \sum_n (m + n) c_{m,n} \\
&= \sum_m \sum_n m c_{m,n} + \sum_m \sum_n n c_{m,n} \\
&= \sum_m m P(X = m) + \sum_n n P(Y = n) \\
&= E(X) + E(Y)
\end{aligned}
$$

This completes the proof.

Let us compute some expectations for the Bernoulli process. First we compute the "hard way," directly from the definition, then we compute using the Addition Rule.

Consider S_n, the number of heads in the first n tosses. The distribution for S_n is $p_k = \binom{n}{k} p^k q^{n-k}$. So

$$E(S_n) = \sum_k k p_k = \sum_{k=0}^{\infty} k \binom{n}{k} p^k q^{n-k}$$

Unfortunately, we cannot simplify this very easily. On the other hand, $S_n = X_1 + X_2 + \cdots + X_n$. Hence,

$$E(S_n) = E(X_1) + E(X_2) + \cdots + E(X_n) = np$$

since all of these have the same expectation: p. In addition, we have shown that

$$\sum_{k=0}^{\infty} k \binom{n}{k} p^k q^{n-k} = np$$

a fact that is not very easy to prove.

Next, consider the waiting time for the kth head, W_k. The distribution for W_k is $p_n = \binom{n-1}{k-1} p^k q^{n-k}$. Therefore,

$$E(W_k) = \sum_n n p_n = \sum_{n=k}^{\infty} n \binom{n-1}{k-1} p^k q^{n-k}$$

Again, there is no easy way to compute this infinite sum. However, $W_k = T_1 + T_2 + \cdots + T_k$. Hence, $E(W_k) = E(T_1) + E(T_2) + \cdots + E(T_k)$. But all the $T_1, \ldots, T_k$ are equidistributed so, in particular, they all have the same expectation. Therefore, $E(W_k) = kE(T_1)$, and we need only compute one expectation: the expectation of $T_1 = W_1$. We shall have to resort to some trickery, but it still is not too difficult.

$$E(T_1) = \sum_n n P(T_1 = n) = \sum_n n q^{n-1} p = p \sum_n n q^{n-1}$$

Now $n q^{n-1}$ should be a familiar expression: it is the derivative of q^n with respect to q. Therefore,

$$E(T_1) = p \sum_n \frac{d}{dq}(q^n) = p \frac{d}{dq}\left(\sum_n q^n\right) = p \frac{d}{dq}\left(\frac{q}{1-q}\right) = p \frac{1}{(1-q)^2} = p \frac{1}{p^2} = \frac{1}{p}$$

Intuitively, it is quite reasonable that $E(T_1) = 1/p$, for if p is large, we do not expect to wait very long for a head to occur, while if p is small, we expect to wait a long time. As before, we get the added bonus of an identity:

$$E(W_k) = \sum_{n=k}^{\infty} n \binom{n-1}{k-1} p^k q^{n-k} = k/p$$

a fact that is quite hard to believe otherwise.

3.5 The Inclusion-Exclusion Principle

Imagine that we have a well-shuffled deck of cards and that we turn the cards over one at a time. While doing this, we call out the names of the cards in their *unshuffled* order (as in bridge), beginning with the deuce of clubs and ending with the ace of spades. What is the probability that *none* of the cards turned over match the name called out when it is turned over? The answer (to an accuracy of 10^{-15}) is $\frac{1}{e}$. This is strange for two reasons: it depends on the number e which should not appear in a finite counting problem, and it does not depend on the number of cards in the deck.

We shall prove this result and several others by a useful formula called the *inclusion-exclusion principle*. The proof of this principle will follow easily from the formalism of random variables. The abstract setting for the principle is the computation of the probability of the union of events in terms of the probabilities of the events and of their intersections. For example, if we have two events A and B, then we know that

$$P(A \cup B) = P(A) + P(B) - P(A \cap B)$$

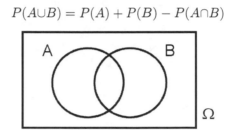

If we refer to the diagram, it is clear what this means: $P(A) + P(B)$ "counts" $P(A \cap B)$ twice. Thus, we "include" $P(A)$ and $P(B)$ and then "exclude" $P(A \cap B)$.

For three events A, B and C we must include, exclude and then include once again:

$$\begin{aligned} P(A \cup B \cup C) \;=\; & P(A) + P(B) + P(C) \\ & - P(A \cap B) - P(A \cap C) - P(B \cap C) \\ & + P(A \cap B \cap C) \end{aligned}$$

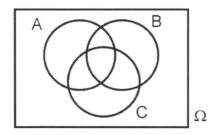

It is quite easy to think through the proof of this directly. However, for the general case it will take a bit more work. Here is the general formula:

Inclusion-Exclusion Principle.

$$P(A_1 \cup A_2 \cup \cdots \cup A_n) = \sum_i P(A_i) - \sum_{i<j} P(A_i \cap A_j) + \sum_{i<j<k} P(A_i \cap A_j \cap A_k)$$
$$- \cdots + (-1)^{n+1} P(A_1 \cap A_2 \cap \cdots \cap A_n)$$

Note that the second sum is really a double sum over both i and j but subject to $i<j$, the third is a triple sum and so on.

Proof. To prove this principle, we introduce a special kind of integer random variable called an *indicator*. If A is an event, the *indicator* of A is the integer random variable I_A corresponding to the question "Did A happen?" More precisely, for any sample point $\omega \in \Omega$,

$$I_A(\omega) = \begin{cases} 1 & \text{if } \omega \in A \\ 0 & \text{if } \omega \notin A \end{cases}$$

One sometimes also sees the notation χ_A for the indicator. We have already encountered such a random variable before. In the Bernoulli process, H_n is the event "the nth toss is heads" and its indicator I_{H_n} is the random variable X_n.

The probability distribution of the indicator I_A is

$$p_n = \begin{cases} P(\overline{A}) & \text{if } n = 0 \\ P(A) & \text{if } n = 1 \\ 0 & \text{otherwise} \end{cases}$$

Therefore, the expectation of I_A is $E(I_A) = 0 \cdot P(\overline{A}) + 1 \cdot P(A) = P(A)$. As a result of this, we see that all probabilities may be reduced to the computation of expectations; one could dispense with sample spaces and events altogether and develop probability theory using only random variables and expectations.

We now consider what happens when we add and multiply indicators. The easy operation is multiplication: $I_A I_B = I_{A \cap B}$ should be obvious. Addition, however, is not so easy because the sum of indicators need not be an indicator: $I_A + I_B$ takes value 2 on $A \cap B$. However, if we put in a correction term, we get an identity: $I_A + I_B = I_{A \cup B} + I_{A \cap B}$. So while multiplication corresponds to intersection, addition does not quite correspond to union.

The last operation we consider is the complement. Here the result is clear: $I_{\overline{A}} = 1 - I_A$. This suggests what we should do in general to compute $I_{A_1 \cup A_2 \cup \cdots \cup A_n}$ in terms of the A_i's: convert to an intersection by using the de Morgan law. Thus:

$$I_{A_1 \cup A_2 \cup \cdots \cup A_n} = 1 - I_{\overline{A_1} \cap \overline{A_2} \cap \cdots \cap \overline{A_n}}$$
$$= 1 - (1 - I_{A_1})(1 - I_{A_2}) \cdots (1 - I_{A_n})$$

We now multiply out the last expression as in high school algebra:

$$
\begin{aligned}
&= 1 - \left[1 - \sum_i I_{A_i} + \sum_{i<j} I_{A_i} I_{A_j} - \cdots \right] \\
&= \sum_i I_{A_i} - \sum_{i<j} I_{A_i} I_{A_j} + \cdots + (-1)^{n+1} I_{A_1} I_{A_2} \cdots I_{A_n} \\
&= \sum_i I_{A_i} - \sum_{i<j} I_{A_i \cap A_j} + \cdots + (-1)^{n+1} I_{A_1 \cap A_2 \cap \cdots \cap A_n}
\end{aligned}
$$

Finally, we take the expectation of this expression using the Addition Rule for Expectations. The result is the inclusion-exclusion principle.

We now return to our first question. Think of the situation as follows. Start with a new unshuffled deck and then shuffle it. The result is a random permutation of the unshuffled deck. What is the probability that no card is in the same position in both the unshuffled and the shuffled decks? To be more precise, consider the integers $1, \ldots, n$ instead of the 52 cards. The sample space is the set Ω of all permutations of $1, \ldots, n$. Thus $|\Omega| = n!$ The notation for permutations is

$$
\begin{array}{ccccc}
1 & 2 & 3 & \ldots & n \\
i_1 & i_2 & i_3 & \ldots & i_n
\end{array}
$$

where one should think of the top row as the unshuffled integers and the lower row as the shuffled ones. A *fixed point* of a permutation is a number j for which $i_j = j$, i.e., the same number j appears twice in one column in our notation. For example, let $n = 3$. There are 6 permutations with number of fixed points as follows:

permutation	number of fixed points
1 2 3 1 2 3	3
1 2 3 2 1 3	1
1 2 3 2 3 1	0
1 2 3 3 2 1	1
1 2 3 3 1 2	0
1 2 3 1 3 2	1

A permutation with no fixed points is called a *derangement*. In the case of the permutations of 3 elements, 2 out of the 6 permutations are derangements, so the probability of a completely random permutation being a derangement is $1/3$ in this case.

Let F be the event "There is at least one fixed point." We want to compute $P(\overline{F})$. Counting F directly is not very easy, but we can write F as the union of events that we can count quite easily. Let A_i be the event "Position i is a fixed point." Then $F = A_1 \cup A_2 \cup \cdots \cup A_n$. Since the A_i are not disjoint, we must apply the inclusion-exclusion principle:

$$P(F) = \sum_i P(A_i) - \sum_{i<j} P(A_i \cap A_j) + \cdots$$

Now an element of A_1 has 1 as a fixed point so it is just a permutation of $\{2, \ldots, n\}$. Therefore, $|A_1| = (n-1)!$ and similarly $|A_i| = (n-1)!$ Any element of $A_1 \cap A_2$ has two fixed points so it is a permutation of $\{3, \ldots, n\}$. So $|A_1 \cap A_2| = (n-2)!$ More generally $|A_1 \cap A_2 \cap \cdots \cap A_k| = (n-k)!$ If we divide by $n!$ we get the probabilities, e.g., $P(A_1 \cap A_2 \cap \cdots \cap A_k) = \frac{(n-k)!}{n!}$. Substituting these into our formula for $P(F)$ gives us:

$$
\begin{aligned}
P(F) &= \binom{n}{1}\frac{(n-1)!}{n!} - \binom{n}{2}\frac{(n-2)!}{n!} + \binom{n}{3}\frac{(n-3)!}{n!} - \cdots \\
&= \frac{n!}{1!(n-1)!}\frac{(n-1)!}{n!} - \frac{n!}{2!(n-2)!}\frac{(n-2)!}{n!} + \cdots \\
&= \frac{1}{1!} - \frac{1}{2!} + \frac{1}{3!} - \cdots + (-1)^{n+1}\frac{1}{n!}
\end{aligned}
$$

From calculus you should immediately recognize this expression as the beginning of the expansion for $1 - e^x$ when $x = -1$. This expansion converges extremely rapidly, and it is essentially $1 - e^{-1}$ when, say, n is larger than 7. We conclude that

$$P(\overline{F}) = \frac{1}{e} \text{ with high accuracy when } n > 7$$

We now consider an application of the inclusion-exclusion principle to surface physics. Suppose we have a molecular beam firing molecules at a target crystal. Assume that a molecule adheres to the crystal if it strikes an unoccupied lattice site and rebounds (and is lost) if it strikes a previously occupied site. If we assume that the molecules are fired at random at the crystal sites, how long must we wait until all the crystal sites are covered? We model this problem using an idealized collection of objects labeled $1, 2, 3, \ldots$ to represent the molecules being fired at the target. The target is represented as n containers into which the molecules are being dropped sequentially. The question is to determine the length of time one needs to wait until every container contains at least one molecule.

The answer to our question will of course be a probability distribution. Let W be the waiting time until all the containers are occupied. We want to compute $P(W \leq k)$. This is the probability that if we place k molecules into n containers, then all the containers are occupied. Let A_i be the event "the ith container is empty." Then $(W \leq k) = \overline{A_1} \cap \overline{A_2} \cap \cdots \cap \overline{A_n}$. By the inclusion-exclusion principle,

$$P(W \leq k) = P\left(\overline{A_1} \cap \overline{A_2} \cap \cdots \cap \overline{A_n}\right)$$

$$= 1 - P(A_1 \cup A_2 \cup \cdots \cup A_n)$$

$$= 1 - \sum_i P(A_i) + \sum_{i<j} P(A_i \cap A_j) - \cdots$$

Now the sample space Ω consists of all placements of k molecules into n containers. Thus $|\Omega| = n^k$. The event A_1 consists of all placements of k molecules into the last $n-1$ containers. Thus $|A_1| = (n-1)^k$. Similarly, $|A_1 \cap A_2| = (n-2)^k$ and so on. So the probability of A_i is $\frac{(n-1)^k}{n^k} = (1 - \frac{1}{n})^k$, that of $A_i \cap A_j$ is $\frac{(n-2)^k}{n^k} = (1 - \frac{2}{n})^k$, and so on. Therefore:

$$P(W \leq k) = 1 - \binom{n}{1}\left(1 - \frac{1}{n}\right)^k + \binom{n}{2}\left(1 - \frac{2}{n}\right)^k - \cdots$$

3.6 Exercises

Exercise 3.1 The thirteen diamonds are taken from a deck of cards and are thoroughly shuffled. One diamond is drawn at random and scored as follows: two through ten score as their rank, face cards score ten, and the ace scores eleven. Let S be the score. Describe the sample space and probability measure used in this problem. Write out S explicitly as a function on the sample space. Write out the probability measure P explicitly as a function. Do S and P have the same domain?

Exercise 3.2 In San Francisco, a drunk leaves a bar and every ten seconds staggers either one yard down the street with probability 3/4 or one yard up the street with probability 1/4. Where is the drunk after one minute? After two minutes? What is the most likely location in each case? How is the most likely location varying in time?

Exercise 3.3 A machine that produces screws is subject to occasional surges in its power supply. These occur independently during each second of time with 90% probability, and the machine produces one screw every second. In one version of the machine, there is a fuse that shuts off the machine permanently when a power surge occurs. We wish to know how many screws the machine produces after it is turned on. Which random variable in the Bernoulli process corresponds to this question? Answer the question.

Exercise 3.4 Another version of the machine in Exercise 3.3 has a temporary circuit breaker so that during a power surge the functioning of the machine is interrupted only for one second. We run the machine for one minute and wish to know how many screws are produced. Which random variable in the Bernoulli process corresponds to this question? Answer the question.

Exercise 3.5 In a bridge game, the deck is thoroughly shuffled and dealt.

You are dealt a hand containing four spades. How many spades was your partner dealt?

Exercise 3.6 Three office workers take a coffee break. They choose one of their number at random to pay for the coffee as follows. All three flip a coin simultaneously and the one having a different outcome pays for the coffee. If all coins come up the same, they flip the coins again. How long does it take to determine who pays for the coffee?

Exercise 3.7 If one has a coin with a bias $p \neq 1/2$, one can nevertheless use it to synthesize a fair coin by the following trick. Flip the coin twice. If the two tosses come out different, we can say that we got heads if the first toss was heads and tails otherwise. If the two tosses are the same, we toss the coin two more times and proceed as above. Show that this produces a fair coin toss. How many tosses of the biased coin are required to produce one "fair toss?"

Exercise 3.8 Given any bias p between 0 and 1 and a fair coin, one can synthesize a biased coin toss with bias p as follows. Write the binary expansion of $q = 1 - p$. This is just a sequence of zeroes and ones after the binary point (i.e., the binary equivalent of the decimal point). Now start tossing the fair coin. When we get a head write down a 1 and for a tail write a 0. Compare the sequence we obtain with the binary expansion of q. Continue tossing until the first time that the two sequences differ. At this point we stop and record what happened on the last toss. Show that what we record is equivalent to a biased coin toss with bias p. How long does it take to complete such a toss? Does it depend on p?

Exercise 3.9 In Exercise 3.2, is the position of the drunk after one minute independent of his position after two minutes?

Exercise 3.10 Prove that if X and Y are independent random variables and if $f(x)$ and $g(y)$ are two functions, then $f(X)$ and $g(Y)$ are also independent random variables.

Exercise 3.11 What is the distribution of S in Exercise 3.1? What is its average value?

Exercise 3.12 Compute the average position of the drunk in Exercise 3.2 after one minute and after two minutes. How is the drunk's average position changing in time? How do these questions differ from the questions asked in Exercise 3.2?

Exercise 3.13 In the game of Chuck-A-Luck, three dice are agitated in a cage shaped like a hourglass. A player may wager upon any of the outcomes 1 through 6. If precisely one die exhibits that value, the player wins at even odds; if two dice show that value, the player wins at 2 to 1 odds; if all three dice show the player's choice, the payoff is 3 to 1. If none of the three dice show the player's choice, the player loses. Compute the expected value of the player's winnings for a bet of one dollar on "2." Is the game fair? If not, suggest payoff odds that would make the game fair.

Exercise 3.14 What is the average number of dots shown by a die tossed once at random? You wish to maximize the value shown by the die. If you are allowed to throw the die a second time, when should you do so? What is the expected value shown by a die when one rethrowing is allowed?

Exercise 3.15 James Bond is imprisoned in a cell from which there are three obvious ways to escape: an air-conditioning duct, a sewer pipe and the door (the lock of which does not work). The air-conditioning duct leads agent 007 on a two-hour trip, whereupon he falls through a trap door onto his head, much to the amusement of his captors. The sewer pipe is similar but takes five hours to traverse (it takes longer to swim than to crawl even for James Bond). Each fall produces temporary amnesia, and he is returned to the cell immediately after each fall. Assume that he always immediately chooses one of the three exits from the cell with probability 1/3. On the average, how long does it take before he notices that the door lock does not work?

Exercise 3.16 As new engines are coming off the assembly line in Detroit, they are then tested to determine the maximum deliverable horsepower. In a lot of 50 engines, 49 deliver a maximum of 200 horsepower, while 1 of them does not work at all, thereby delivering a maximum of 0 horsepower. What is the average maximum horsepower of the engines in the lot? Is the average a reasonable description of the maximum horsepower of the engines in the lot?

Exercise 3.17 A gambler hits upon what seems to be a foolproof system. He begins with a one-dollar bet playing the game of black-or-red in roulette, and each time that he loses, he doubles the amount bet over the previous bet until he wins *once*, at which point he quits. In this way, he stands to recoup his losses when he finally does win. He realizes that there is a small chance that he will lose everything he has ($1023), but he considers this probability to be small enough that he can ignore it. The probability of winning on a given trial is 18/38, in which case he wins an amount equal to what he originally bet, otherwise he loses his bet. What is the probability that he eventually wins and what is his net gain when he does? What is the probability that he loses all and how much does he lose? What is his average net gain using this system? Is it really foolproof? Is the risk he is taking a reasonable one?

Exercise 3.18 Compute the average length of a Craps game. For the rules of the game, see Exercise 2.19.

In the remaining problems of this chapter it may be necessary to approximate the harmonic series using Euler's approximation:

$$1 + \frac{1}{2} + \frac{1}{3} + \cdots + \frac{1}{n} \approx \log(n) + \gamma$$

where $\gamma = 0.57721\ldots$ is the Euler-Mascheroni constant.

Exercise 3.19 (Coupon collector's problem) A young baseball fan wants to collect a complete set of 262 baseball cards. The baseball cards are available in a completely random fashion, one per package of chewing gum, which she buys twice a day. How long on the average does it take her to get the complete set?

Exercise 3.20 A superpower has 262 missiles stored in well-separated silos. An enemy is considering a sneak attack. However, for the attack to succeed *every* one of the missiles must be destroyed. We will consider this problem again in Exercise 3.21, but for now we consider the following simple model. Assume each attacking warhead hits one of the enemy missiles in its silo with each enemy missile being equally likely to be the one that is hit. On average, how many warheads will be needed to ensure the destruction of every enemy missile?

Exercise 3.21 The analysis in Exercise 3.20 is overoptimistic for several reasons. There is a significant probability that a given warhead will hit none of the silos. Furthermore, we want not the average number of warheads required but rather the number of warheads needed to ensure with a very high probability (say 99%) that all the enemy missiles have been destroyed. Compute the number of warheads needed if each attacking warhead has probability 0.75 of hitting its target. Even this is optimistic inasmuch as the shock waves from nuclear explosions are such that one cannot expect the various warheads converging on one target or on nearby targets to be independent. However, it gives one an idea of just how foolhardy so-called preemptive warfare can be.

Exercise 3.22 A molecular beam is firing metal ions toward the face of a crystal. If an ion strikes an unoccupied site on the crystal, it promptly occupies that site, otherwise it bounces away and is lost. Every ion hits the crystal somewhere, with each site being equally likely. If there are 10^{16} crystal sites, on average how many ions must the beam fire at the crystal in order to fill every site?

Exercise 3.23 Guests arrive at random at a party, and the hosts seats them as they arrive successively one at a time around a large circular table. Twenty guests arrive, ten single men and ten single women. On average, how many of the twenty adjacent pairs around the table will consist of a man and a woman?

Exercise 3.24 The host in Exercise 3.23 invites twenty couples to a cocktail party. As the couples do not know each other, the host decides to mix his guests by assigning each man to one of the women in such a way that every possible arrangement is equally likely. On average, how many couples find themselves assigned to each other?

Exercise 3.25 Return to Exercise 2.8. On average, how many people call out their birthdays before a match is found, assuming that one is eventually found? How does it compare with the observed value?

Exercise 3.26 The Polish mathematician Stefan Banach kept two match boxes, one in each pocket. Each box initially contained n matches. Whenever he wanted a match he reached into one of his pockets completely at random. When he found that the box he chose was empty, how many matches were in the other box? How many were there on average?

Exercise 3.27 Jacob Bernoulli proposed the following dice game. The player pays one dollar and throws a single die. He then throws a set of n dice, where n is the number shown by the first die. The total number of dots shown by the n dice is then used to determine the payoff. If the number is less than twelve he loses the bet, if the number equals twelve his dollar is returned, while if the number exceeds twelve he receives two dollars. Find the expected number of dots shown by the n dice. Is the game favorable to the player?

Exercise 3.28 Nicolas Bernoulli proposed the following coin-tossing game which has since been called the St. Petersburg paradox. A player pays an entrance fee of E rubles to the casino. A fair coin is then tossed until it comes up heads. If it requires n tosses to get the first head, the player is paid 2^n rubles, for a net gain of $2^n - E$ rubles. What is the player's expected net gain? The paradox arises from the fact that one is placing no limit on the resources of the casino. If the casino possesses a total of $R = 2^N$ rubles, compute the net expected gain of the player. For the game to be fair, what should E be? For example, if the casino has resources of 33.55 million rubles, what entrance fee would be fair?

Exercise 3.29 What is the expected duration of the St. Petersburg game for the casino mentioned at the end of Exercise 3.28?

Exercise 3.30 What is the probability that in n tosses of a fair coin two heads never occur in a row, i.e., no run of two or more heads ever occurs?

Exercise 3.31 The generalization of Chevalier de Méré's first problem (Exercise 2.25) is called the problem of points. The problem concerns a game between two players that was interrupted before its conclusion. Suppose that N points are required to win the game, that player A has $N - a$ points and that player B has $N - b$ points. In a given trial, A wins with probability p and B with probability $q = 1 - p$. How should the stakes be divided? The problem was first solved by de Montmort. Can you solve it also?

Exercise 3.32 Generalize Exercise 3.14 to produce a kind of analog, for dice throwing, of draw poker. The player throws five dice. The player then has the option to choose a subset of the dice for rethrowing. This subset can be empty but cannot consist of all the dice. The process is then repeated for the rethrown dice, continuing until no more dice may be rethrown. The object is to maximize the total number of dots showing on the dice. Devise a strategy and calculate the expected outcome for this strategy. The optimal strategy will produce an expected outcome of about $24\frac{4}{9}$.

3.7 Answers to Selected Exercises

Answer to 3.1. A sample point in this case is the drawing of a diamond from a deck of cards. There are thirteen cards in a suit, so the sample space Ω consists of the thirteen diamonds that can be drawn from a deck of cards:

$$\Omega = \{A\Diamond, 2\Diamond, 3\Diamond, 4\Diamond, 5\Diamond, 6\Diamond, 7\Diamond, 8\Diamond, 9\Diamond, 10\Diamond, J\Diamond, Q\Diamond, K\Diamond\}$$

An event is a subset of Ω so there are a total of $2^{13} = 8192$ events in all. Obviously, it would be tedious to list all of the events, but it is easy to give some examples:

$$
\begin{array}{rl}
\text{"A face card is drawn"} & \{J\diamondsuit, Q\diamondsuit, K\diamondsuit\} \\
\text{"An even numbered card is drawn"} & \{2\diamondsuit, 4\diamondsuit, 6\diamondsuit, 8\diamondsuit, 10\diamondsuit\} \\
\text{"A five of diamonds is drawn"} & \{5\diamondsuit\} \\
\text{"Nothing is drawn"} & \emptyset \\
\text{"Something is drawn"} & \Omega \\
\text{"A ten-point card is drawn"} & \{10\diamondsuit, J\diamondsuit, Q\diamondsuit, K\diamondsuit\}
\end{array}
$$

The probability measure P implied by the phrase "thoroughly shuffled" is the "equally likely probability." It means that each card is as likely to be drawn as any other card. In practice, it is difficult to achieve the equally likely probability with a physical deck of cards. However, it is now common to shuffle electronic decks in online games, and these decks can be shuffled such that the equally likely probability is actually achieved.

The random variable S and the probability measure P are both mathematical functions. They are, however, defined on completely different domains and mean very different things. The domain of S is the sample space Ω. It has these values:

sample point	$A\diamondsuit$	$2\diamondsuit$	$3\diamondsuit$	$\cdots$	$8\diamondsuit$	$9\diamondsuit$	$10\diamondsuit$	$J\diamondsuit$	$Q\diamondsuit$	$K\diamondsuit$
value of S	11	2	3	$\cdots$	8	9	10	10	10	10

If we number the cards from 1 to 13, starting with $A\diamondsuit$ and ending with $K\diamondsuit$, then the random variable S can be specified as:

```
> cards <- 1:13
> S <- cards
> S[1] <- 11
> S[11:13] <- 10
```

Note the use of indexing to change some of the elements of a vector. The scores are the same as the card numbers with the exceptions that $A\diamondsuit$ has score 11 and the face cards have score 10. The graph is then obtained using plot(cards, S):

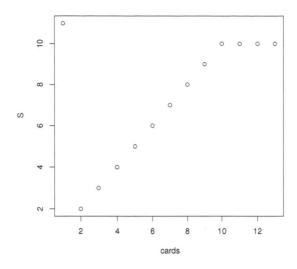

The domain of P, on the other hand, is the collection of all events, so P is defined on all 8192 events. Giving a tabulation like that for S is unreasonable. Instead, we give some examples:

Question: How probable is...	Answer		
drawing a face card?	$P(\{J\Diamond, Q\Diamond, K\Diamond\})$	$=$	$\frac{3}{13}$
drawing an even numbered card?	$P(\{2\Diamond, 4\Diamond, 6\Diamond, 8\Diamond, 10\Diamond\})$	$=$	$\frac{5}{13}$
drawing a five of diamonds?	$P(\{5\Diamond\})$	$=$	$\frac{1}{13}$
drawing nothing?	$P(\emptyset)$	$=$	0
drawing something?	$P(\Omega)$	$=$	1
drawing a ten-point card?	$P(\{10\Diamond, J\Diamond, Q\Diamond, K\Diamond\})$	$=$	$\frac{4}{13}$

Answer to 3.2. The drunk executes a nonsymmetric random walk. The probability distribution after six steps (one minute) is

$$P(S_6' = x) = \binom{6}{3 + \frac{x}{2}} \left(\frac{1}{4}\right)^{3 + \frac{x}{2}} \left(\frac{3}{4}\right)^{3 - \frac{x}{2}}$$

where x is measured in yards and "up" the street is the positive x-direction. To draw the graph of this probability distribution, we use the R function `dbinom` that computes the binomial probability distribution. The first parameter is the vector of values we wish to compute. In this case, we want the entire range of values from 0 to 6, so we set `t` equal to this range by assigning it to the vector `0:6`. The second parameter is the number of steps, i.e., the n in S_n. The third parameter is the bias, which is $1/4$ in this case. However, we

want S'_n, not S_n. For this, we simply change the values on the x-axis from the
range 0 to 6 to the range -6 to 6, and plot using the new range of values:

```
> t <- 0:6
> p <- dbinom(t, 6, 1/4)
> x <- 2*t-6
> plot(x, p)
```

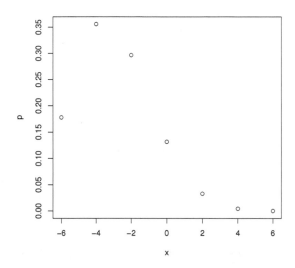

Next, we wish to compute the most likely location for the drunk. The maximum value of the probability is easily found using `max(p)`. To get the position
where this maximum occurs, we select it from the `x` vector using:

```
> x[p==max(p)]
```

The result is -4. Note that there can be more than 1 location where the
maximum probability occurs. For example, after the first 3 steps, the positions
-3 and -1 both have the probability 27/64. So one should actually speak of
the most likely *locations*. R has no problem with this, and `x[p==max(p)]` will
be a vector of elements in general.

To get the distribution after 2 minutes, we could repeat all of this again
with 6 replaced by 12, but it is easier just to do this once by writing an R
function for it as follows:

```
> drunk <- function(n) {
+    t <- 0:n
+    p <- dbinom(t, n, 1/4)
+    x <- 2*t-n
+    plot(x, p)
+    x[p==max(p)]
+ }
> drunk(12)
```

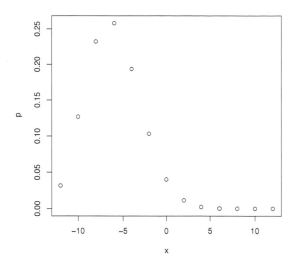

The most likely location is moving down the street at a rate of roughly 3 yards per minute. One can actually observe the movement of the most likely location by applying the **drunk** function repeatedly:

```
> ignore <- sapply(0:60, drunk)
```

However, this will probably run so quickly that it is almost not visible, so we add a call to **Sys.sleep** to slow it down. This function causes a pause for a specified number of seconds. Another problem is that the scales for the 2 axes will be continually changing. To ensure that the scales remain the same for each graph, we set the **xlim** and **ylim** parameters. The **xlim** parameter specifies the lower and upper limits for the x-axis. Notice the use of a 2-element vector to specify the limits. The **ylim** is for the y-axis. Here is the complete program:

```
> drunk <- function(n) {
+     t <- 0:n
+     p <- dbinom(t, n, 1/4)
+     x <- 2*t-n
+     plot(x, p, xlim=c(-60,60), ylim=c(0,1))
+     Sys.sleep(0.2)
+ }
> ignore <- sapply(0:60, drunk)
```

Answer to 3.9. Intuitively, the position of the drunk after one minute certainly influences the position after two minutes, but let us check this. For S'_6 and S'_{12} to be independent we must have:

$$P((S'_6 = x_1) \cap (S'_{12} = x_2)) = P(S'_6 = x_1) \, P(S'_{12} = x_2)$$

for all positions x_1 and x_2. For example, let $x_1 = -6$ and $x_2 = 12$. Clearly, if the drunk is six yards *down* the street after one minute, it is not possible to reach a position twelve yards *up* the street in the next minute. Thus,

$$(S'_6 = -6) \cap (S'_{12} = 12) = \emptyset$$

Hence,

$$P((S'_6 = -6) \cap (S'_{12} = 12)) = 0$$

But both $P(S'_6 = -6)$ and $P(S'_6 = 12)$ are nonzero. This contradicts independence. Of course, one could check this using many other choices of x_1 and x_2.

Answer to 3.11. In Exercise 3.1, S takes values from 2 to 11. The answer to the question "What is the score of the drawn card?" is a probability distribution. Now in the answer to Exercise 3.1 we saw how to define this using R. To get the probability distribution we need to compute how many times S takes each of the possible scores 2 to 11. For example, we would like to compute how many times S takes value 11. The vector of when this happens is S==11. Notice that this is almost the same as the event notation $(S = 11)$. The number of elements in this set is then computed using sum(S==11). The probability will then be sum(S==11)/13. We need to do this for each possible value of the random variable. As in the answer to Exercise 3.2 we put this computation in a function for convenience:

```
> P <- function(s) sum(S==s)/13
```

We now have the probability distribution as a function; namely, for each score s, the R expression P(s) is the same as the probabilistic expression $P(S = s)$. We then convert it to a vector by using the sapply function:

```
> p <- sapply(1:11, P)
```

Notice that we used 1:11, not 2:11, even though 1 is not a possible score. The subscripts for vectors in R always start from 1. We can then compute the expectation by multiplying the probabilities with the scores and by summing them together:

```
> sum((1:11)*p)
[1] 7.307692
```

All of this could have been done in a single command:

```
> sum((1:11) * sapply(1:11, function(s) sum(S==s)/13))
```

Answer to 3.12. We first compute the expectations of X'_n and S'_n in the Bernoulli process: random walk. The expectation of S'_n is easy to compute because of the Addition Rule for Expectations:

$$
\begin{aligned}
E(X'_n) &= (-1) \cdot q + 1 \cdot p = p - q \\
E(S'_n) &= E(X'_1 + X'_2 + \cdots + X'_n) \\
&= E(X'_1) + E(X'_2) + \cdots + E(X'_n)
\end{aligned}
$$

$$= n(p - q)$$

In Exercise 3.2, $p = 1/4$ and $q = 3/4$, so the average position of the drunk after n steps is $n(p-q) = -n/2$. After one minute this is three yards down the street, and after two minutes this is six yards down the street. The average position is moving exactly three yards/min down the street. Compare this with Exercise 3.2, where we computed the most likely position of the drunk. The most likely position will generally not be the same as the average position. A more important point is that the answer to the question "Where is the drunk after one minute?" is the *whole* probability distribution S'_6. It is from this answer that one may derive such information as the most likely position and the average position.

Answer to 3.18. Let L be the length of a Craps game. The game ends after only one roll if 2, 3, 7, 11 or 12 is rolled. These have probability $1/3$ so $P(L = 1) = 1/3$. For later rolls, the length depends on the point. In a fashion similar to the solution to Exercise 2.19, suppose the point has probability p of being rolled during a single roll. For this point, the game will then last n rolls with probability equal to:

$$p\left(1 - p - \frac{1}{6}\right)^{n-2}\left(p + \frac{1}{6}\right)$$

where the first p is for "getting" the point, and the last $\left(p + \frac{1}{6}\right)$ is for ending the game at the nth roll. The "contribution" of this point to the average length of the game is:

$$\sum_{n=2}^{\infty} np\left(1 - p - \frac{1}{6}\right)^{n-2}\left(p + \frac{1}{6}\right)$$

Let $r = 1 - p - \frac{1}{6}$. Then the contribution of the point is:

$$\frac{p\left(p + \frac{1}{6}\right)}{r} \sum_{n=2}^{\infty} nr^{n-1}$$

Now use the trick we used in Section 3.4 for computing the expectation of the first waiting time W_1 in the Bernoulli process:

$$\sum_{n=2}^{\infty} nr^{n-1} = \sum_{n=2}^{\infty} \frac{d}{dr} r^n = \frac{d}{dr}\left(\sum_{n=2}^{\infty} r^n\right) = \frac{d}{dr}\left(\frac{r^2}{1 - r}\right)$$

$$= \frac{2r(1 - r) + r^2}{(1 - r)^2} = \frac{2r - r^2}{(1 - r)^2} = \frac{r(2 - r)}{(1 - r)^2}$$

Therefore, the contribution of this point is:

$$\frac{p\left(p + \frac{1}{6}\right)}{r} \times \frac{r(2 - r)}{(1 - r)^2} = \frac{p\left(p + \frac{1}{6}\right)\left(1 + p + \frac{1}{6}\right)}{\left(p + \frac{1}{6}\right)^2} = p\left(p + \frac{7}{6}\right)\left(p + \frac{1}{6}\right)^{-1}$$

Finally, we must sum this expression over all six possible points. The expectation $E(L)$ is this sum plus the probability $1/3$ for the first roll:

$$E(L) = \frac{1}{3} + \frac{5}{12} + \frac{23}{45} + \frac{235}{396} + \frac{235}{396} + \frac{23}{45} + \frac{5}{12} = 3\frac{62}{165} \approx 3.3758$$

Answer to 3.25. Let B_n be the number of students in a classroom of n students who must call out their birthdays before a match is found. Now $(B_n>b)$ is the event that *none* of the first b students has the same birthday as anyone else in the class. The sample space Ω is the set of all placements of n objects (i.e., the students) into 365 containers. For a placement to be in $(B_n>b)$ it must have the first b objects placed in b distinct containers and the remaining $n-b$ objects placed (arbitrarily) in the remaining $365-b$ containers. By our counting rules, there are

$$\frac{365!}{(365-b)!}(365-b)^{n-b}$$

ways to do this. Hence,

$$P(B_n>b) = \frac{365!(365-b)^{n-b}}{(365-b)!(365)^n}$$

This formula will produce numbers that are much too large for R to compute, so we take logarithms and then exponentiate the result:

```
> birthday <- function(n, b)
+ exp(lfactorial(365)+(n-b)*log(365-b)
+    - lfactorial(365-b)-n*log(365))
```

Consider the case of a class of $n = 80$ students. We would like to graph the probability as a function of b, but our **birthday** function has two parameters, so we cannot use **sapply**. In this case, we use the **mapply** function. The m should be taken to mean "multiple," while the s in **sapply** means "single." The first parameter of **mapply** is the function, while all the rest of the parameters are the vectors to be applied to the first parameter. Note that **sapply** is a special case of **mapply** except that the function is the second parameter of **sapply**.

```
> plot(1:80, mapply(birthday, 80, 1:80))
```

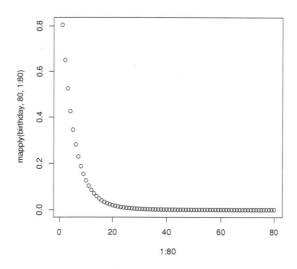

In the case of 12 students,

```
> birthday(80,12)
[1] 0.08578222
```

so there is only an 8.6% chance that more than 12 students in a class of 80 will be able to call out their birthdays without having a match. It is extremely unlikely for all the students to call out their birthdays without a match because

```
> birthday(80,80)
[1] 8.566805e-05
```

The average value of B_n can be computed (at least for reasonably large n so that B_n is nearly certain to have a value) by means of the above formula for $P(B_n > b)$. We omit the details. One uses the same trick we used in Section 3.4 for computing the expectation of the first waiting time W_1 in the Bernoulli process. The expectation is:

$$E(B_n) = \left(1 - \exp\left(\frac{0.5 - n}{365}\right)\right)^{-1}$$

For $n = 80$, this is:

```
> 1/(1-exp((0.5-80)/365))
[1] 5.109331
```

The median value of B_n is between 3 and 4. So more often than not, a class of 80 will require 3 or fewer students to call out their birthdays before a match is found!

We will see a distribution of this kind in Section 6.3, where we will give it a name: the *exponential distribution*. Finding this distribution here is no surprise in light of the ubiquity of this distribution in nature. In fact, in contrast to the calculations in Exercise 3.26, which are not very obvious, the calculations in this exercise are very easy once we have learned about the "memorylessness" property that characterizes the exponential distribution. See Section 7.1.

Answer to 3.26. There are many ways to solve this problem. In the spirit of this chapter, we convert the question to a waiting time. The stochastic process we use is the Bernoulli process with a fair coin. The heads and tails correspond to choosing one or the other matchbox. Let V_n be the waiting time for either $n+1$ heads or $n+1$ tails to occur. This is the time when Stefan Banach first discovers an empty matchbox. When he does, he has used $V_n - (n+1)$ of the matches from the other box so

$$n - (V_n - (n+1)) = 2n + 1 - V_n$$

are left. Therefore, the probability distribution of $M_n = 2n + 1 - V_n$ is the answer to our question.

The event $(V_n = a)$ means that one of the following disjoint events has occurred:

$(S_{a-1} = n) \cap (X_a = 1)$ This means that n heads have occurred between time 1 and time $a - 1$, and a head occurred at time a.

$(S_{a-1} = a - 1 - n) \cap (X_a = 0)$ This event is that n tails (and hence $a - 1 - n$ heads) have occurred between time 1 and time $a - 1$, and a tail occurred at time a.

Since $(V_n = a)$ is the union of these two events,

$$
\begin{aligned}
P(V_n = a) &= P(S_{a-1} = n)\,P(X_a = 1) + P(S_{a-1} = a - 1 - n)\,P(X_a = 0) \\
&= \frac{1}{2}P(S_{a-1} = n) + \frac{1}{2}P(S_{a-1} = a - 1 - n) \\
&= \frac{1}{2}\binom{a-1}{n}\frac{1}{2^{a-1}} + \frac{1}{2}\binom{a-1}{a-1-n}\frac{1}{2^{a-1}} \\
&= \binom{a-1}{n}\frac{1}{2^{a-1}}
\end{aligned}
$$

because $\binom{a-1}{n} = \dfrac{(a-1)!}{n!(a-1-n)!} = \binom{a-1}{a-1-n}$. Finally, we convert this to the answer to the original question:

$$
\begin{aligned}
P(M_n = b) &= P(2n + 1 - V_n = b) = P(V_n = 2n + 1 - b) \\
&= \binom{2n-b}{n}2^{b-2n}
\end{aligned}
$$

The expectation of M_n is:

$$E(M_n) = \sum_{b=0}^{n} b\binom{2n-b}{n}2^{b-2n}$$

This expression can be simplified although it is not easy to do. We omit the details. The result is

$$E(M_n) = (2n+1)\binom{2n}{n}2^{-2n} - 1$$

Here is the original sum written using R:

```
> matchbox <- function(n,b) choose(2*n-b,n)*2^(b-2*n)
> plot(0:30, mapply(matchbox, 30, 0:30))
```

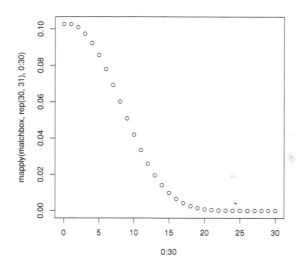

Finally, we compute the expectation both using the sum and using the simplified formula:

```
> bmatchbox <- function(n,b) b*matchbox(n,b)
> EM <- function(n) sum(mapply(bmatchbox, n, 0:n))
> EM(30)
[1] 5.257269
> EM <- function(n) (2*n+1)*choose(2*n,n)*2^(-2*n) - 1
> EM(30)
[1] 5.257269
```

Answer to 3.28. We model this with the waiting time W_1 for the first head, when the bias is $p = 1/2$. The player's expected net gain is

$$\sum_{n=1}^{\infty}(2^n - E)\left(\frac{1}{2}\right)^n = \sum_{n=1}^{\infty}\left(1 - E\left(\frac{1}{2}\right)^n\right)$$
$$= \sum_{n=1}^{\infty}(1) - \sum_{n=1}^{\infty}E\left(\frac{1}{2}\right)^n$$

$$= \sum_{n=1}^{\infty} (1) - E$$

The first term above diverges to infinity, so the expected net gain is infinite no matter how large the entrance fee E is! If the casino only possesses $R = 2^N$ rubles, then model is not quite W_1 because the game is stopped if the number of tails is N. At this point, the casino must give the player the R rubles it possesses. With this in mind, the expected net gain for the player is now

$$\sum_{n=1}^{N} (2^n - E)\left(\frac{1}{2}\right)^n + (2^N - E)\left(\frac{1}{2}\right)^N \;=\; \sum_{n=1}^{N}\left(1 - E\left(\frac{1}{2}\right)^n\right) + 1 - E2^{-N}$$

$$= \; N - E(1 - 2^{-N}) + 1 - E2^{-N}$$

$$= \; N + 1 - E$$

For the game to be fair, E should be $N+1$ rubles. Since 33.55 million is 2^{25}, if the casino has this many rubles, the entrance fee should be at least 26 rubles.

CHAPTER 4

General Random Variables

Jacob Bernoulli was one of the eight prominent mathematicians in the Bernoulli family. He was born in 1654, coincidentally the same year that Pascal and Fermat began their correspondence which was to create the mathematical theory of probability. Jacob is best known for the work *Ars Conjectandi* (The Art of Conjecture), published eight years after his death in 1713 by his nephew Nicholas. In this work he described the known results in probability theory and in enumeration, often giving alternative proofs of known results, and systematically applied probability theory to games of chance. He both discovered and proved the Law of Large Numbers, the first significant theorem of probability. In his *Ars Conjectandi,* he said, "We define the art of conjecture, or stochastic art, as the art of evaluating as exactly as possible the probabilities of things, so that in our judgments and actions we can always base ourselves on what has been found to be the best, the most appropriate, the most certain, the best advised; this is the only object of the wisdom of the philosopher and the prudence of the statesman." This summation of the purpose of probability theory continues to be relevant today.

So far we have considered only integer random variables. We now allow random variables to take any real values. Unfortunately, technical difficulties will appear that did not occur with integer random variables. We begin with an example so that we can gradually work our way out of the difficulties.

Consider the process of dropping a point on the interval $[0, a]$. Intuitively, the point is just as likely to fall on one part of $[0, a]$ as another. For example, it should be just as probable for the point to fall on the left half of the interval as to fall on the right half. More generally, the probability that the point falls in any given subinterval should be proportional to the length of that subinterval. Unfortunately, this leads to the inescapable conclusion that the probability of the point taking any one particular value x is *zero*.

So we see that the intuitive concept of an integer random variable, i.e., of a variable which takes its values with certain probabilities, is inadequate for describing the phenomenon of a general random variable. In fact there is an intriguing philosophical paradox here: how can the point land anywhere at all if the probability of its landing in any one place is zero? We will avoid such seeming paradoxes by decreeing that the probabilistic structure of a random variable is given by the probabilities that it takes values in *intervals*. More precisely, if X is a random variable, the probabilistic structure of X is given by the probability that X is between c and d for any real numbers $c \leq d$. We write $P(c \leq X \leq d)$ for this probability. For example, if X is the random variable corresponding to a point dropped at random on $[0, a]$, then for any pair of real numbers $c < d$ in $[0, a]$,

$$P(c \leq X \leq d) = \frac{d - c}{a}$$

There is a neat way to express the probabilistic structure of random variables in general: the **probability distribution function**, also called the **cumulative probability distribution**. We define this to be the function

$$F(x) = P(X \leq x)$$

To compute probabilities on "half-open" intervals we use the fact that:

$$P(c < X \leq d) = P(X \leq d) - P(X \leq c) = F(d) - F(c)$$

For other kinds of event, we use Boolean operations and limits to express the probability using the above formula. Therefore, the probabilistic structure of a random variable is completely determined by its distribution function.

If the probability distribution function F is continuous, then the probability of X taking any particular value is zero. In this case, $F(x) = P(X \leq x) = P(X < x)$, and we can ignore the distinction between $\leq$ and $<$ in computations. On the other hand, if $P(X = x) > 0$ for some number x, then F will have a discontinuous jump at x, and the distinction does matter. We will see an example of this in Section 6.4. Fortunately, in practice, most distribution functions are continuous.

Consider once again the random variable X corresponding to dropping a point at random on $[0, a]$. The distribution function of X is

$$F(x) = P(X \leq x) = \begin{cases} 0 & \text{if } x < 0 \\ x/a & \text{if } 0 \leq x \leq a \\ 1 & \text{if } x > 0 \end{cases}$$

When a random variable has this distribution function we shall say that it is **uniformly distributed** on $[0, a]$.

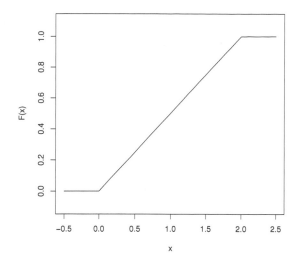

We see that the probabilistic meaning of dropping a point at random is that we have a random variable X uniformly distributed on $[0, a]$. We might also say that we are "choosing" or "sampling" a point at random from $[0, a]$. The process of sampling a sequence of n points at random from $[0, a]$ is called the **Uniform process**. More precisely, a Uniform process of sampling n points from $[0, a]$ is a sequence of n independent random variables $X_1, X_2, \ldots, X_n$ uniformly distributed on $[0, a]$. It is the continuous analog of the finite sampling process in Chapter 2. Note that we do not have to distinguish between sampling with or without replacement because the probability that any two of the sampled points coincide is zero.

To simulate the Uniform process, use the `runif` function. The three parameters specify the number of points to be sampled and the interval to be sampled from. One can specify any interval. On this graph we show 100 points sampled uniformly from the interval $[0, 10]$, and then the same points sorted in order. To distinguish the two plots, we use the `pch` parameter. This specifies the plotting symbol to be used. Many plotting symbols are available. One can specify either a number or a character. In this case, we used "u" for the original points and "o" for the points in order.

```
> points <- runif(100,0,10)
> plot(1:100, points, pch="u")
> points(1:100, sort(points), pch="o")
```

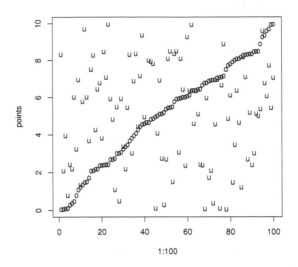

Most graphs use the numbered symbols rather than the characters. To see what these look like, plot them with a command like this:

```
> plot(1:25,1:25,pch=1:25)
```

which shows the numbered plot symbols.

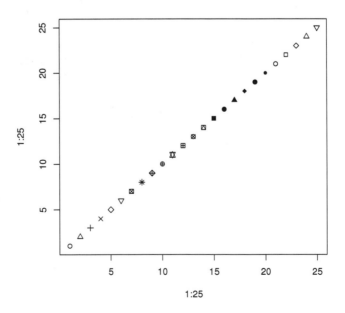

4.1 Order Statistics

Sorting and ordering are a natural way to organize our world. When presented with data, one of the first actions most people will do is to arrange the information in order. Moreover, people have a special fascination for the extremes: the biggest, the smallest, the highest or the lowest. In the Uniform process, the points are dropped on the interval independently at random. So they need not be in order. If we sort the points, we obtain a new set of random variables called the **order statistics**. If we have random variables $X_1, X_2, \ldots, X_n$, the sorted values are written $X_{(1)}, X_{(2)}, \ldots, X_{(n)}$. Each of the order statistics answers a probabilistic question about the original values. For example, in the Uniform process, the random variable $X_{(1)}$ answers the question "which point is the smallest?" The pronunciation of $X_{(1)}$ is "X order 1." The formula for $X_{(1)}$ is

$$\min(X_1, X_2, \ldots, X_n)$$

To compute this for a specific sequence x of numbers, use the `min` function:

```
> min(x)
```

Similarly, to compute the last order statistic, use the `max` function. More generally, to compute all of the order statistics for a specific vector x, use the `sort` function. For example, to compute the third order statistic of a vector x use:

```
> sort(x)[3]
```

While it is certainly useful to be able to compute the order statistics for a particular sample, it does not answer the probabilistic question of what the smallest value is. For this we need to find the probability distribution of $X_{(1)}$. To determine the answer to this question in the Uniform process, note that the event $(X_{(1)} > t)$ is the same as saying that *all* the X_i are greater than t. Hence:

$$
\begin{aligned}
P\big(X_{(1)} > t\big) &= P((X_1 > y) \cap (X_2 > t) \cap \cdots \cap (X_n > t)) \\
&= P(X_1 > y)\, P(X_2 > t) \cdots P(X_n > t) \\
&= \left(\frac{a - t}{a}\right)^n
\end{aligned}
$$

Therefore, the distribution function of $X_{(1)}$ is

$$F_{(1)}(x) = P\big(X_{(1)} \le x\big) = 1 - P\big(X_{(1)} > x\big) = 1 - \left(\frac{a - x}{a}\right)^n$$

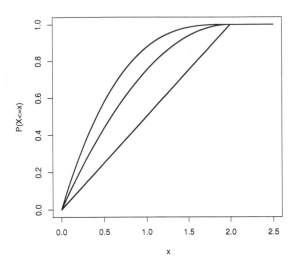

The distribution function is more and more "concentrated" near 0 as n increases; the more points one drops, the more likely that the minimum is small.

We need a way to express more clearly the fact that the distribution of $X_{(1)}$ is more concentrated near 0 for larger n. Indeed, as we shall see, the distribution of a random variable is not a very good way to visualize the behavior of the random variable. A better way is to examine the rate at which the probability is increasing. If the distribution function $F(x)$ of a random variable X is continuous and has a derivative, then this derivative is called the **density** of X:

$$f(x) = \frac{d}{dx}F(x) \qquad \int_{-\infty}^{x} f(u)du = F(x)$$

For example, the density of X_1 in the Uniform process is

$$f_1(x) = \begin{cases} 1/a & 0 \leq x \leq a \\ 0 & x < 0 \text{ or } x > a \end{cases}$$

Using density we see much more clearly why X_1 is said to be uniformly distributed on $[0, a]$: its density is constant on $[0, a]$.

On the other hand, the density of $X_{(1)}$ is

$$f_{(1)}(x) \quad = \quad \frac{n(a - x)^{n-1}}{a^n} \qquad \text{for } 0 \leq x \leq a$$

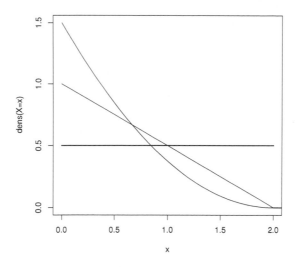

Notice how the density is sharply peaked at $x = 0$ just as we intuitively would expect.

4.2 The Concept of a General Random Variable

We are now ready to give a formal definition of the intuitive ideas we have just introduced.

Definition. A **random variable** X is a function that maps sample points in a sample space Ω to real numbers, with the property that the subsets $(X \leq x) = \{\omega \in \Omega \mid X(\omega) \leq x\}$ are events of Ω for all real numbers x. The **probability distribution function** of a random variable X is the function

$$F(x) = P(X \leq x)$$

As similarly noted for integer random variables, the technical assumption that the subsets $(X \leq x)$ are events will never bother us. We state it for purely grammatical reasons.

Integer Random Variables

Integer random variables are characterized by the fact that their distribution functions are constant except at integers, where they have discontinuous jumps. Here is a typical example:

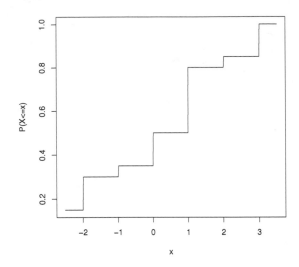

Being discontinuous, the distribution function of an integer random variable is rather unpleasant to deal with. As a result, one generally considers instead the probability distribution $p_n = P(X = n)$. It is unfortunate that $F(x)$ and p_n are both referred to as the distribution of an integer random variable. This is why one will sometimes add the adjective "cumulative" to emphasize that $F(x)$ is the accumulated probability *up to* x, not the probability *at* x.

We can now appreciate the naming convention used by R for the functions associated with each probability distribution, and the reason for the confusion that can result from it. The letter **d** was used for the density function, whether it is for a continuous random variable or an integer random variable. It is not actually a density for integer random variables, but it is certainly analogous to a density, so the name is reasonably appropriate. Similarly, the letter **p** was used for the probability distribution function for both continuous and integer random variables. With practice, one gets used to the notation.

Continuous Random Variables

A random variable X is a **continuous random variable** if its distribution function $F(x)$ is continuous and piecewise differentiable. The derivative $f(x) = F'(x)$ is called the **density** of X. It is the continuous analog of the probability distribution p_n of an integer random variable. This can be made quite precise using infinitesimals: the probability that X takes a given value x is the infinitesimal $f(x)dx$. In other words, the probability that X takes a value in a very small interval $[x, x + h]$ is close to $f(x)h$, and the smaller the interval, the better the approximation.

This suggests that the probability for a continuous random variable X to take a given value x is not quite zero but rather infinitesimal, if $f(x) \neq 0$. So although $(X = x)$ is an unlikely event, it is not impossible. We will write $\text{dens}(X = x)$ for the density $f(x)$ of X at x. However, one should take caution when using the notation for the density, since $\text{dens}(X = x)$ does not act like a probability $P(X = x)$. To give a concrete example, let X be uniformly distributed on $[0, 1]$. Then $2X$ is uniformly distributed on $[0, 2]$. Therefore, $\text{dens}(X = x) = 1$, but $\text{dens}(2X = 2x) = 1/2$, even though the events $(X = x)$ and $(2X = 2x)$ are obviously the same. In general, before performing any calculations involving densities, one should first convert them to probabilities. For example,

$$\text{dens}(X = x) = \frac{d}{dx}P(X \le x) = \frac{d}{dx}P(2X \le 2x)$$

$$\text{dens}(2X = 2x) = \frac{d}{d(2x)}P(2X \le 2x) = \frac{1}{2}\frac{d}{dx}P(2X \le 2x)$$

therefore, $\text{dens}(X = x) = 2\text{dens}(2X = 2x)$.

The density of X acts precisely like a mass density on the real line, a familiar concept in calculus. Thus, for example, to compute $P(a < X \le b)$ we must integrate the density:

$$P(a \le X \le b) = \int_a^b f(x)dx$$

In the case of an integer random variable, we get a sum:

$$P(m \le X \le n) = \sum_{i=m}^n p_i$$

The integral is the continuous analog of a sum.

Independence

The concept of the independence of two arbitrary random variables ought to be obvious, given the definition in the integer case. Namely, two random variables X and Y are **independent** if the events $(X \le x)$ and $(Y \le y)$ are independent for any pair of real numbers x and y:

$$P((X \le x) \cap (Y \le y)) = P(X \le x)\,P(Y \le y)$$

Properties of Densities and Distributions

The distribution function $F(x)$ of an arbitrary random variable satisfies:

1. if $x \le y$, then $F(x) \le F(y)$
2. $\lim_{x \to -\infty} F(x) = 0$
3. $\lim_{x \to \infty} F(x) = 1$
4. F is right continuous, i.e., $\lim_{\substack{y \to x \\ y > x}} F(y) = F(x)$

All these are obvious consequences of the definition of the distribution function. It is an interesting exercise to show the converse: any function $F(x)$ satisfying (1) to (4) is the distribution function of some random variable X on some sample space.

When X is a continuous random variable, its density $f(x)$ satisfies properties analogous to those of the distribution p_n of an integer random variable; namely,

1. $f(x) \geq 0$

2. $\displaystyle\int_{-\infty}^{\infty} f(x)dx = 0$

4.3 Joint Distribution and Joint Density

Just as we did for integer random variables, we measure the correlation of two arbitrary random variables by using a **joint distribution function**. The joint distribution function of random variables X and Y is

$$F(x,y) = P((X \leq x) \cap (Y \leq y))$$

If X and Y are continuous, then they also have a **joint density**:

$$\text{dens}(X = x, Y = y) = f(x,y) = \frac{\partial}{\partial x}\frac{\partial}{\partial y}F(x,y)$$

In terms of infinitesimals, the probability that X takes the value x and Y takes the value y is $f(x,y)dxdy$. As with ordinary densities, be careful not to treat $\text{dens}(X = x, Y = y)$ as a probability.

Suppose that F_X, F_Y and f_X, f_Y denote the distribution functions and densities of the continuous random variables X and Y, respectively. We can recover the densities from the joint densities by integrating:

$$\int_{-\infty}^{\infty} f(x,y)dy = f_X(x)$$

$$\int_{-\infty}^{\infty} f(x,y)dx = f_Y(y)$$

We call these the **marginal densities** or **marginals**. One can also recover the distributions from the joint distribution:

$$F_X(x) = \lim_{y \to \infty} F(x,y)$$

$$F_Y(x) = \lim_{x \to \infty} F(x,y)$$

In terms of the joint distribution and joint density, two random variables, X, Y are independent if and only if either

$$F(x,y) = F_X(x)F_Y(y)$$

or

$$f(x,y) = f_X(x)f_Y(y)$$

4.4 Mean, Median and Mode

Continuous probability distributions are analogous to the physical concept of a mass distribution. As we saw in Section 3.4, we can think of a probability distribution as a mass distributed along a long, thin rod. In the case of integer random variables, the mass consists of weights evenly spaced along the rod. Continuous probability distributions, on the other hand, correspond to a mass distributed anywhere along the rod. As in the case of integer random variables, the rod will normally have a *center of mass* where the rod will balance. This is the expectation of the random variable. For an integer random variable X, the expectation is the sum

$$E(X) = \sum_n np_n$$

For a continuous random variable X, the expectation is the continuous analog:

$$E(X) = \int_{-\infty}^{\infty} xf(x)dx$$

if it exists.

When X is a random variable that has a lower bound b on its support (for example, when X is a positive random variable), there is another formula for the expectation that is sometimes handy

$$
\begin{aligned}
E(X) &= b + \int_b^{\infty} P(X>x)\, dx \\
&= b + \int_b^{\infty} (1 - F(x))dx
\end{aligned}
$$

To get this formula, just apply integration by parts to the definition of the expectation. We leave the details as an exercise.

The expectation of a continuous random variable also satisfies the property we found very useful for integer random variables:

Continuous Addition Rule for Expectations. For any two continuous random variables X and Y,

$$E(X + Y) = E(X) + E(Y)$$

Proof. This is essentially the same proof as in the integer case.

$$
\begin{aligned}
E(X + Y) &= \int_{-\infty}^{\infty}\int_{-\infty}^{\infty} (x + y)f(x,y)dxdy \\
&= \int_{-\infty}^{\infty}\int_{-\infty}^{\infty} xf(x,y)dxdy + \int_{-\infty}^{\infty}\int_{-\infty}^{\infty} yf(x,y)dxdy \\
&= \int_{-\infty}^{\infty} xf_X(x)dx + \int_{-\infty}^{\infty} yf_Y(y)dy \\
&= E(X) + E(Y)
\end{aligned}
$$

While the expectation is the most important measure of the center of a random variable, there are other measures that are useful in statistics. A **median** of a random variable X is any number μ such that

$$P(X \leq \mu) \geq \frac{1}{2} \qquad \text{and} \qquad P(X \geq \mu) \geq \frac{1}{2}$$

We leave it as Exercise 4.4 to prove that the median always exists, but need not be unique. By contrast, the expectation is unique but need not exist. Many random variables deal with time, especially waiting times for some occurrence. The median for a temporal random variable is called the *half-life* of the random variable.

The *mode* of a continuous random variable with density $f(x)$ is a value of x for which $f(x)$ attains a maximum. A mode need not exist and need not be unique. The term "mode" is sometimes also used for a local maximum of the density function $f(x)$. A random variable whose density function has several local maximums is said to be *multimodal*. Modes are useful mainly for describing what the density function looks like graphically.

4.5 The Uniform Process

We now make a detailed investigation of the Uniform process in order to illustrate the concepts we have just introduced. Recall that the Uniform process of sampling n points from $[0, a]$ is the same as a sequence of independent random variables $X_1, X_2, \ldots, X_n$, each being uniformly distributed on the interval $[0, a]$. For example, these random variables might be the measurements of the heights of a random sample of n people. If we wish to ignore the order in which the people are measured, we simply write down the heights in increasing order. We call this new sequence the **order statistics** of the original sample. In effect, we "forget" what the order of sampling was and consider only the *set* of n measurements.

To be more precise, we introduce the following notation:

$$X_{(1)} \;=\; min(X_1, \ldots, X_n)$$

$$X_{(2)} \;=\; \text{next larger point after } X_{(1)}$$

$$\cdots$$

$$X_{(n)} \;=\; max(X_1, \ldots, X_n)$$

How are the order statistics distributed? What are their joint distributions? What are the distributions of the gaps between successive order statistics? Unlike the gaps in the Bernoulli process, these are not independent; for if one is big, the others must be small. What are the joint distributions of the gaps? We shall now answer these and other questions.

Let $F_{(k)}(x)$ and $f_{(k)}(x)$ denote the distributions and density of the kth order statistic $X_{(k)}$. Thus $F_{(k)}(x) = P(X_{(k)} \leq x)$. Now $(X_{(k)} \leq x)$ is the event "at least k of the n points fall in the interval $[0, x]$." In other words, there could be k points, $(k+1)$ points, and so on. We know the probability for each of these, so we just add them up to get the distribution function:

$$F_{(k)}(x) = P\big(X_{(k)} \le x\big) = \binom{n}{k}\left(\frac{x}{a}\right)^k \left(\frac{a-x}{a}\right)^{n-k}$$

$$+ \binom{n}{k+1}\left(\frac{x}{a}\right)^{k+1}\left(\frac{a-x}{a}\right)^{n-k-1}$$

$$+ \cdots$$

$$+ \binom{n}{n}\left(\frac{x}{a}\right)^n$$

For example, the first summand is the probability that *exactly* k points fall in $[0, x]$, and hence exactly $n - k$ fall in $(x, a]$. Similarly for the other summands. Needless to say, this expression is awkward.

Consider now the density $f_{(k)}(x)$. For this we use the event $A =$ "The kth order statistic $X_{(k)}$ is in $(x, x + h]$, and no other point falls in $(x, x + h]$." The probability $P(A)$ is the product of three factors:

1. One of the n points falls in $[x, x + h]$. This has probability $n\dfrac{h}{a}$.

2. Next, $k - 1$ of the remaining $n - 1$ points fall in $[0, x]$. This has probability $\binom{n-1}{k-1}\left(\dfrac{x}{a}\right)^{k-1}$.

3. Finally, the remaining $n - k$ points fall in $[x + h, a]$. There are no choices to be made, so the probability is $\left(\dfrac{a-x-h}{a}\right)^{n-k}$.

Combining all of these gives us the probability of A:

$$P(A) = \binom{n-1}{k-1}\left(\frac{x}{a}\right)^{k-1} n\frac{h}{a}\left(\frac{a-x-h}{a}\right)^{n-k}$$

Unfortunately, what we really want is $P\big(x < X_{(k)} \le x + h\big)$, not $P(A)$. This is starting to look very complicated.

However, there is no reason to panic! We never really have to compute this expression for the following reason. If more than one of the $X_{(j)}$ fall in $[x, x + h]$, the resulting probability involves a factor of $\left(\dfrac{h}{a}\right)^2$ or an even higher power of $\dfrac{h}{a}$. In other words,

$$P\big(x < X_{(k)} \le x + h\big) = P(A) + \left(\frac{h}{a}\right)^2 \text{(CE)}$$

where CE stands for "complicated expression." Now divide by h and take the limit as $h \to 0$:

$$f_{(k)}(x) = \lim_{h \to 0} \frac{P\big(x < X_{(k)} \le x + h\big)}{h}$$

$$= \lim_{h \to 0} \binom{n-1}{k-1} \left(\frac{x}{a}\right)^{k-1} n\frac{1}{a} \left(\frac{a-x-h}{a}\right)^{n-k} + \frac{h}{a^2} \text{(CE)}$$

$$= \lim_{h \to 0} n \binom{n-1}{k-1} \left(\frac{x^{k-1}(a-x-h)^{n-k}}{a^n}\right) + \frac{h}{a^2} \text{(CE)}$$

$$= n \binom{n-1}{k-1} \left(\frac{x^{k-1}(a-x)^{n-k}}{a^n}\right)$$

We never have to compute the complicated expression because no matter what it is, it disappears when we let h go to zero. We shall use this trick repeatedly. In fact, it is precisely because we can make this kind of simplification that the density is so much more computable than the distribution.

Next, we consider the joint distribution of two order statistics. For example, how does the tenth point influence the twentieth? This is an important question in biostatistics, because of the necessity of biologists to rely on small samples. Order statistics allow one to detect deviations from randomness in a relatively small sample.

As with the above computation, it is much more convenient to compute the joint density rather than the joint probability distribution. Let $X_{(j)}$, $X_{(k)}$ be two of the order statistics, where $j < k$. Then the joint distribution is

$$F_{(j,k)}(x, y) = P\big((X_{(j)} \leq x) \cap (X_{(k)} \leq y)\big)$$

and the density is

$$f_{(j,k)}(x, y) = \frac{\partial}{\partial x} \frac{\partial}{\partial y} F_{(j,k)}(x, y)$$

Again, as with the computation above, we need only compute the probability of the event "$X_{(j)}$ falls in $[x, x+h]$, $X_{(k)}$ falls in $[y, y+\epsilon]$, and no other points fall in these intervals." We think of these two intervals as dividing $[0, a]$ into five containers into which we place n objects with occupancy numbers: $j-1, 1, k-j, 1$, and $n-k$. There are $\binom{n}{j-1, 1, k-j-1, 1, n-k}$ ways to place the n objects with these occupancy numbers. Therefore, the event in question has the probability

$$\binom{n}{j-1, 1, k-j-1, 1, n-k} \left(\frac{x}{a}\right)^{j-1} \frac{h}{a} \left(\frac{y-x-h}{a}\right)^{k-j-1} \frac{\epsilon}{a} \left(\frac{a-y-\epsilon}{a}\right)^{n-k}$$

Dividing by $h\epsilon$ and letting $h \to 0$ and $\epsilon \to 0$ gives the joint density:

$$f_{(j,k)}(x, y) = \begin{cases} \frac{n!}{(j-1)!(k-j-1)!(n-k)!} \frac{x^{j-1}(y-x)^{k-j-1}(a-y)^{n-k}}{a^n} & \text{if } 0 \leq x \leq y \leq a \\ 0 & \text{otherwise} \end{cases}$$

Finally, we consider the joint density of all n order statistics. Let $x_1 < x_2 < \ldots < x_n$ be real numbers in $[0, a]$. Now

$$P\big((x_1 < X_{(1)} \leq x_1 + h_1) \cap \ldots \cap (x_n < X_{(n)} \leq x_n + h_n)\big) = n! \frac{h_1}{a} \cdots \frac{h_n}{a}$$

because there are $n!$ ways of placing the n points in the intervals $[x_1, x_1 + h_1]$,

..., $[x_n, x_n + h_n]$. The h_i's are chosen so small that there is no overlap. There-fore, the joint density of all n order statistics is:

$$f(x_1, \ldots, x_n) = \begin{cases} \dfrac{n!}{a_n} & \text{if } 0 \leq x_1 < x_2 < \cdots < x_n \leq a \\ 0 & \text{otherwise} \end{cases}$$

The **gaps** of the Uniform process are the distances between successive points in increasing order. The gap between 0 and $X_{(1)}$ is written L_1, the gap between $X_{(j)}$ and $X_{(j+1)}$ is written L_{j+1}, and the gap between $X_{(n)}$ and a is L_{n+1}. The order statistics may be written in terms of the gaps:

$$X_{(k)} = L_1 + L_2 + \cdots + L_k$$

The gaps are not independent: if one is large the others must be small. But the gaps are nevertheless equidistributed! When we have conditional probabilities, we will be able to prove this rigorously. However, one can prove this probabilis-tically. Since one of our main objectives is to learn to think probabilistically, this kind of proof is actually preferable.

Imagine that we drop $n + 1$ points on a circle of circumference a. It is in-tuitively obvious by symmetry that the gaps (measured along the circumfer-ence) so obtained are all equidistributed. On the other hand, this experiment is stochastically equivalent to the following experiment. Fix one point (call it 0) on the circle and then drop n more points at random. If we cut the circle at 0 and stretch it out over the interval $[0, a]$, then the gap distributions on $[0, a]$ are the same as those on the circle (the probability that another of the n points falls at the same place as 0 is zero). Therefore, the gap distributions on $[0, a]$ are all the same. This completes the proof.

In particular, all the gaps are distributed the same as $L_1 = X_{(1)}$. We already computed the density of $X_{(1)}$ so the density of any gap L_i is

$$f(x) = \frac{n}{a^n}(a - x)^{n-1} \text{ on } [0, a]$$

The expectation of L_i is given by:

$$E(L_i) = \int_0^a x \frac{n}{a^n}(a - x)^{n-1} dx$$

but there is a much easier way to compute this. Since

$$E(L_1) = E(L_2) = \cdots = E(L_{n+1})$$

and

$$L_1 + L_2 + \cdots + L_{n+1} = a$$

it follows that

$$E(L_i) = \frac{a}{n + 1}$$

by the Addition Rule. We can now appreciate the power of the Addition Rule, for the L_i's are not independent.

Similarly, we can compute the expectation $E(X_{(k)})$ quite easily. For

$$X_{(k)} = L_1 + \ldots + L_k$$

which implies that

$$E(X_{(k)}) = E(L_1) + \ldots + E(L_k) = \frac{ka}{n+1}$$

While this is certainly what one would intuitively expect, a direct computation would be tedious.

Consider now a seeming paradox. Suppose we label a reference point g on a circle of circumference a, then we drop n points at random. What is the expected length of the gap that includes the reference point g? The answer is $\frac{2a}{n+1}$, *not* $\frac{a}{n}$, as one might intuitively expect. The paradox lies not in any contradiction but rather in having a false intuition. Think of the experiment in reverse order: drop n points at random and *then* choose a reference point g. The reference point g is more likely to fall in a longer gap simply *because* it is longer. The seeming paradox comes from the impression that one is performing the following quite different experiment: drop n points at random on a circle and then pick a *gap* at random (i.e., any gap is as likely to be chosen as any other). This second experiment does indeed have the expectation a/n.

4.6 Table of Probability Distributions

Random variables are a central concept in the theory of probability. For example, we saw that the Uniform process is simply the study of n independent, uniformly distributed random variables. One could regard probability theory abstractly as the study of certain functions on sample spaces, which satisfy certain laws. However, this would miss the point, because it is the examples that make the theory, and we can only learn probability theory by carefully studying the examples, especially the important ones.

Random variables are classified by their distributions. And when one speaks of a distribution one usually has a standard model in mind. Learning probability theory therefore requires learning not just the distribution but also the natural phenomena that give rise to them. We will now make a list of distributions and models. We will add to our list in subsequent chapters.

Bernoulli distribution. X has a *Bernoulli distribution* if X is an integer random variable which takes just two values, usually taken to be 0 and 1. This distribution depends on one parameter, $p = P(X = 0)$. The standard model for this distribution is a toss of a biased coin, X_i, during the Bernoulli process.

Binomial distribution. X has the *binomial distribution* if X is an integer random variable, and

$$P(X = k) = \binom{n}{k} p^k q^{n-k}$$

The binomial distribution depends on two parameters: n and p. The standard model is S_n, the number of heads in the first n tosses of the Bernoulli process. Here p measures the "bias" of the coin.

Geometric distribution. X has the *geometric distribution* if X is an integer random variable, and

$$P(X = n) = q^{n-1}p$$

The standard models are the waiting time W_1 for the first head in the Bernoulli process, and the gap T_k between the $(k-1)^{st}$ and k^{th} occurrences of heads in the Bernoulli process.

Negative Binomial (Pascal) distribution. X has this distribution if X is an integer random variable, and

$$P(X = n) = \binom{n-1}{k-1} q^{n-k} p^k$$

This distribution has one parameter k. The standard model is the kth waiting time W_k of the Bernoulli process.

Uniform distribution. X has this distribution if X is a continuous random variable with density

$$f(x) = 1/a, \qquad \text{on } [0, a]$$

The standard model is any X_i, "dropping a point at random," in the Uniform process. Here the parameter a is the length of the interval on which one is dropping (or sampling) points.

Distributions of order statistics. These are called the **beta distributions**. X has a beta distribution if it is a continuous random variable, and its density is

$$f(x) = (k+l-1)\binom{k+l-2}{k-1}\frac{x^{k-1}(a-x)^{l-1}}{a^{k+l-2}}, \qquad \text{on } [0, a]$$

There are three parameters a, k and l, although in most treatments, the parameter a is set to 1, so only two are used. These parameters are called the *shape parameters*, and are commonly denoted by the letters $\alpha = k$ and $\beta = l$. Because the factorial function is meaningful for nonintegers, the beta distribution parameters can be any positive real numbers. The standard model is the kth order statistic $X_{(k)}$ of $n = k + l - 1$ points dropped at random on $[0, a]$. The parameter l is the position of the order statistic if we number the points in descending order, starting from the maximum and continuing to the minimum. The standard model is defined whenever the beta distribution parameters are integers.

Distribution	Type	Parameter(s)	Model(s)
Bernoulli	integer	p	X_i in Bernoulli
Binomial	integer	n, p	S_n in Bernoulli also X_i when $n = 1$
Geometric	integer	p	W_1 or any Bernoulli T_k
Negative Binomial	integer	k	W_k in Bernoulli
Uniform	continuous	a	X_i in Uniform
Beta	continuous	a, n, k	$X_{(k)}$ in Uniform also L_i when $k = 1$

Important Probability Distributions

Type	Distribution or density	Expectation
Bernoulli	$p_0 = q = 1 - p$, and $p_1 = p$	p
Binomial	$p_k = \binom{n}{k} p^k q^{n-k}$	np
Geometric	$p_n = q^{n-1} p$	$1/p$
Negative Binomial	$p_n = \binom{n-1}{k-1} q^{n-k} p^k$	k/p
Uniform	$f(x) = 1/a$ on $[0, a]$	$a/2$
Beta	$f(x) = n \binom{n-1}{k-1} x^{k-1} (a - x)^{n-k} a^{-n}$	$ka/(n+1)$

Expectations of Important Distributions

4.7 Scale Invariance

Anyone who has looked at population data for cities can see that there will be a few very large cities and a large number of much smaller ones. When the data is plotted using logarithmic scales, it tends to be approximately a straight line. An impressive variety of other phenomena also have this characteristic. The first extensive study was published by Zipf in 1949.

These distributions share this property: when one conditions on the subset of the entities with the highest ranks (e.g., the largest cities), the resulting conditional distribution looks the same as the entire distribution. In other words, whatever is causing the disparity among the entities, it is the same no matter what scale we use to look at the phenomenon. We model this probabilistically with a continuous random variable S supported on an interval $[0, a]$, for which

$$P(S \leq st/a \mid S \leq s) = P(S \leq t)$$

for all t, s in $(0, a]$. The factor s/a is the unique factor which linearly *rescales* (or *magnifies* or *zooms*) from the interval $[0, a]$ to the interval $[0, s]$. A random variable with this property is known by a variety of names such as **self-similar**, **scale-invariant** and **Pareto**. The term **fractal** has been used for phenomena of this kind, so this term could also be used to describe this distribution. One can show that if a random variable S is scale-invariant, then it has a beta distribution whose first parameter α can be any positive number, and whose second parameter β is equal to 1. See Section B.2 for a proof. When α is an integer, the standard model of the beta distribution is the last order statistic, i.e., the maximum of a set of α points dropped on $[0, a]$.

One example of a scale-invariant distribution is the Pareto principle or 80/20 rule. This is an observation about economies, but it has since been applied to a great variety of other phenomena. The Pareto principle is concerned with the allocation of resources to individuals in a population. If one sorts individuals by the amount of resources that they own, then the 80/20 rule holds when 80% of the resources are concentrated in 20% of the individuals. Moreover, this same degree of concentration holds at every scale. Thus, among the wealthiest 20% of the population, 80% of their wealth (which is 64% of the wealth of the whole population) is concentrated in the wealthiest 20% (i.e., the wealthiest 4% of the entire population). It is easy to compute the exponent α for the 80/20 rule by solving for α in the equation $0.20^{\alpha} = 0.80$, obtaining `log(0.80)/log(0.20) = 0.1386469`. Thus, $\alpha \approx 1/7$, and the Pareto distribution satisfies

$$P(S \leq t) = \sqrt[7]{t}$$

Other Pareto-like distributions can be defined using other "rules," such as a 90/10 rule. Each such rule has a corresponding value of α.

The Pareto distribution, and the beta distributions in general, are supported in an interval $[0, 1]$. As a result, the values may be regarded as probabilities. Random variables whose values are probabilities occur in a number of contexts, most notably the Shannon Coding Theorem in Section 9.2. When a probability distribution has this property, we can view the inverse of the distribution function as being a distribution function. For example, the R function `qbeta`, which is the quantile function for the beta distribution, can itself be regarded as a distribution function. In the case of the Pareto distribution, the inverse distribution reverses the roles of the individuals and the resources. The resources now have the primary role, and they satisfy the 20/80 rule, the reverse of the 80/20 rule; namely, 80% of the resources are owned by 20% of the individuals. The 20/80 distribution is the beta distribution with shape parameter $\alpha = $ `log(0.20)/log(0.80) = 7.212567`. In other words, it is as if the resources are randomly selecting their owners by first uniformly at random selecting 7 potential owners, and then choosing the wealthiest among the 7. As another example, when one sorts the sizes of cities, one finds that it approximately satisfies a Pareto distribution. If this distribution is the 90/10 rule, then it has the beta distribution with $\alpha = $ `log(0.90)/log(0.10) = 0.04575749`. The inverse distribution has $\alpha = $ `log(0.10)/log(0.90) = 21.85435`. So it is as if each person first uniformly at random selects 22 potential cities from among all possible cities, and then moves to the largest one among the 22.

The standard model for the scale-invariant distribution is the maximum value of a set of points dropped on an interval. Thus scale-invariance is a manifestation of the result of sorting a collection of uniformly distributed random numbers. The Pareto distribution is also a manifestation of sorting, since the distribution appears only if one sorts the population by some criterion such as wealth, size or frequency of use. So it should not be too surprising that the same distribution occurs in each case.

Suppose that S is scale-invariant on $[0, a]$, and consider the random variable $X = 1/S$. This is a random variable on $[b, \infty)$, where $b = 1/a$. It is easy to see, by inverting all of the variables in the scale-invariance condition, that X satisfies the condition

$$P(X > xy/b \mid X > x) = P(X > y)$$

for all x, y in $[m, \infty)$. This is exactly the same as the original scale-invariance condition, except that the comparisons are in the opposite direction, i.e., one is using the complementary probability distribution rather than the probability distribution. By the characterization of scale-invariant distributions, if X satisfies the condition above, then for some positive number α, we have:

$$P(X > x) = \left(\frac{x}{b}\right)^{-\alpha}$$

The literature on the Pareto distribution is usually stated in terms of the random variable $X = 1/S$, and the emphasis is on the *long-tail* that the distribution exhibits. Unfortunately, this formulation loses the connection with the beta distribution.

In any case, much has been said about the ubiquity of the scale-invariant distributions, going back to the work of Zipf. Explanations for its ubiquity are seldom very convincing. Yet, when one realizes that the scale-invariant distributions are just special cases of the beta distributions, its ubiquity becomes far less mysterious, since the beta distributions maximize entropy, as we will see in Section 9.3.

4.8 Exercises

Exercise 4.1 Show that if X is a random variable on $[b, \infty)$, then

$$E(X) \;=\; b + \int_b^\infty P(X > x)\, dx$$

Exercise 4.2 A boy makes a date with his girl friend. They are to meet at some time between 6 PM and 7 PM, but since both are absent-minded they forget which time they had agreed upon. As a result, each arrives at a random moment between 6 and 7. Each waits for 10 minutes and if the other fails to appear, he or she promptly leaves in a blue funk. What is the probability that true love prevails (at last this one evening)?

Exercise 4.3 When five points are chosen uniformly at random from the interval $[1, 2]$, what is the distribution of the natural logarithm of the smallest point?

Exercise 4.4 Prove that the median of any random variable exists. Does it have to be unique?

Exercise 4.5 After grading an examination, a teacher arranges the papers in order by grade. The **sample median** is the middle grade if there are an odd number of papers, and is the average of the two middle grades otherwise. Give a definition of the sample median in the Uniform process and compute its distribution.

Exercise 4.6 How far apart are the largest and the smallest points in the Uniform process of sampling n points from $[0, a]$? We call this the *spread*. Compare the spread with the second largest order statistic, $X_{(n-1)}$.

Exercise 4.7 Show that any function satisfying the four properties of a distribution function is in fact the distribution of some random variable on some sample space.

Exercise 4.8 One can also develop a theory of discrete order statistics. For these random variables, the interval $[0, a]$ is replaced by the set of integers $\{1, 2, \ldots, A\}$, each of which is equally likely to be chosen, and a given integer may be chosen more than once. The formulas one gets are quite complicated. It should be clear, however, that when A is large compared to n, the number of points chosen, we may approximate the discrete order statistics with the continuous ones. The principle that the gaps are equidistributed holds both for the discrete and for the continuous cases. Compute the distribution of the first order statistic. Note that this is an *integer* random variable.

Exercise 4.9 During World War II, the Allies estimated the number of tanks that had been produced by German industry by collecting the serial numbers of abandoned tanks. There are actually two questions one can ask here. One can ask for the most likely number of tanks that have been produced, or one can ask for the most reasonable rough estimate of the number. The former question would be most appropriate if we placed a very high value on getting the *exact* number, nearby numbers being useless. The latter question is clearly more appropriate in the context of this problem.

To answer these questions, we must rephrase them in the language of probability. Assume n serial numbers have been collected, the largest of which is $X_{(n)}$. The first question should read: what is the number A of tanks such that when n numbers are chosen uniformly from $\{1, \ldots, A\}$ the probability that we get the actually observed values is as large as possible? The answer is $X_{(n)}$ itself. Prove this. We call this the **maximum likelihood estimator** of A. See Exercise 2.22 for another example of such an estimator. For the second question, we want an estimate of A such that if one makes many estimates of the same number A by this method, we will, on average, be close to the correct value. We will consider this question later in Exercise 4.18.

Exercise 4.10 A biologist is studying organelles in a cell. The organelles in question are spheres of equal but unknown radius r within a given type of cell, and they are distributed randomly throughout the cell. The biologist estimates r by observing a cross-section of the cell and by measuring the radii of the visible granules. Suppose that n granules are observed and that the largest observed radius is $R_{(n)}$. Find the maximum likelihood estimator of r. The measurements $R_1, \ldots, R_n$ of the n radii will not be uniformly distributed. However, the random variables

$$\sqrt{r^2 - R_1^2}, \ldots, \sqrt{r^2 - R_n^2}$$

will be uniformly distributed.

Exercise 4.11 Compute directly, without first finding densities, the joint distribution function of the two order statistics $X_{(j)}$ and $X_{(k)}$ in the Uniform process.

Exercise 4.12 Suppose that $X_1, X_2, \ldots, X_n$ are independent uniformly distributed random variables on the intervals $[0, a_1], \ldots, [0, a_n]$, respectively. Compute the densities and the joint densities of the order statistics

$$X_{(1)} \leq X_{(2)} \leq \cdots \leq X_{(n)}$$

Exercise 4.13 In Exercise 4.12 above, compute the distributions of the gaps.

Exercise 4.14 Compute the average value of the natural logarithm of the smallest point among five chosen, uniformly from $[1, 2]$. Is this the same as the natural logarithm of the average value of the smallest point? Explain this. See Exercise 4.3.

Exercise 4.15 Compute the average of the median of the set of order statistics. See Exercise 4.5.

Exercise 4.16 An enzyme randomly breaks each of 24 identical (and very long) DNA molecules into 2 pieces. How long is the shortest piece produced? What is the average length of the shortest piece?

Exercise 4.17 Shuffle a deck of cards and turn up cards one at a time until the first spade appears. On average, how many cards, including the spade, do we expose?

More generally, if we are looking for one of n cards in the deck, on average, how many cards must we expose until we find one of them?

Exercise 4.18 Return to Exercise 4.9. To answer the second question we require a random variable with the property that its expectation is A. The maximum likelihood estimator will not do because $E(X_{(n)}) \neq A$. What should one use?

Exercise 4.19 Return to Exercise 4.10. Find a random variable R for this problem such that $E(R) = r$.

Exercise 4.20 A long DNA molecule is broken into N pieces. Find the average length of the ith longest piece produced, $1 \leq i \leq N$. Use probabilistic reasoning as follows. Let $n = N - 1$ so that our model is the Uniform process of sampling n points from $[0, a]$, where a is the length of the DNA molecule. The problem is to compute the expectations of $L_{(1)}, L_{(2)}, \ldots, L_{(n+1)}$, i.e., of the order statistics of the gaps. Here is how to compute $L_{(1)}$. First find $P(L_{(1)} > t)$. Since the event $(L_{(1)} > t)$ is the same as

$$(L_1 > t) \cap (L_2 > t) \cap \cdots \cap (L_{n+1} > t),$$

when this event occurs we can remove a subsegment of length t from each of the gaps. The resulting process is the Uniform process of sampling n points from $[0, a - (n + 1)t]$. Geometrically, the event $(L_{(1)} > t)$ is an n-dimensional cube whose side has length $a - (n+1)t$. Thus, $P(L_{(1)} > t) = \left(\frac{a - (n+1)t}{a} \right)^n$, and dens$(L_{(1)} = t) = n(n+1)\frac{(a+(n+1)t)^{n-1}}{a^n}$. It is now easy to calculate $E(L_{(1)})$.

Similarly, to compute $L_{(2)}$ we simply note that when we remove a subsegment of length $L_{(1)}$ from each gap (which we can do since $L_{(1)}$ is the smallest gap), what remains is the Uniform process of sampling $n - 1$ points from $[0, a - (n + 1)L_{(1)}]$. Moreover, $L_{(2)}$ is the sum of $L_{(1)}$ and the length of the smallest gap in this smaller process. This gives $E(L_{(2)})$. Continuing by induction we can compute $E(L_{(k)})$ for all k. Although the above reasoning is not, strictly speaking, rigorous, we will show how to make it completely rigorous in Chapter 6. See Exercise 6.24.

Exercise 4.21 A biologist uses an enzyme to break a DNA molecule into 10 pieces. The original molecule was 10,000 base pairs long. Upon examining the pieces, the biologist finds that the smallest is only 10 base pairs long! How probable is it that the smallest of 10 pieces could be this short or shorter? Use the results of Exercise 4.20. Does the biologist have a case for believing that the enzyme attacked the DNA molecule in a nonrandom fashion?

Exercise 4.22 A gambler is playing a sequence of games. For each trial he can choose to bet either on heads or on tails of a toss of a fair coin. If he bets on heads he gains or loses \$1 depending on whether the coin shows heads or tails, respectively. Similarly, if he bets on tails he gains or loses \$2 if the coin shows tails or heads, respectively. In each trial the gambler chooses one or the other bet at random betting on heads with probability p. Let A be the event that he bets on heads and let B be the event that the coin shows heads. Write his net gain in one trial using indicators.

Exercise 4.23 The President of the U.S. holds frequent news conferences. The journalists who attend these conferences are usually the same group, more or less. Let us suppose that in the first 2 years of his term the President answers 400 questions put to him by the 100 regular journalists. During this time 4 of the journalists have never been accorded recognition. These 4 get together and complain that they are being discriminated against, arguing that the probability that none of them was ever recognized is only 8×10^{-8}. On the other hand, the President's Press Secretary argues that the probability for 4 or more of the journalists to be ignored is really about 11%, which is not significant evidence for discrimination. Who is right? Formulate the 2 models being used and calculate the required probabilities. Refer to Exercise 2.16 for 1 model. For the other use the finite sampling process of sampling 400 journalists with replacement from among 100. In Chapter 5 we will develop techniques for approximating the answer. See Exercise 5.30.

Exercise 4.24 In the game of Treize, popular in seventeenth century France, 13 balls labeled from 1 through 13 were placed in an urn and drawn out one at a time at random without replacement. The players bet on the waiting time until either the nth ball drawn was labeled n or else the urn was emptied. Compute the distribution of this waiting time. Generalize to N balls.

Exercise 4.25 An obvious modification of the game of Treize would be to allow players to bet on the number of times that the nth ball drawn was labeled n. Compute the distribution of this number. Generalize to N balls. What is the average number of matches? See Exercise 3.24.

Exercise 4.26 A sociologist claims that he can determine a person's profession by a single glance. A psychologist decides to test his claim. She makes a list of 13 professions and chooses photographs (all in a standard pose) of 13 individuals, 1 in each profession. She then asks the sociologist to match the photos with professions. The sociologist identifies only 5 correctly. What do you think of his claim? Note that this exercise is closely related to Exercise 4.25.

Exercise 4.27 Prove the inclusion-exclusion principle for the *max* and *min* functions. Hint: $\int_{c}^{\infty} I_{(-\infty,x)}dy = x - c$, provided that $x > c$.

Exercise 4.28 Return to the molecular beam in Exercise 3.22. Assume the beam fires 10^{14} ions per second. Compute the distribution of the waiting time until the crystal is totally covered. Give a formula. Do not try to evaluate it. In Exercise 5.31 we will show how to compute an accurate approximation for the value of this expression. Do the same computation as above for the baseball fan problem (Exercise 3.19) and for the preemptive nuclear attack problem (Exercise 3.20).

Exercise 4.29 The standard card deck used by parapsychology experimenters is called the Zener deck. It has 25 cards, 5 each of 5 kinds. A typical test consists of the experimenter in 1 room and the subject in another. The experimenter shuffles the deck thoroughly and then turns the cards over 1 at a time at a fixed rate. Simultaneously, the subject is trying to perceive the sequence of cards. In order to test whether the subject's perceived sequence could have been simply a random guess, we must calculate the distribution of the number of matches occurring in a random permutation of the deck relative to some standard ordering of the deck.

Exercise 4.30 In an ancient kingdom, the new monarch was required to choose his queen by the following custom. One hundred prospective candidates are chosen from the kingdom, and once a day for one hundred days, one of the candidates chosen at random was presented to the monarch. The monarch had the right to accept or reject each candidate on the day of her presentation. When a given candidate is rejected she leaves, and the monarch cannot change his mind about that candidate. Assume that the preferences of the monarch can be expressed in a linear order (from the best to the worst) and that the monarch wants the best candidate, second best will not do. What strategy should he employ? What is the probability that he succeeds in his quest for perfection?

Suppose that instead of simply ranking the candidates, the monarch rates each of them on a 0 to 10 scale. In other words, using some sort of objective criteria, he can determine a real number between 0 and 10 for each candidate. Assume that the ratings are uniformly distributed on $[0, 10]$. Again, the monarch wants the best candidate among the one hundred. What strategy should he employ now? What is the probability of success?

This problem goes by a great many names, such as the *secretary problem* and the *optimal stopping problem*.

4.9 Answers to Selected Exercises

Answer to 4.3. Let $X_1, X_2, \ldots, X_5$ be five points chosen uniformly and independently from $[1, 2]$. What is $\log(X_{(1)})$? The answer is the probability distribution

$$P\big(\log(X_{(1)}) \leq t\big) = P\big(X_{(1)} \leq e^t\big)$$

Now one must be careful: the interval $[1, 2]$ is not of the form $[0, a]$. The distribution $X_{(1)}$ for this interval is, by the change of variables $x \to x - 1$, given by

$$F_{(1)}(x) = \begin{cases} 0 & \text{if } x < 1 \\ 1 - (2 - x)^5 & \text{if } 1 \leq x \leq 2 \\ 0 & \text{if } x > 2 \end{cases}$$

Hence,

$$P\big(\log(X_{(1)}) \leq t\big) = F_{(1)}\big(e^t\big) = \begin{cases} 0 & \text{if } t < 0 \\ 1 - (2 - e^t)^5 & \text{if } 0 \leq t \leq \log(2) \\ 0 & \text{if } t > \log(2) \end{cases}$$

Differentiating, we find the density

$$f(t) = \begin{cases} 5(2 - e^t)^4 e^t & \text{if } 0 \leq t \leq \log(2) \\ 0 & \text{otherwise} \end{cases}$$

We now graph the densities of $X_{(1)}$ and $\log(X_{(1)})$ both on $[0, \log(2)]$. However, one must be a little careful here. The second $X_{(1)}$ was originally on $[1, 2]$, so the two $X_{(1)}$'s are not the same random variable. We have also introduced the **plot** parameters for specifying axis labels. Up to now, we have used the expressions that define the points being plotted. These are not appropriate for most publications. One should normally specify both the x-axis label with `xlab` and the y-axis label with `ylab`, and the graph as a whole with `main`.

```
> X1 <- function(x) 5*(log(2)-x)^4/(log(2))^5
> lX1 <- function(x) 5*(2-exp(x))^4*exp(x)
> trange <- (0:(log(2)*100))/100
> plot(trange, sapply(trange, X1), type="l",
+        xlab="x",ylab="First order statistic",
+        main="Comparison of X(1) and log(X(1))")
> points(trange, sapply(trange, lX1), type="l")
```

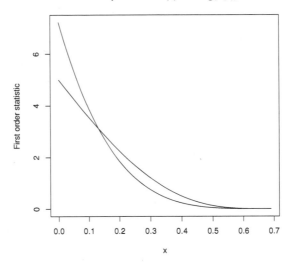

The density of $X_{(1)}$ is much more concentrated at 0 than $\log(X_{(1)})$. One can intuitively see why this should be so. Observe that $\log(x)$ tends to "spread out" the early part of $[1, 2]$ and to "bunch up" the later part. For example, 1.5 is halfway between 1 and 2, but $\log(1.5) \approx 0.4$ is over 58% of the way from 0 to $\log(2)$. As a result, the smallest point among the $\log(X_i)$'s will tend to be larger than the smallest point among the X_i's. Notice also that $\operatorname{dens}(\log(X_{(1)}) = t)$ is *not* obtained by substituting $e^t - 1$ into $f_{(1)}(x)$. When changing variables, one must use the distribution function, not the density.

Answer to 4.4. Let X be any random variable. The distribution function $F(x) = P(X \leq x)$ is an increasing function, and we know that $\lim_{x \to -\infty} F(x) = 0$, and $\lim_{x \to \infty} F(x) = 1$. Therefore, there will be values of x for which $F(x) < \frac{1}{2}$, and values of x for which $F(x) \geq \frac{1}{2}$. So the set $\{x \mid F(x) \geq \frac{1}{2}\}$ is a nonempty set of real numbers, which has a lower bound. By the properties of real numbers, there is a "limit" μ, which is a real number such that $F(x) < \frac{1}{2}$ for all $x < \mu$ and $F(x) \geq \frac{1}{2}$ for all $x > \mu$. At μ itself we can say nothing yet.

By part (a) of Exercise 1.13,

$$P(X<\mu) = \lim_{\substack{x \to \mu \\ x < \mu}} P(X \leq x)$$

This probability can be at most $\frac{1}{2}$ because $P(X \leq x) < \frac{1}{2}$ for all $x < \mu$. Similarly,

$$P(X>\mu) = \lim_{\substack{x \to \mu \\ x > \mu}} P(X>x) = \lim_{\substack{x \to \mu \\ x > \mu}} (1 - P(X \leq x))$$

is at most $\frac{1}{2}$ because $P(X \leq x) \geq \frac{1}{2}$ for all $x > \mu$. Therefore,

$$P(X \leq \mu) = 1 - P(X>\mu) \geq \frac{1}{2}$$

$$P(X \geq \mu) = 1 - P(X<\mu) \geq \frac{1}{2}$$

We conclude that μ is a median of X as desired.

While the median exists, it need not be unique. For example, let X take values ± 1 with equal probability. Then any real number μ in the interval $[-1, 1]$ is a median for X. Probability distributions need not have a mean. We will see an example in Section 5.7. The fact that the median exists, even when a mean may not, makes it a more useful statistical quantity in many applications.

Answer to 4.6. The spread S is

$$S = X_{(n)} - X_{(1)} = a - \left(L_{(1)} + L_{(n+1)}\right)$$

Now in Section 4.5 we computed the joint density of the order statistics. So in particular, the joint density of $X_{(1)}$ and $X_{(n)}$ is

$$\text{dens}\left(X_{(1)} = x_1, X_{(n)} = x_n\right) = \begin{cases} n(n-1)\frac{(x_n - x_1)^{n-2}}{a^n} & \text{if } 0 \leq x_1 \leq x_n \leq a \\ 0 & \text{otherwise} \end{cases}$$

To compute the distribution function of the spread, we consider the event $\left(X_{(n)} - X_{(1)} \leq s\right) = \left(X_{(n)} \leq s + X_{(1)}\right)$. The values of x_1 and x_n for which this event occurs are those pairs in a trapezoidal region. We illustrate this region in the case $a = 5$ and $s = 3$:

```
> plot(c(0,0,2,5,0), c(0,3,5,5,0), type="l",
+      xlab="x5", ylab="x1")
```

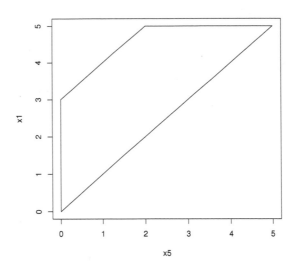

Therefore,

$$P\big(X_{(n)} - X_{(1)} \leq s\big) = \int_0^{a-s} \int_{x_1}^{x_1+s} \frac{n(n-1)}{a^n}(x_n - x_1)^{n-2} dx_n dx_1$$

$$+ \int_{a-s}^a \int_{x_1}^a \frac{n(n-1)}{a^n}(x_n - x_1)^{n-2} dx_n dx_1$$

$$= \frac{n(n-1)}{a^n} \int_0^{a-s} \left[\frac{(x_n - x_1)^{n-1}}{n-1}\right]_{x_1}^{x_1+s} dx_1$$

$$+ \frac{n(n-1)}{a^n} \int_{a-s}^a \left[\frac{(x_n - x_1)^{n-1}}{n-1}\right]_{x_1}^a dx_1$$

$$= \frac{n}{a^n}\left(\int_0^{a-s} s^{n-1} dx_1 + \int_{a-s}^a (a - x_1)^{n-1} dx_1\right)$$

$$= \frac{n}{a^n}\left(\left[s^{n-1}x_1\right]_0^{a-s} + \left[-\frac{(a-x_1)^n}{n}\right]_{a-s}^a\right)$$

$$= \frac{n}{a^n}\left(s^{n-1}(a-s) + \frac{s^n}{n}\right)$$

The density of the spread is then

$$\operatorname{dens}(S = s) = \frac{n(n-1)}{a^n}(a-s)s^{n-2}$$

which is the same as the density of the $(n-2)^{nd}$ order statistic! The reason for this surprising fact is that the L_i's are not only equidistributed but in fact have the much stronger property of **exchangeability** that we discuss in Chapter 6. Using this property, one can compute the distribution of the spread in a much easier way. See Exercise 6.33.

Answer to 4.7. Let $F(t)$ be a function satisfying the four properties of a distribution function. Let the sample space Ω be the real line $(-\infty, \infty)$. Define the elementary events to be the half-open intervals $(a, b]$. The events of Ω are all those obtainable by Boolean operations from the elementary events. We define the probability measure P on elementary events by $P((a, b]) = F(b) - F(a)$. The four properties of a distribution function ensure that P is a probability measure. Finally, we define a random variable on Ω by $X(\omega) = \omega$ for all $\omega \in \Omega$. Clearly $F(t)$ is the distribution function of X.

Answer to 4.16. Assume that each DNA molecule has a preferred direction or orientation. We do this just to have a way of referring to a position on the molecule. As a result, the point where the enzyme breaks the molecule may be viewed as a point in $[0, q]$. If this point is completely random, and if every molecule is independently broken, then we may use the Uniform process to model this phenomenon. Here the number of points is 24 and the length of the interval is the length of each DNA molecule. The shortest piece produced is either $X_{(1)}$ or $a - X_{(24)}$, whichever is shorter. That is, we wish to compute the distribution of the random variable $L = \min\left(L_{(1)}, L_{(25)}\right)$. We will, in fact, compute the distribution in general for $L = \min\left(L_{(1)}, L_{(n+1)}\right)$.

The first step is to find the joint density of $L_{(1)}$ and $L_{(n+1)}$. We use the procedure of Section 4.5. To compute $\mathrm{dens}\left(L_{(1)} = x, L_{(n+1)} = y\right)$, consider the event

$$\left(x < L_{(1)} \leq x + h\right) \cap \left(y < L_{(n+1)} \leq y + \epsilon\right) =$$
$$\left(x < X_{(1)} \leq x + h\right) \cap \left(a - y - \epsilon \leq X_{(n+1)} < a - y\right)$$

which can be represented pictorially like this:

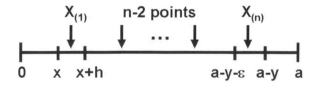

Except for terms that disappear in the limit, the probability of this event is

$$\binom{n}{1, n-2, 1} \frac{h}{a} \left(\frac{a - y - \epsilon - x - h}{a}\right)^{n-2} \frac{\epsilon}{a}$$

Dividing by $h\epsilon$ and letting $h, \epsilon \to 0$, we have

$$\mathrm{dens}\left(L_{(1)} = x, L_{(n+1)} = y\right) = \begin{cases} \binom{n}{1, n-2, 1} \dfrac{(a-y-x)^{n-2}}{a^n} & \text{if } 0 \leq x + y \leq a \\ 0 & \text{otherwise} \end{cases}$$

The restrictions on x and y are important. Notice that

$$\binom{n}{1, n-2, 1} = \binom{n}{1}\binom{n-1}{n-2}\binom{1}{1} = n(n-1)$$

We are now ready to compute the distribution of L. We use the same technique that we used in Section 4.1 for computing the distribution of $X_{(1)} = \min(X_1, X_2, \ldots, X_n)$. But now, of course, it is more complicated because $L_{(1)}$ and $L_{(n+1)}$ are not independent. First note that the values of L range from 0 to $a/2$. Thus, if l is in the interval $[0, a/2]$, then

$$
\begin{aligned}
P(L>l) &= P\big((L_{(1)}>l)\cap(L_{(n+1)}>l)\big) \\[2mm]
&= \int_{y>l}\int_{x>l \text{ and } x+y\leq a} \mathrm{dens}\big(L_{(1)} = x, L_{(n+1)} = y\big)\, dx dy \\[2mm]
&= \int_{l}^{a-l}\int_{l}^{a-x} \frac{n(n-1)}{a^n}(a-y-x)^{n-2} dy dx \\[2mm]
&= \frac{n(n-1)}{a^n}\int_{l}^{a-l}\left[-\frac{(a-y-x)^{n-1}}{n-1}\right]_{y=l}^{y=a-x} dx \\[2mm]
&= \frac{n}{a^n}\int_{l}^{a-l}(a-l-x)^{n-1} dx \\[2mm]
&= \frac{n}{a^n}\left[-\frac{(a-l-x)^n}{n}\right]_{l}^{a-l} \\[2mm]
&= (a-2l)^n a^{-n}
\end{aligned}
$$

Written in R this is

```
> minpiece <- function(a,n,l) ((a-2*l)/a)^n
```

The density of L is therefore

$$
\mathrm{dens}(L = l) = \begin{cases} \dfrac{2n}{a}\left(\dfrac{a-2l}{a}\right)^{n-1} & \text{if } 0\leq l\leq \frac{a}{2} \\[3mm] 0 & \text{otherwise} \end{cases}
$$

The expectation of L is, by an easy computation,

$$E(L) = \frac{a}{2(n+1)}$$

For our special case of $n = 24$, the average size of the smallest piece produced is $E(L) = \frac{a}{50}$.

One can devise a test of the randomness of the breakage by this enzyme, using this calculation. Suppose that the DNA molecule is 220 million base pairs long. The probability distribution looks like this:

```
> pts <- 10000
> a <- 220*10^6
```

```
> l <- (1:m)*(a/2)/m
> p <- mapply(minpiece, a, 24, l)
> plot(l, p, type="l")
```

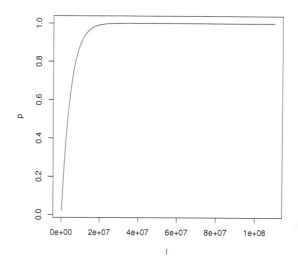

Now compute the values of l for which $P(L{\leq}l)$ is equal to 0.025 and 0.975. Call them l_1 and l_2. Then

$$P(l_1{<}L{\leq}l_2) = P(L{\leq}l_2) - P(L{\leq}l_1) = 0.975 - 0.025 = 0.95$$

To compute l_1 with R we look for the first value of l where p exceeds 0.025:

```
> min(l[p>0.025])
[1] 121000
```

Similarly to compute l_2 we look for the first value of l where p exceeds 0.975:

```
> min(l[p>0.925])
[1] 15675000
```

If the length of the smallest piece falls within these bounds, we conclude that nothing unusual has happened. If L falls outside these bounds, then we suspect that the enzyme may not be breaking the DNA molecule completely at random. It may be, for example, that the enzyme prefers the ends of the DNA molecule. Such a conclusion, however, would require further study. See Section 5.4 for more examples of such tests.

Answer to 4.17. Note that this is *not* correctly modeled by W_1 in the Bernoulli process when $p = 1/4$ because the sampling is not being done with replacement. To see why W_1 is not the correct model, just note that in the Bernoulli process there is no maximum number of tosses that may be necessary to get the first head. Whereas, in this problem, the maximum number of cards one can turn up before getting a spade is 39. The first case is that a spade is turned up immediately at the start. This has probability $1/4$. For the rest of the turns, one turns up a succession of nonspades followed by a spade. For the case of n nonspades, the term in the computation of the expectation is:

$$(n+1)\frac{39\cdots(39-n+1)}{52\cdots(52-n+1)}\frac{13}{52-n}$$

The first factor above ranges from 2 to 40, the numerators of the second factor range from 39 down to 1, the denominators range from 52 down to 14, and the denominator of the last factor ranges from 51 down to 13. For the second factor, the factors accumulate. The cumprod function does the accumulation, so the expected number of cards is

```
> 1/4+sum((2:40)*cumprod((39:1)/(52:14))*13/(51:13))
[1] 3.785714
```

If one multiplies all of these terms out manually, the exact answer is $53/14$. The analysis is similar for the case of one of n cards rather than one of 13, and the average number of cards that we must expose in general is $\frac{53}{n+1}$.

Answer to 4.18. Since $E(X_{(n)}) = \dfrac{nA}{n+1}$, the random variable with the desired property is $\dfrac{n+1}{n}X_{(n)}$.

Answer to 4.20. The expectation of $L_{(1)}$ is

$$E(L_{(1)}) = \frac{a}{(n+1)^2}$$

The expectation of $L_{(2)}$ is

$$
\begin{aligned}
E(L_{(2)}) &= E(L_{(1)}) + E\left(\frac{a-(n+1)L_{(1)}}{n^2}\right)\\
&= \frac{a}{(n+1)^2} + \frac{a}{n^2} - \frac{(n+1)a}{n^2(n+1)^2}\\
&= \frac{a}{n+1}\left(\frac{1}{n+1}+\frac{1}{n}\right)
\end{aligned}
$$

In general,

$$E(L_{(k)}) = \frac{a}{n+1}\left(\frac{1}{n+1}+\cdots+\frac{1}{n-k+2}\right)$$

Answer to 4.21. Applying the results of Exercise 4.20, the probability that the smallest of 10 pieces could be this short or shorter is 8.6%. Consequently, the occurrence is somewhat unusual, but not particularly surprising.

CHAPTER 5

Statistics and the Normal Distribution

Abraham de Moivre was a French mathematician best known today for de Moivre's formula, which links complex numbers and trigonometry. A French Huguenot, de Moivre was jailed in 1685, but was released shortly thereafter and fled to England. He had a considerable reputation as a mathematician, with many distinctions and honors such as election to the Royal Society of London, as well as to the Berlin and Paris academies. He was also a close friend of both Isaac Newton and Edmond Halley. Yet he never succeeded in securing a permanent position, and made his living precariously as a tutor and consultant. Among his many contributions to probability theory, he was responsible for the definitions of statistical independence as well as the normal distribution. He discovered Stirling's formula, now incorrectly attributed to James Stirling, and used it to prove that the binomial distribution converges to the normal distribution, a fact that we now recognize as an important special case of the Central Limit Theorem.

The normal distribution arises whenever we make a succession of imperfect measurements of a quantity that is supposed to have a definite value. If all the students in a class take the same test, we may think of their grades as being imperfect measurements of the average capability of the class. In general, when we make a number of independent measurements, the average is intuitively going to be an approximation of the quantity we are trying to find.

On the other hand, the various measurements will tend to be more or less spread out on both sides of the average value. We need a measure for how far individual measurements are spread out around the quantity being measured. This will tell us, for example, how many measurements must be made in order to determine the quantity to a certain accuracy. It will also make it possible to formulate statistical tests to determine whether or not the data in an experiment fit the model we have proposed for the experiment.

5.1 Variance

The variance of a random variable X is a measure of the spread of X away from its mean.

Definition. Let X be a random variable whose mean is $E(X) = m$. The **variance** of X is $Var(X) = E\big((X - m)^2\big)$, if this expectation converges. The square root of the variance is called the **standard deviation** of X and is written $\sigma(X) = \sqrt{Var(X)}$. We sometimes write $\sigma^2(X)$ for $Var(X)$.

If X is an integer random variable having probability distribution $p_n = P(X = n)$, then

$$Var(X) = \sum_n (n - m)^2 p_n$$

If X is a continuous random variable whose density is $f(x) = \text{dens}(X = x)$, then

$$Var(X) = \int_{-\infty}^{\infty} (x - m)^2 f(x)dx$$

In the continuous case we can imagine that $f(x)$ is the density at x of a thin rod. This rod has a total mass of 1 and balances at the mean m. The moment of inertia of this rod about its balance point is precisely the variance. If we rotate the rod about the balance point, it would have the same angular momentum if all the mass were concentrated at a distance $\sigma(X)$ from the point of rotation.

It is possible for a random variable not to have a mean. We give an example in Section 5.7. It is also possible for a random variable to have a mean but not to have a finite variance. We shall see examples in the exercises. However, in a great many physical processes it is reasonable to assume that the random variables involved do have a finite variance (and hence also a mean). For example, on an exam if the possible scores range from 0 to 100, the measurement of someone's exam score is necessarily going to have a finite variance.

A useful formula for the variance is the following:

$$Var(X) = E\big(X^2\big) - E(X)^2$$

This is an easy formula to verify. The crucial step is that the expectation is additive, even when the random variables involved are not independent.

$$
\begin{aligned}
Var(X) &= E\big((X - m)^2\big) \\
&= E\big(X^2 - 2mX + m^2\big) \\
&= E\big(X^2\big) - 2mE(X) + m^2 \\
&= E\big(X^2\big) - 2m^2 + m^2 \\
&= E\big(X^2\big) - m^2
\end{aligned}
$$

As we just remarked, the expectation is additive. In general, it is not multiplicative; that is, $E(XY)$ need not be $E(X)E(Y)$. The variance is a measure of the extent to which the expectation is not multiplicative when $X = Y$, for in this case $Var(X) = E(X \cdot X) - E(X)E(X)$. The **covariance** of X and Y in general is the difference

$$Cov(X,Y) = E(XY) - E(X)E(Y)$$

We will not be using covariance except in a few optional exercises. Covariance is often used as a measure of the independence of random variables because of the following important fact:

Product Rule for Expectation. If X and Y are independent random variables, then $E(XY) = E(X)E(Y)$, or equivalently $Cov(X,Y) = 0$.

Proof. We will only consider the case of integer random variables. The proof for the continuous case requires annoying technicalities that obscure the basic idea. We leave this as an exercise.

We compute the distribution of the product XY in terms of the distributions of X and of Y. The event $(XY = n)$ is the disjoint union of the events $(X = j) \cap (Y = k)$ for all integers j and k whose product is n. Hence,

$$P(XY = n) = \sum_{jk=n} P((X = j) \cap (Y = k))$$

Now X and Y are independent, so

$$P(XY = n) = \sum_{jk=n} P(X = j)\,P(Y = k)$$

Next, we use this formula to compute the expectation of XY; namely,

$$
\begin{aligned}
E(XY) &= \sum_n nP(XY = n) \\
&= \sum_n n \sum_{jk=n} P(X = j)\,P(Y = k) \\
&= \sum_n \sum_{jk=n} jkP(X = j)\,P(Y = k)
\end{aligned}
$$

Now summing over n and then summing over all j's and k's whose product is n is the same as simply summing over all j's and k's with no restrictions. Thus,

$$
\begin{aligned}
E(XY) &= \sum_j \sum_k jkP(X = j)\,P(Y = k) \\
&= \sum_j \sum_k jP(X = j)\,kP(Y = k) \\
&= \sum_j jP(X = j) \sum_k kP(Y = k) \\
&= E(X)E(Y)
\end{aligned}
$$

This gives the Product Rule. Note that, unlike the Addition Rule for Expectations, we must have that X and Y are independent for the product rule to work.

We add that it is possible for nonindependent random variables X and Y to satisfy $E(XY) = E(X)E(Y)$. As a result, the covariance is not a true measure of independence. If $Cov(X, Y) = 0$, one says that X and Y are *uncorrelated*.

We now derive the most important consequence of the Product Rule for Expectations; namely, the variance is additive for independent random variables. In fact, this property uniquely characterizes the variance among all possible measures of the "spread" of a random variable about its mean.

Addition Rule for Variance. If X and Y are independent, then

$$Var(X + Y) = Var(X) + Var(Y)$$

Proof. To see why this is so, we simply calculate $Var(X+Y)$ using the Product Rule for Expectations:

$$
\begin{aligned}
Var(X + Y) &= E((X + Y)^2) - (E(X + Y))^2 \\
&= E(X^2 + 2XY + Y^2) - (E(X) + E(Y))^2 \\
&= E(X^2) + 2E(XY) + E(Y^2) \\
&\quad - E(X)^2 - 2E(X)E(Y) - E(Y)^2 \\
&= E(X^2) - E(X)^2 + E(Y^2) - E(Y)^2 \\
&= Var(X) + Var(Y)
\end{aligned}
$$

In terms of the standard deviations, $\sigma(X + Y) = \sqrt{\sigma^2(X) + \sigma^2(Y)}$; the standard deviations, $\sigma(X)$ and $\sigma(Y)$, of independent random variables act like the components of a vector whose length is $\sigma(X + Y)$.

There are two more properties of the variance that are important for us. Both are quite obvious:

$$Var(cX) = c^2 Var(X) \qquad \text{and} \qquad Var(X + c) = Var(X)$$

The first expresses the fact that the variance is a quadratic concept. The second is called **shift invariance**. It should be obvious that merely shifting the value of a random variable X by a constant only changes the mean and not the spread of the measurement about the mean.

(0)	$Var(X) = E(X^2) - E(X)^2$	$\sigma(X) = \sqrt{Var(X)}$
(1)	If X and Y are independent, then:	
	$Var(X + Y) = Var(X) + Var(Y)$	$\sigma(X + Y) = \sqrt{\sigma^2(X) + \sigma^2(Y)}$
(2)	$Var(cX) = c^2 Var(X)$	$\sigma(cX) = c\sigma(X)$
(3)	$Var(X + c) = Var(X)$	$\sigma(X + c) = \sigma(X)$

Basic Properties of Variance and Standard Deviation

We now compute the variances of some of the random variables we have encountered so far. Consider a single toss of a biased coin. This is described by any random variable X_n of the Bernoulli process. Recall that

$$X_n = \begin{cases} 0 & \text{if the } n^{th} \text{ toss is tails} \\ 1 & \text{if the } n^{th} \text{ toss is heads} \end{cases}$$

Since $1^2 = 1$ and $0^2 = 0$, X_n^2 is the same as X_n. Therefore, the variance of any X_n is

$$Var(X_n) = E(X_n^2) - E(X_n)^2 = E(X_n) - E(X_n)^2 = p - p^2 = pq$$

The variance of X_n as a function of the bias p is

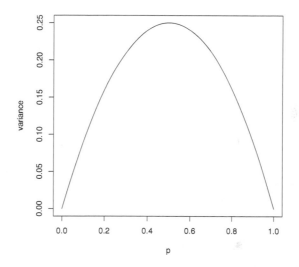

Notice that the largest variance corresponds to a fair coin ($p = 1/2$). We intuitively think of a fair coin as having the most "spread out" distribution of all biased coins; the more biased a coin is, the more its distribution is "concentrated" about its mean.

Next, consider the number of heads S_n in n tosses of a biased coin. Since $S_n = X_1 + X_2 + \cdots + X_n$ is the sum of n independent random variables all of whose variances are the same,

$$Var(S_n) = nVar(X_1) = npq$$

If we try to compute $Var(S_n)$ directly from the definition, we get

$$Var(S_n) = E((S_n - np)^2) = \sum_{k=0}^{n} (k - np)^2 \binom{n}{k} p^k q^{n-k}$$

That this is npq is far from obvious.

We leave the computation of the variances of the gaps T_k and the waiting times W_k as exercises.   The results are:

$$Var(T_k) = \frac{q}{p^2} \qquad Var(W_k) = \frac{kq}{p^2}$$

Now consider a point X dropped at random uniformly on $[0, a]$. Clearly, the average value of the point is $a/2$, the midpoint of $[0, a]$. The variance is

$$
\begin{aligned}
Var(X) &= E(X^2) - E(X)^2 \\
&= \int_0^a x^2 \mathrm{dens}(X = x)\, dx - \left(\frac{a}{2}\right)^2 \\
&= \int_0^a x^2 \left(\frac{1}{a}\right) dx - \left(\frac{a}{2}\right)^2 \\
&= \frac{1}{a}\left[\frac{x^3}{3}\right]_0^a - \frac{a^2}{4} \\
&= \frac{1}{a}\left(\frac{a^3}{3}\right) - \frac{a^2}{4} \\
&= \frac{a^2}{12}
\end{aligned}
$$

So the standard deviation is

$$\frac{a}{\sqrt{12}} = \frac{a}{2\sqrt{3}} \approx 0.2887a$$

We can think of this in the following way. Given a uniform bar of length a, its midpoint is the center of mass. If the bar were set spinning around its center of mass, the angular momentum would be the same if the mass were all at a distance $\dfrac{a}{2\sqrt{3}}$ from the center of rotation.

We leave it as an exercise to compute the variances of the gaps and the order statistics of the Uniform process. Notice that we cannot use the fact that $X_{(k)} = L_1 + L_2 + \ldots + L_k$, because the gaps are *not* independent. The variances are:

$$
\begin{aligned}
Var(L_k) &= \frac{a^2 n}{(n+1)^2(n+2)} \\
Var(X_{(k)}) &= \frac{a^2 k(n - k + 1)}{(n+1)^2(n+2)}
\end{aligned}
$$

Distribution	Process Model	Mean	Variance
Bernoulli	X_i in Bernoulli	p	pq
Binomial	S_n in Bernoulli	np	npq
Geometric	W_1 or any T_k in Bernoulli	$\dfrac{1}{p}$	$\dfrac{q}{p^2}$
Negative Binomial	W_k in Bernoulli	$\dfrac{k}{p}$	$\dfrac{kq}{p^2}$
Uniform	X_i in Uniform	$\dfrac{a}{2}$	$\dfrac{a^2}{12}$
Beta	$X_{(k)}$ in Uniform also L_i when $k = 1$	$\dfrac{ka}{n+1}$	$\dfrac{a^2 k(n-k+1)}{(n+1)^2(n+2)}$

Important Expectations and Variances

Standardization

If we shift a random variable X by a constant, replacing X by $X + c$, or if we make a scale change, multiplying X by a nonzero constant, we have not altered X in a significant way. We have only reinterpreted a measurement of X by a linear change of variables. The idea of standardization is to choose a single "standard" random variable among all those related to one another by a linear change of variables. Then, in order to determine if two random variables are "essentially" the same, we should compare their standardized versions.

Definition. A random variable X is **standard** if $E(X) = 0$ and $Var(X) = 1$. If X has mean m and finite variance σ^2, then

$$\frac{X - m}{\sigma}$$

is standard. We call $(X - m)/\sigma$ the **standardization** of X. A physicist would say that $(X - m)/\sigma$ expresses X in "dimensionless units."

The covariance of the standardizations of two random variables X and Y is called the **coefficient of correlation** and is written $\rho(X, Y)$. It is easy to prove that $|\rho(X, Y)| \leq 1$ and that $\rho(X, Y) = Cov(X, Y)/(\sigma(X)\sigma(Y))$. We leave these as exercises. If $|\rho(X, Y)| \approx 1$, then X and Y are nearly the same, except for rescaling and shifting.

We call $\sigma(X)$ the *standard* deviation because of its appearance in the standardization. We think of $\sigma(X)$ as being the natural unit for measuring how far a given observation of X deviates from the mean.

5.2 Bell-Shaped Curve

In this section we introduce one of the most important distributions in probability: the normal distribution. The traditional explanation for the importance of the normal distribution relies on the Central Limit Theorem, which we will discuss in Section 5.3. However, we feel that the explanation using entropy and information, given in Chapter 9, is better because it provides a context which explains the ubiquity not only of the normal distribution but of several other important distributions as well.

Definition. A continuous random variable X is said to have the **normal distribution** or with mean m and variance σ^2 if

$$\text{dens}(X = x) = \frac{1}{\sigma\sqrt{2\pi}} \exp\left(\frac{-(x-m)^2}{2\sigma^2}\right)$$

For brevity, we will write simply "X is $N(m, \sigma^2)$." Some authors write $N(m, \sigma)$ — instead of $N(m, \sigma^2)$ — one should beware.

Another name for the normal distribution is the *Gaussian distribution*, especially in older texts and articles. It would make much more sense, from a historical point of view, to name the normal distribution after de Moivre, who introduced it, if it is to be named after any person. Unlike most distributions, the formula for the normal density comes with the mean and variance already specified. We should, however, verify that m and σ^2 really are the mean and the variance. In fact, it is not obvious that this formula actually defines the density of a continuous random variable. To verify these facts we use the following basic formula, which everyone ought to have seen in a calculus course at some time:

$$\int_{-\infty}^{\infty} e^{-x^2}\,dx = \sqrt{\pi}$$

To prove this, let $A = \displaystyle\int_{-\infty}^{\infty} e^{-x^2}\,dx$. Then, since x is a dummy variable, $A = \displaystyle\int_{-\infty}^{\infty} e^{-y^2}\,dy$ also. Therefore, $A^2 = \displaystyle\int_{-\infty}^{\infty} e^{-x^2}\,dx \int_{-\infty}^{\infty} e^{-y^2}\,dy$. Now we switch to polar coordinates and integrate:

$$\begin{aligned}
A^2 &= \int_{-\infty}^{\infty}\int_{-\infty}^{\infty} e^{-(x^2+y^2)}\,dxdy. \\
&= \int_{0}^{2\pi}\int_{0}^{\infty} e^{-r^2} r\,dr d\theta \\
&= \int_{0}^{2\pi}\left[-\frac{1}{2}e^{-r^2}\right]_{0}^{\infty} d\theta \\
&= \int_{0}^{2\pi}\frac{1}{2}d\theta = \pi
\end{aligned}$$

Hence, $A = \sqrt{\pi}$.

We use this formula to show that the normal density really defines a density. We first change variables to $y = \dfrac{x - m}{\sigma\sqrt{2}}$ so that $\sigma\sqrt{2}\,dy = dx$. Then

$$\int_{-\infty}^{\infty} \frac{1}{\sigma\sqrt{2\pi}} \exp\left(-\frac{(x-m)^2}{2\sigma^2}\right) dx \;=\; \int_{-\infty}^{\infty} \frac{1}{\sigma\sqrt{2\pi}} e^{-y^2} \sigma\sqrt{2}\,dy$$

$$= \frac{1}{\sqrt{\pi}} \int_{-\infty}^{\infty} e^{-y^2}\,dy$$

$$= 1$$

Next, we compute $E(X)$ when X is $N(m, \sigma^2)$. Again, we use the change of variables $y = \dfrac{x - m}{\sigma\sqrt{2}}$. Note that $x = \sigma\sqrt{2}y + m$.

$$
\begin{aligned}
E(X) &= \int_{-\infty}^{\infty} x \frac{1}{\sigma\sqrt{2\pi}} \exp\left(-\frac{(x-m)^2}{2\sigma^2}\right) dx \\[6pt]
&= \int_{-\infty}^{\infty} \left(\sigma\sqrt{2}y + m\right) \frac{1}{\sigma\sqrt{2\pi}} e^{-y^2} \sigma\sqrt{2}\,dy \\[6pt]
&= \frac{1}{\sqrt{\pi}} \int_{-\infty}^{\infty} \left(\sigma\sqrt{2}y + m\right) e^{-y^2}\,dy \\[6pt]
&= \frac{\sigma\sqrt{2}}{\sqrt{\pi}} \int_{-\infty}^{\infty} e^{-y^2}\,dy + \frac{m}{\sqrt{\pi}} \int_{-\infty}^{\infty} e^{-y^2}\,dy \\[6pt]
&= \frac{\sigma\sqrt{2}}{\sqrt{\pi}} \left[\frac{-1}{2} e^{-y^2}\right]_{-\infty}^{\infty} + \frac{m}{\sqrt{\pi}} \sqrt{\pi} \\[6pt]
&= 0 + m
\end{aligned}
$$

Finally, we leave it as an exercise to show that $Var(X) = \sigma^2$. It can be done using integration by parts.

Although these computations look messy, we are inescapably forced to consider this density function because this is the distribution corresponding to the concept of *total randomness* or complete *randomness*. In Chapter 9, we will make this concept more precise. So it is important that one has an intuitive idea of what it means for a random variable to be normal.

The normal density function is symmetric about the mean m and its mode is the same as the mean. Beyond $3.5\,\sigma$ units from m, the value of the normal density is essentially zero. The natural unit for measuring deviations from the mean is the standard deviation. When x is the deviation from the mean measured in this natural unit, then we get the standard normal density. Two other commonly used properties of the normal distribution are:

- The area within 1.96 standard deviations of the mean is 0.95.
- The area within 2.58 standard deviations of the mean is 0.99.

These will be important when we compute significance levels.

Occasionally, one will see tables of the **error function**, $erf(t)$. This function is closely related to the normal distribution although it is not the same:

$$erf(t) = P(|Y| \leq t) = \frac{1}{\sqrt{\pi}} \int_{-t}^{t} e^{-y^2} dy$$

where Y has the distribution $N(0, 1/2)$. If X has the standard normal distribution, then

$$P(-x \leq X \leq x) = erf\left(\frac{x}{\sqrt{2}}\right)$$

and

$$P(X \leq x) = \frac{1}{2}\left(1 + erf\left(\frac{x}{\sqrt{2}}\right)\right)$$

5.3 The Central Limit Theorem

The traditional explanation for the importance of the normal distribution relies on the Central Limit Theorem. Briefly, this theorem states that the average of a sequence of independent, equidistributed random variables tends to the normal distribution no matter how the individual random variables are distributed. The explanation for the ubiquity of the normal distribution then goes as follows. Suppose that X is the random variable representing the measurement of a definite quantity but which is subject to chance errors. The various possible imperfections (minute air currents, stray magnetic fields, etc.) are supposed to act like independent equidistributed random variables whose sum is the total error of the measurement X. Unfortunately, this explanation fails to be very convincing because there is no reason to suppose that the various contributions to the total error are either independent or equidistributed. We will have to wait until Chapter 9 to find a more fundamental reason for the appearance of the normal distribution. The explanation given there uses the concepts of entropy and information.

Intuitively, the sum of independent, equidistributed random variables is progressively more disordered as we add more and more of them. As a result, the standardization of the sum necessarily approaches having a normal distribution as $n \to \infty$. This tendency to become disordered is exhibited even when the random variables are not quite independent and equidistributed. It is this tendency that accounts for the ubiquity of the normal distribution.

Suppose that $X_1, X_2, \ldots, X_n, \ldots$ are independent equidistributed random variables whose common mean and variance are $m = E(X_i)$ and $\sigma^2 = Var(X_i)$. Let S_n be the sum $X_1 + X_2 + \cdots + X_n$. Then the mean of S_n is

$$E(S_n) = E(X_1) + E(X_2) + \cdots + E(X_n) = nm$$

and its variance is

$$Var(S_n) = Var(X_1) + Var(X_2) + \cdots + Var(X_n) = n\sigma^2$$

since the X_i's are independent. Therefore, the standard deviation of S_n is

$\sqrt{Var(S_n)} = \sigma\sqrt{n}$, and the standardization of S_n is

$$Y_n = \frac{S_n - nm}{\sigma\sqrt{n}} = \frac{X_1, X_2, \ldots, X_n - nm}{\sigma\sqrt{n}}$$

This is an important formula to remember. The Central Limit Theorem then says that Y_n tends to the normal distribution as $n\to\infty$.

Central Limit Theorem. If $X_1, X_2, \ldots, X_n, \ldots$ are independent, equidistributed random variables with mean m and variance σ^2, then

$$P(Y_n \leq t) = P\left(\frac{X_1, X_2, \ldots, X_n - nm}{\sigma\sqrt{n}} \leq t\right) \to \frac{1}{\sigma\sqrt{2\pi}}\int_{-\infty}^{t} e^{-\frac{x^2}{2}}\,dx \quad \text{as } n\to\infty$$

We present a proof of the Central Limit Theorem in Section 8.5. For example, in the Bernoulli process, the random variable S_n is the sum of independent, equidistributed random variables X_i whose common mean is $m = p$ and whose common variance is $\sigma^2 = pq$. Then $Y_n = \dfrac{S_n - np}{\sqrt{npq}}$ tends toward the standard normal distribution. That is, S_n is approximately distributed according to $N(np, npq)$. This approximation is surprisingly accurate even for small values of n. Here is the distribution of S_4 in the Bernoulli process using a fair coin, with the normal distribution $N(2, 1)$ superimposed:

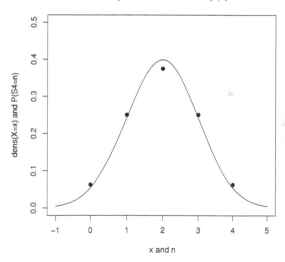

Comparison of S4 with N(2,1)

For example, we know that $P(1 \leq S_4 \leq 3) = 1 - \frac{2}{16} \approx 0.875$. On the other hand, when X is $N(2, 1)$, $P(0.5 \leq X \leq 3.5) \approx 0.8664$. You can see that the fit is quite close. We now see why we chose the scaling in our animation in Section 3.1.

The two most common manifestations of the Central Limit Theorem are the following:

- As $n\to\infty$, the sum S_n "tends" to the distribution $N(nm, n\sigma^2)$.

- As $n \to \infty$, the **sample average** (or **sample mean**) $\overline{m} = \dfrac{S_n}{n}$ "tends" to the distribution $N(m, \sigma^2/n)$.

Note that the sample average $\overline{m}$ is a random variable. It is easy to see that its expectation is the mean:

$$E(\overline{m}) = E\left(\frac{S_n}{n}\right) = \frac{nm}{n} = m$$

Hence, the sample mean is an approximation to the true mean m. The variance of the sample mean depends on σ^2:

$$Var(\overline{m}) = Var\left(\frac{S_n}{n}\right) = \frac{Var(S_n)}{n^2} = \frac{n\sigma^2}{n^2} = \frac{\sigma^2}{n}$$

Intuitively, it is clear that as n increases, the sample average will be a better and better approximation to the mean m. The Central Limit Theorem tells us precisely how good an approximation it is. These are illustrated in the following graphs for which the variance takes several values for the same mean $m = 0$.

Various normal distributions with mean 0

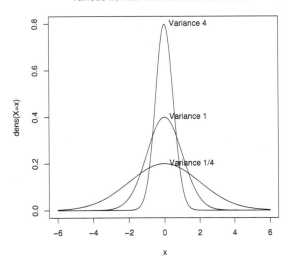

We remark that the independence of the random variables is essential in the Central Limit Theorem. For example, the gaps L_i of the Uniform process are equidistributed, but their sum, $L_1 + L_2 + \cdots + L_{n+1}$, is the length of the interval, which we know with certainty.

Statistical Measurements

Suppose we make n measurements $X_1, X_2, \ldots, X_n$ of the same quantity. Implicitly, we are assuming that these measurements are equidistributed and independent random variables. Each measurement has a distribution whose mean is the quantity we wish to measure. But the measurements are imperfect and so tend to be spread to a certain extent on both sides of the mean. Statisticians refer to this situation as a "random sample."

Definition. A **random sample** of size n is a set of n independent, equidistributed random variables $X_1, X_2, \ldots, X_n$.

In Sections 5.4 and 5.5, we will consider the problem of measuring the mean of a distribution using random samples. In particular, we would like to know how small a random sample is sufficient for a given measurement. If we wish to determine the average number of cigarettes smoked per day by Americans, it would be highly impractical to ask every American for this information. Statistics enables one to make accurate measurements based on surprisingly small samples.

In addition to the measurement problem, we will also consider the problem of using a random sample as a means for making predictions of the future. The predication will, of course, be a probabilistic one: with a certain probability the next measurement will lie within a certain range. For all the statistical problems we will study, we will assume only that the variance of each measurement X_i is finite. In most cases, this is a reasonable assumption especially if the measurements lie in a finite interval. For example, the number of cigarettes smoked by one individual in one day is necessarily between 0 and 10^6.

Main Rule of Statistics. In any statistical measurement we may assume that the individual measurements are distributed according to the normal distribution $N(m, \sigma^2)$.

To use this rule, we first find the mean m and variance σ^2 from information given in our problem or by using the sample mean $\overline{m}$ and/or sample variance $\overline{\sigma^2}$ defined below. We then compute using either the **pnorm** or the **qnorm** function.

As stated, the main rule says only that our results will be "reasonable" if we assume that the measurements are normally distributed. We can actually assert more. In the absence of better information, we *must* assume that a measurement is normally distributed. In other words, if several models are possible, we must use the normal model unless there is a significant reason for rejecting it.

When the mean m and/or the variance σ^2 of the measurements X_i are not known, the following random variables may be used as approximations:

sample mean	$\overline{m} = (X_1 + X_2 + \cdots + X_n)/n$
sample variance	$\overline{\sigma^2} = \dfrac{(X_1 - \overline{m})^2 + (X_2 - \overline{m})^2 + \cdots + (X_n - \overline{m})^2}{n - 1}$

For example, an exam graded on a scale of 0 to 100 is given in a class of 100 students. The sample mean is found to be 81 with sample variance 100 (standard deviation 10). Based on this data, we can predict that if the exam is given to another student, the student will score between 61 and 100 with probability 0.95 (within $2\overline{\overline{\sigma}}$ of $\overline{m}$). In actual exam situations, the distribution of an individual exam score is more complicated than the normal distribution, but in the absence of any better information we follow the Main Rule.

When the mean m is known but the variance is not, there is a slightly better approximation to the variance.

Definition. The **sample variance when m is known** is:

$$\overline{\overline{\sigma^2}} = \frac{(X_1 - m)^2 + (X_2 - m)^2 + \cdots + (X_n - m)^2}{n}$$

Note that the denominator differs for the two approximations to the variance. The reason for the different denominators in the two expressions is subtle. We leave it as an exercise to show that the expectations of the random variables are both equal to the true variance. The distributions of the random variables $\overline{m}$, $\overline{\sigma^2}$, and $\overline{\overline{\sigma^2}}$ are very important in statistics and we will undertake to compute them in some cases, leaving the rest as exercises.

5.4 Significance Levels

Let us begin with an example. We are presented with a coin having an unknown bias p. We are told that the coin is fair, but we are suspicious and would like to check this assertion. So we start tossing the coin. After 100 tosses we get only 41 heads. Do we have a reason to suspect that the coin is not fair?

In such an experiment, we carefully examine the model we have postulated in order to determine what kind of behavior is consistent with the model. If the observed behavior is consistent with the model, we have no reason to suspect that the coin is unfair. In this case, the postulated model is the Bernoulli process with bias $p = 1/2$. The postulated model is called the **hypothesis**. Determining whether the postulated model is consistent with the observed behavior is called **hypothesis testing**.

The average value of S_{100} is 50 for our postulated model. We are interested in the possible deviation of S_{100} from its mean value 50, because very large deviations are unlikely in the model but would not be if the coin is unfair. The usual statistical procedure in such a case is to determine precisely how large a deviation from the mean is reasonable in the model. We actually observed 41. This is 9 away from the mean 50. To determine whether this should be considered a large deviation, we compute the probability that an observation will be 9 or more away from the mean. We can do this either directly with the binomial distribution or approximately with the normal distribution. Using the `pbinom` function we compute

```
> pbinom(41,100,1/2) + 1 - pbinom(59,100,1/2)
[1] 0.072757
```

Alternatively, using the `pnorm` function we have

```
> pnorm(41,50,5) + 1 - pnorm(59,50,5)
```

[1] 0.07186064

With either one the probability of the event is about 7%. While this probability is somewhat small, it is not convincingly so, and we must conclude that our suspicions about the unfairness of the coin are groundless. The probability we just computed is called the **significance** of our test or the **p-value**. The **significance level** is the maximum probability level at which the test can be labeled as **having significance** or as being a **significant test** at the significance level. The most commonly used significance levels are 0.05, 0.01 and 0.001. In our case, the significance is 0.07. In statistical language, one would say, "the experiment has no significance at the 0.05 level." Notice that we say *no* significance. Statistically speaking, a significant result occurs only when a postulated model is *rejected*.

Looking at the reasoning a bit more carefully, we have said the following. Assuming that the coin is fair, about 95% of the time we will get between 40 and 60 heads when we toss the coin 100 times. But 5% of the time we will not be within this range.

> The significance level represents the probability that we will reject the postulated model even though this model is correct.

Notice the indirectness of this kind of reasoning. We say nothing about whether or not the coin is *really* fair or unfair, or even that it is fair or unfair with a certain probability. Statistics never tells one anything for certain, even in the weak sense of probabilistic certainty. All we can do is to devise tests for determining at some significance level whether or not the data we have collected are *consistent* with the model. Because of the abbreviated terminology that statisticians and scientists frequently use when discussing the result of an experiment, one should be careful not to ascribe properties to statistical statements, which they do not possess.

The 0.05 significance level is so commonly used by statisticians and scientists that this level is assumed when no significance level is specified. The 0.01 significance level is also common, and an experiment is said to be *very significant* if this level is being used. For example, in our coin tossing test we found that getting 41 heads was not (statistically) significant. On the other hand, getting 39 heads would be significant but would not be very significant, while getting 35 heads would be very significant. The 0.001 is also sometimes used, and a test that is significant at this level would be called *extremely significant*.

Let us consider another example. We are given a die, and we wish to test whether it is loaded. We decide to consider whether "3" is special, and we choose to work at the 0.05 significance level. Our experiment consists of rolling the die 120 times, and we find that "3" comes up 25 times. Our postulated model is now the Bernoulli process with bias $p = 1/6$. The mean and variance of a single roll are $m = p = 1/6$ and $\sigma^2 = pq = 5/36$. The expected number of 3's is 20, so we compute the probability that the number of 3's would be 5 or more away from 20:

```
> pbinom(15,120,1/6) + 1 - pbinom(25,120,1/6)
[1] 0.2254459
```

Therefore, the experiment is *not* significant, and we have no reason to suspect that the die is loaded. How many 3's would we have to roll to get a significant result at the 0.05 level? For this we need to invert the probability distribution function. The inverse function for the binomial distribution is `qbinom`. To be significant we need to roll either a small number of 3's or a large number of 3's. The total probability for the 2 cases should be 0.05, so we allocate 0.025 to each of these cases:

```
> qbinom(0.025,120,1/6)
[1] 12
> qbinom(0.975,120,1/6)
[1] 28
```

So we would have to roll fewer than 12 or more than 28 3's to have a significant result.

A quick rule of thumb for testing the Bernoulli process (to be used only if one is in a hurry) is the following.

Bernoulli Rule of Thumb. If one tosses n times a coin with bias p, then the result is significant if the number of heads lies outside $np \pm 2\sqrt{npq}$, and very significant if the number of heads lies outside $np \pm 3\sqrt{npq}$.

The first rule above is quite accurate, at least for reasonably large samples where the normal distribution is a good approximation. The second rule is actually between very significant and extremely significant. The accurate range for a test being very significant is $np \pm 2.6\sqrt{npq}$, while the range for a test being extremely significant is $np \pm 3.3\sqrt{npq}$.

5.5 Confidence Intervals

The concept of a confidence interval is a variation on the statistical themes we have just been describing. Instead of testing a hypothesis, one is interested in the accuracy of a measurement or in the prediction of the future.

Let us consider a very simple example of the prediction of the future. Suppose we have two competing airlines on a given route, both having the same departure time. Suppose that every day exactly 1000 passengers show up and that each one chooses one or the other airline independently of the other passengers. Both airlines want to be able to accommodate as many passengers as possible. They could do this, of course, by providing 1000 available seats. Needless to say, this would be prohibitively expensive, particularly since the probability that all 1000 seats would ever be needed is essentially zero. By providing 1000 seats we would have the absolute certainty that there will never be an overflow; if we are willing to accept a 5% chance of an overflow, the number of seats we must provide decreases dramatically.

To compute this, we again use the Bernoulli process with the bias $p = 1/2$.

The number of passengers choosing one particular airline is S_{1000}. We wish to compute the number of passengers y so that $P(Y \leq y) = 0.95$:

```
> qbinom(0.95,1000,1/2)
[1] 526
```

In other words, we need only provide 526 seats to have 95% **confidence** of not having an overflow. This is quite a dramatic drop from 1000 seats. Even for 99% confidence we need only a few more seats:

```
> qbinom(0.99,1000,1/2)
[1] 537
```

We could also have used the normal distribution. The mean is 500 and the variance is $1000pq = 250$, so the two confidence intervals are:

```
> qnorm(0.95, 500, sqrt(250))
[1] 526.0074
> qnorm(0.99, 500, sqrt(250))
[1] 536.7828
```

which are almost exactly the same as we computed using the Bernoulli process.

We speak of the interval $[0, 526]$ as being a 95% **confidence interval** for S_{1000}. In general, any interval $[a, b]$ for which $P(a \leq X \leq b) = 0.95$ is called a 95% confidence interval for the random variable X. When X is normally distributed (or approximately normally distributed) with distribution $N(m, \sigma^2)$, we generally use either a **one-sided confidence interval** or a **two-sided confidence interval**. A one-sided interval is of the form $(-\infty, t]$ or of the form $[t, \infty)$. A two-sided interval is chosen to be symmetric about the mean: $[m - t, m + t]$. When testing statistical hypotheses, one uses either a one-sided or a two-sided confidence interval. The corresponding tests are then referred to as a **single-tail test** or a **double-tail test**, respectively.

Now we consider the problem of the accuracy of statistical measurements. Suppose we wish to determine the percentage of adult Americans who smoke. To find out this number we randomly sample n persons. How many persons do we have to sample in order to determine the percentage of smokers to two decimal-place accuracy? Of course, we can determine the percentage to this accuracy with absolute certainty only by asking virtually the whole population, because there is always the chance that those not asked will all be smokers. Therefore, we must choose a confidence level. The usual level is 95% so we will use this. Up to now, our computations all involved determining either the probability given an observation or the confidence interval given the probability. In each case, the sample size n was known. Now we want to compute the sample size. One could either write a small program that will try a sequence of values until one works, or one could solve for the number mathematically. We will show how one can compute the sample size using the normal distribution.

The model we are using is the Bernoulli process with bias p, where p is the percentage we are trying to compute. Each person we ask will be a smoker with probability p. If we randomly ask n persons, the number who smoke divided by

n will be an approximation $\bar{p}$ to p. This number is the sample mean $\bar{m} = S_n/n$. We use the Central Limit Theorem in its second manifestation. We find that $\bar{p} = \bar{m}$ has approximately the distribution $N(p, pq/n)$. Therefore, $\dfrac{\bar{m} - p}{\sqrt{pq}/\sqrt{n}}$ is approximately $N(0, 1)$. We require a two-sided interval in this problem:

$$P\left(-1.96 \leq \frac{\bar{m} - p}{\sqrt{pq}/\sqrt{n}} \leq 1.96\right) = 0.95$$

or

$$P\left(|\bar{m} - p| \leq \frac{1.96\sqrt{pq}}{\sqrt{n}}\right) = 0.95$$

We want to choose n so that $|\bar{m} - p| \leq 0.005$ in order to have two-place accuracy. That is, $\dfrac{1.96\sqrt{pq}}{\sqrt{n}} = 0.005$ or $n \approx 154000pq$. Unfortunately, we do not know p. However, we do know that pq takes its largest value when $p = q = 1/2$ because $\sigma^2 = pq = p(1 - p)$ has its maximum at $p = 1/2$. Therefore, $n \leq 154000 \times 0.25 = 38,500$. In other words, to determine the percentage of smokers to two decimal places accuracy with 95% confidence we must sample up to 38,500 persons.

In practice, one would first determine p to one decimal-place accuracy. This requires only a sample of 385 persons. Using this number, one can compute $\sigma^2 = pq$ more precisely. Using this better value of σ, we can determine more precisely how many persons must be sampled in order to find p to two decimal-place accuracy. For example, suppose that with the smaller sample we find that $p = 0.10 \pm 0.05$. The worst case for σ^2 is now $pq = (0.15)(0.85) = 0.1275$. We must then sample $n = 19,635$ persons to determine p to within 0.005.

One must be careful not to confuse the accuracy with the confidence. The accuracy tells us how accurately we think we have measured a certain quantity. The confidence tells us the probability that we are right. To illustrate the distinction between these two concepts we consider the above measurement problem with two accuracies and three confidence levels. In general, improving confidence does not require much more effort while increasing accuracy requires a great deal of additional effort.

Confidence	One-place accuracy	Two-place accuracy
95%	385	38,500
99%	667	66,700
99.9%	1089	108,900

In the exercises, we consider more examples of significance levels and confidence intervals. Some of these have a distinct air of the supernatural about them. How, for example, is it possible to make conclusions about the television preferences of a population of 200,000,000 persons based on a sample of only 400 of them? In fact, the size of the population is irrelevant to the statistical analysis.

5.6 The Law of Large Numbers

The Law of Large Numbers was the first nontrivial theorem in probability theory. While it has been superseded for most applications today by the Central Limit Theorem, it has many uses nonetheless which makes it more than just a bit of historical memorabilia. Intuitively, the Law of Large Numbers says that the average of a sequence of independent, equidistributed random variables tends to the common mean of the random variables. Implicit in this statement is the assumption that the random variables have a mean. Because we are dealing with random variables, the precise statement is that the limit of the average is the mean with probability 1:

Strong Law of Large Numbers. If $X_1, X_2, \ldots$ are a sequence of independent equidistributed random variables with a common mean m,

$$P\left(\lim_{n \to \infty} \frac{X_1 + X_2 + \cdots + X_n}{n} = m \right) = 1.$$

The reason for calling this theorem the *Strong* Law of Large Numbers is to distinguish it from a weaker form of the law in which one only asserts that the sample mean will be within ϵ of the mean m in the limit, for any $\epsilon > 0$. If the X_i's have a finite variance, then the Strong Law of Large Numbers is a consequence of the Central Limit Theorem. We leave the proof as an exercise. The general case is much more difficult.

The Law of Large Numbers was once taken as the justification for defining probability in terms of the frequency of occurrence of an event. This led to an interpretation of probability called the frequentist point of view that had a significant influence at one time. The problem with founding an interpretation of probability on the Law of Large Numbers is that it is a purely mathematical theorem. In order for it to make sense, we must already have the concepts of probability, random variables and expectations. To use the Law as the *definition* of probability leads to circular reasoning.

The Law of Large Numbers is less useful in applications than the Central Limit Theorem because it does not specify how large a sample is required to achieve a given accuracy. However, it does have interesting *theoretical* applications such as the Shannon Coding Theorem in Section 9.2. Another theorem which has great usefulness in probability theory is the Bienaymé-Chebyshev Inequality. Its importance stems primarily from its simplicity.

Bienaymé-Chebyshev Inequality. Let X be a random variable with mean $E(X) = m$ and variance $Var(X) = \sigma^2$, then for all $t > 0$,

$$P(|X - m| \geq t) \leq \frac{\sigma^2}{t^2}$$

Proof. Suppose that X is a continuous random variable with density $f(x)$. The proof in the case of an integer random variable is similar. Clearly, we may assume that m is zero, for if not we just replace X by $X - m$. Now estimate the variance as follows:

$$\sigma^2 \;=\; Var(X) = E(X^2)$$
$$=\; \int_{-\infty}^{\infty} x^2 f(x)\,dx$$
$$\geq\; \int_{|x|\geq t} x^2 f(x)\,dx$$
$$\geq\; \int_{|x|\geq t} t^2 f(x)\,dx$$
$$=\; t^2 \int_{|x|\geq t} f(x)\,dx$$
$$=\; t^2 P(|X|\geq t)$$

The second inequality above is a consequence of the fact that $x^2 \geq t^2$ in the domain of integration. If we now solve for $P(|X|\geq t)$ we get the desired inequality.

The last result we will consider in this section is one of the most astonishing facts about probability: the Kolmogorov Zero-One law. As with the other theorems in this section, it has little practical usefulness, but it has many theoretical applications. The Law of Large Numbers, for example, can be proved using it.

Definition. Suppose that $X_1, X_2, \ldots$ is a sequence of random variables which are independent but not necessarily equidistributed. A **tail event** A is an event such that

1. A is defined in terms of the random variables $X_1, X_2, \ldots$
2. A is independent of any finite set of the X_i's, i.e.,

$$P(A \cap (X_1 = t_1) \cap (X_2 = t_2) \cap \cdots \cap (X_n = t_n)) =$$
$$P(A)\,P(X_1 = t_1)\,P(X_2 = t_2)\cdots P(X_n = t_n)$$

for any n and any set of t_i's.

Kolmogorov Zero-One Law. If A is a tail event, then $P(A) = 0$ or 1.

At first it might seem that there cannot be any tail events except for Ω and $\overline{\Omega}$, because tail events seem both to depend on the X_i's and not to depend on the X_i's. However, there are, in fact, many nontrivial examples. Here is one. Toss a fair coin infinitely often, and define

$$X_i' = \begin{cases} +1 & \text{if the } n^{th} \text{ toss is a head} \\ -1 & \text{if the } n^{th} \text{ toss is a tail} \end{cases}$$

as we did for the Random Walk process. Now let A be the event "The infinite series $\sum_{n=1}^{\infty} \dfrac{X_n'}{n}$ converges." This is a tail event because the convergence or divergence of a series is determined by the terms of the series but is independent

of any finite set of them. We all should know at least two examples from calculus: the harmonic series $\sum_{n=1}^{\infty} \frac{1}{n}$ diverges but $\sum_{n=1}^{\infty} \frac{(-1)^n}{n}$ converges to $\log(2)$.
What we are doing is to change the signs of the harmonic series randomly and independently. $P(A)$ is the probability that a random choice of signs yields a convergent series. The zero-one law tells us that $P(A)$ can only be 0 or 1; there are no other possibilities. In fact, $P(A) = 1$; we leave this as an exercise.

As another example, suppose that a Rhesus monkey is trained to hit the keys of a typewriter and does so at random, each key having a certain probability of being struck each time, independently of all other times. Let A be the event "the Rhesus monkey eventually types out Shakespeare's *Hamlet*." Again this is clearly a tail event and so $P(A) = 0$ or 1. This is easy to see. *Hamlet* has about 2×10^5 characters and could be written with a typewriter having 100 keys. Suppose each key has 1% probability of being typed. The probability of typing *Hamlet* is $p = (.01)^{2 \times 10^5}$ during a given "session" of 2×10^5 keystrokes. The probability of *not* typing *Hamlet* in one session is $q = 1 - p{<}1$. The probability that in *infinitely* many sessions the monkey *never* types out *Hamlet* is $\lim_{n \to \infty} q^n = 0$. Therefore, $P(A) = 1$. On the other hand, the expected waiting time until the monkey types out *Hamlet* is about $10^{400,000}$ keystrokes. If a million monkeys could type one keystroke every nanosecond, the expected waiting time until one of the monkeys types out *Hamlet* is so long that the estimated age of the universe is insignificant by comparison.

Needless to say, this is not a practical method for writing plays. The Kolmogorov zero-one law has little practical usefulness. But it does have theoretical uses, and it shows how counterintuitive probability theory can be.

5.7 The Cauchy Distribution

Imagine that a novice tennis player is standing 10 m from an infinitely long straight wall. The tennis player hits a ball horizontally in a completely random *direction* toward the wall. Where does the ball hit the wall? Where does it hit on the average? The answer to the former question is, of course, the probability distribution. The answer to the latter question is intuitively obvious; the point closest to the tennis player must be the average, by symmetry. However, we will find that the situation is much more complicated than intuition would lead one to expect.

To compute the probability distribution, assume that we measure distances along the wall in meters with the zero point being the point of the wall closest to the tennis player. The tennis player hits in a random direction θ uniformly chosen from $\left(-\frac{\pi}{2}, \frac{\pi}{2}\right)$. The ball hits the wall at the point $Y = 10 \tan(\theta)$. Thus, the distribution is:

$$\begin{aligned} P(Y {\le} y) &= P(10 \tan(\theta) {\le} y) \\ &= P\left(\theta {\le} \arctan\left(\frac{y}{10}\right)\right) \end{aligned}$$

$$= \frac{\arctan(y/10)}{\pi} + \frac{1}{2}$$

Differentiating, the density of Y is:

$$\text{dens}(Y = y) = \frac{1}{\pi\left(1 + \left(\frac{y}{10}\right)^2\right)} = \frac{10}{\pi(100 + y^2)}$$

This distribution is called a **Cauchy distribution**. If we vary the distance from the wall and the position on the wall that is closest to the tennis player, we get the general Cauchy distribution which has density:

$$\text{dens}(Y = y) = \frac{d}{\pi\left(d^2 + (x - m)^2\right)}$$

where m is the point closest on the wall, and d is the distance from the wall. The parameter d is called the *scale parameter*. This class of distributions is known by many names such as Cauchy-Lorentz, Lorentz and Breit-Wigner. It occurs in many physical processes such as forced resonance and spectroscopy, as well as in statistics.

We now attempt to compute the expectation of the distribution. The ball hits the wall *on average* at

$$E(Y) = \int_{-\infty}^{\infty} \frac{10y\,dy}{\pi(100 + y^2)} = \frac{10}{\pi} \int_{-\infty}^{\infty} \frac{y\,dy}{(100 + y^2)}$$

provided this improper integral converges. Now by definition such an improper integral converges if and only if

$$\int_a^{\infty} \frac{y\,dy}{(100 + y^2)} \quad \text{and} \quad \int_{-\infty}^a \frac{y\,dy}{(100 + y^2)}$$

both converge and their sum does not depend on a. For example, let $a = 0$. Then, using the substitution $x = y^2$, we find that

$$\int_a^b \frac{y\,dy}{(100 + y^2)} = \frac{1}{2} \int_0^{\sqrt{b}} \frac{dx}{100 + x}$$

$$= \frac{1}{2}[\log(100 + x)]_0^{\sqrt{b}}$$

$$= \frac{1}{2}\left(\log\left(100 + \sqrt{b}\right) - \log(100)\right)$$

So $\int_a^b \frac{y\,dy}{(100 + y^2)}$ goes to ∞ as $b \to \infty$. Therefore, the Cauchy distribution has *no average value!* This is completely contrary to the intuitive belief that $E(Y)$ should be 0 by symmetry.

One can easily simulate the Cauchy distribution by generating a sequence of angles uniformly distributed on $(-\pi/2, \pi/2)$ and then applying $10\tan(\theta)$ to these angles. In R, the Cauchy distribution is directly supported, so one can simply use the rcauchy function to generate random numbers. As an example, we "hit" a ball 1000 times and then graphed where it hits the wall. The `par` command is used for specifying "global parameters" that apply to more than one graph command. In this case, we want to have several graphs on a single image. The `mfrow` specifies the number of rows and columns in the image. The "row" in mfrow means that the graphs are to be drawn one row at a time. The `mfcol` parameter specifies that the graphs are to be drawn one column at a time. These two parameters differ only if there are both multiple rows and multiple columns of graphs in the image, which is not the case here.

```
> balls <- 1:1000
> tennis <- 10*tan(runif(length(balls), -pi/2, pi/2))
> par(mfrow=c(1,2))
> plot(balls, tennis, pch=".")
> plot(balls, tennis, pch=".", ylim=c(-100,100))
```

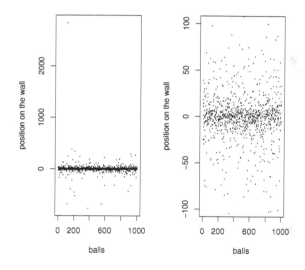

The first graph has no limitation on the y-axis. This gives some idea of why this distribution has no expectation. There will occasionally be points that are extremely far away. The second graph limits the y-axis to 100m in each direction.

Because the Cauchy distribution has no expectation, the Law of Large Numbers does not apply to it. To see what this means in this case, we graphed both the random points and the averages. While the Law of Large Numbers does not apply, it still might be possible that the averages would tend toward the midpoint of the line. This program is similar to previous graphs, except that now we are drawing two graphs on top of each other. The `points` command is used for plotting additional points on the most recent graph. There are many other commands for adding additional graphics to a graph, such as text labels and other axes. The `cumsum` command computes all sums of a vector. If the vector x has elements $x_1, x_2, \ldots x_n$, then `cumsum(x)` has elements x_1, $x_1 + x_2$, $x_1 + x_2 + x_3, \ldots, x_1 + x_2 + \ldots + x_n$. Dividing each of these by the number of summands gives the averages of the vector.

```
> balls <- 1:1000
> tennis <- rcauchy(length(balls), 0, 10)
> plot(balls, tennis, pch=".", ylim=c(-100,100))
> points(balls, cumsum(tennis)/balls, type="l")
```

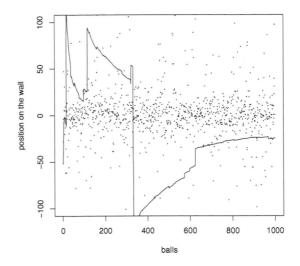

Note that the averages never "settle down." Just as it seems that the averages are approaching 0, an extremely distant ball hits the wall, and the averaging has to start over again. This process never ends. In fact, one can show that the distribution of the sample mean

$$\frac{Y_1 + Y_2 + \cdots + Y_n}{n}$$

of n independent Cauchy random variables has the same distribution as a single Cauchy random variable! See Exercise 8.9.

In the next program, we "hit" 50,000 random balls at the wall and compute the sample means using `cumsum` as we just did above. We then animate the process by drawing the graphs for the balls and sample means in successive ranges. The first range is 1:1000, the second is 100:1100, etc. It is as if we are looking at the process through a window as it speeds by us. We program the animation by writing a function that draws the graph using only the 1000 elements starting at the specified time. The program ends by using the `plot` command to show the entire graph of the sample means. You will need to run the program several times to see some of the more dramatic effects.

```
> balls <- 1:50000
```

```
> tennis <- rcauchy(50000, 0, 1)
> averages <- cumsum(tennis)/balls
> window <- function(time) {
+    duration <- time:(time+999)
+    plot(balls[duration], tennis[duration],
+        pch=".", ylim=c(-5,5))
+    points(balls[duration], averages[duration],
+           type="l")
+    Sys.sleep(0.02)
+ }
> ignore <- sapply((1:490)*100, window)
> plot(balls, averages, type="l")
```

Note that the sample means might be so large at times that the graph of the sample means is not visible in the window. It is worth mentioning that, strictly speaking, rcauchy is not exactly generating Cauchy random numbers. The numbers cannot be larger than the limits imposed by the computer architecture. Because the numbers being generated are bounded, the Law of Large Numbers actually does apply, and the sample means will eventually approach 0, albeit long after round-off errors and other computer limitations have rendered the computations meaningless.

5.8 Exercises

Exercise 5.1 Suppose that X is a random variable whose density is

$$\text{dens}(X = x) = \begin{cases} (\beta - 1)x^{-\beta} & \text{if } x > 1 \\ 0 & \text{if } x \leq 1 \end{cases}$$

where $\beta > 1$. Show that

If $1 < \beta \leq 2$, then X has neither mean nor variance

If $2 < \beta \leq 3$, then X has mean $\frac{\beta-1}{\beta-2}$ but has no variance

If $\beta > 3$, then X has variance $\frac{\beta-1}{(\beta-2)^2(\beta-3)}$

Exercise 5.2 Let X be the random variable S'_n in the symmetric Bernoulli process random walk model. Let $Y = X^2$. Then X and Y are obviously dependent random variables. Show that $Cov(X, Y) = 0$. Hence, the converse of the Product Rule for Expectations does not hold.

Exercise 5.3 Verify the following formula, which holds for arbitrary random variables $X_1, X_2, \ldots, X_n$ (not necessarily independent) so long as both sides exist:

$$Var(X_1 + X_2 + \cdots + X_n) = \sum_i Var(X_i) + 2\sum_{i<j} Cov(X_i, X_j)$$

Use this formula and the exchangeability of the gaps in the Uniform process to compute $Var(X_{(k)})$.

Exercise 5.4 Prove that if X and Y are independent continuous random variables, then $E(XY) = E(X)E(Y)$. To do this one must split X into positive random variables X^+ and X^- such that $X = X^+ - X^-$, where

$$X^+ = \begin{cases} X & \text{if } X \geq 0 \\ 0 & \text{if } X < 0 \end{cases}$$

and similarly for X^-. Do the same for Y.

Exercise 5.5 Prove that for any two random variables X and Y,

$$E(XY) \leq \sqrt{E(X^2)E(Y^2)}$$

This is called the Cauchy-Schwarz inequality. Use this to prove that the correlation coefficient of X and Y satisfies $|\rho(X,Y)| \leq 1$. Also show that:

$$\rho(X,Y) = \frac{Cov(X,Y)}{\sigma(X)\sigma(Y)}$$

Exercise 5.6 Let $X_1, X_2, \ldots, X_n$ be a set of independent random variables, not necessarily equidistributed, but all having the same mean and variance σ^2. Prove that the sample mean has expectation m and that both sample variances have expectation σ^2. Also compute the variance of the sample mean. Can you compute the variance of either sample variance?

Notice that the denominator *must* be different for the two sample variances. This denominator is called the **degrees of freedom** by statisticians. Intuitively, each estimation of a parameter of the unknown distribution causes a loss of one degree of freedom in the random sample.

Exercise 5.7 In Exercise 3.10, we saw that if X and Y are independent, then $g(X)$ and $h(Y)$ are also independent for any two functions g and h. It follows that $g(X)$ and $h(Y)$ are also uncorrelated. Prove the converse: if $g(X)$ and $h(Y)$ are uncorrelated for *every* pair of functions g and h, then X and Y are independent.

Exercise 5.8 Use the normal distributions indicated to compute:

(a)	$P(-.5 \leq X \leq .5)$,	where	X	is	$N(0,1)$
(b)	$P(X \leq -2)$.	where	X	is	$N(0,1)$
(c)	$P(Y \geq 5)$,	where	Y	is	$N(0,4)$
(d)	$P(1 \leq Y \leq 4)$,	where	Y	is	$N(-2,9)$

Exercise 5.9 In each of the following cases, find a number α such that

(a)	$P(X \geq \alpha)$	=	0.03,	where	X is	$N(0,1)$
(b)	$P(-\alpha \leq X \leq \alpha)$	=	0.08,	where	X is	$N(0,1)$
(c)	$P(-2 - \alpha \leq X \leq \alpha - 2)$	=	0.10,	where	Y is	$N(-2,9)$
(d)	$P(Y \geq \alpha)$	=	0.98,	where	Y is	$N(0,4)$

Exercise 5.10 Show explicitly that the normal distribution $N(m, \sigma^2)$ really does have mean m and variance σ^2.

Exercise 5.11 A prestigious scientific journal announces as part of its editorial policy that only results significant at the .01 level will be acceptable for publication (and conversely any result significant at the .01 level is acceptable). They reason that by doing so their readership will have the confidence that at most 1% of the published results will be incorrect. Discuss the fallacy of this policy.

Exercise 5.12 In a scientific paper, you read the following: "In four of our experiments the data are significant at the .05 level. The fifth experiment, however, is significant at the .01 level!" What is misleading about this?

Exercise 5.13 A population scientist believes that roughly 50% of the population is female, but does not want to be too hasty. So he decides to be cautious and to test whether or not at least 45% of the population is female. To do this, he takes a random sample of 100 persons. If he discovers that only 40 of them are female, does he have sufficient evidence to reject the model that (at least) 45% of the population is female? Use a Bernoulli process with $p = .45$ and a one-sided significance test.

Exercise 5.14 A statistician wonders just how careful the scientist in Exercise 5.13 was when he made his random sample. The actual percentage is known from census data to be 51%. Is 40 significantly different from the expected value of 51? Is 40 significantly *smaller* than 51? What do you think of the sampling technique of the scientist?

Exercise 5.15 You own a company that produces medium quality left-handed screws. About 1% of the screws produced by one machine are defective. As the screws are produced, the defective ones are found and discarded. A count is kept of the number of defective screws produced each hour. The machine is readjusted whenever the number of defective screws produced is significantly greater than 1%. You may regard this as a Bernoulli process. The machine makes 10,000 screws per hour. Describe a procedure for determining when the machine is out of adjustment at the 0.05 level and at the 0.01 level.

Exercise 5.16 A congressman wishes to vote according to the "will of the people" on a certain bill. Now in this case, one wishes to know whether the percentage p of his constituency in favor of the bill is above or below 50%. Clearly if p is close to $1/2$ a rather large sample will be required to distinguish between the two possibilities. How is this reflected in a statistical test? For example, suppose that a poll is made soliciting the opinion of a certain number of voters chosen at random. Use a .05 significance level to decide what the congressman should do in each of the following cases:

Number of Pollees	Number of Pollees in Favor of the Bill
100	54
100	41
500	267
500	225
1000	534

Note that the congressman has three choices in each case: (a) vote *for* the bill, (b) vote *against* the bill or (c) order a larger sample be taken.

Exercise 5.17 A company wishes to test the effectiveness of a new magazine advertising campaign. It decides that the campaign is effective if the proportion of subscribers to the magazine who use their product is twice as large as the proportion who do so in the general population. A 10% significance level is agreed upon. It is known that 15% of the general population use the firm's product. A sample of 50 subscribers is ordered and it is found that 10 use the product. What does this suggest about the advertising campaign?

Exercise 5.18 A study has shown that in a certain profession the women are receiving only 88% as much, on average, as their male counterparts receive. However, the study is several years old and a women's organization wants to determine if the women in this profession are losing relative to the men. It is known that the men in the profession now receive 138% of the pay they received when the first study was conducted. A random sample of the female professionals is made. The average pay of these women was found to be the following (as a percentage of the current average pay of men in the profession): 70%, 78%, 80%, 83%, 84%, 86%, 87% and 96%. Compute the sample mean and sample standard deviation. Use a single-tail test at the .05 significance level to determine if the women in this profession are losing ground relative to the men.

Exercise 5.19 The Food and Drug Administration (FDA) suspects that a drug company is producing a certain pill with a purity less than that required by law. The law allows at most 5 parts per million (ppm) of a certain impurity. An FDA laboratory tests a random sample of 50 pills. They find that the pills have a sample mean impurity of 5.4 ppm and a sample variance of 4.38 ppm^2. Can they assert that the pills do not comply with the law? We will return to this question in Exercise 6.49.

Exercise 5.20 A paper company was a major polluter of a small river for many years. When antipollution laws were enacted, it reacted slowly at first but later made a major effort to control its pollution. Unfortunately, the firm suffered from its earlier recalcitrance by acquiring a public image as a major polluter. Indeed, a very large sample revealed that close to 90% considered the firm to be a major polluter. To counter this, they began a public relations campaign. After the campaign, a random sample of 200 individuals were asked whether the company was still a major polluter. It was found that 174 felt this way. Did the campaign have a significant effect?

Exercise 5.21 (Hans Zeisel) Dr. Benjamin Spock, author of a famous book on baby care, and others, were initially convicted of conspiracy in connection with the draft during the Vietnam war. The defense appealed, one ground being the sex composition of the jury panel. The jury itself had no women, but chance and challenges could make that happen. The defense might have claimed that the jury lists (from which the jurors are chosen) should contain 55% women, as in the general population. However, the defense did not. Instead they complained that six judges in the court averaged 29% women in their jury lists, but the seventh judge, before whom Spock was tried, had fewer, not just on this occasion but systematically. The last nine jury lists for that judge contained the following counts:

Women	Men	Total	Proportion Women
8	42	50	0.16
9	41	50	0.18
7	43	50	0.14
3	50	53	0.06
9	41	50	0.18
19	110	129	0.15
11	59	70	0.16
9	91	100	0.09
11	34	45	0.24
Grand totals 86	511	597	0.144

Did the jury lists for this judge have a significantly smaller percentage of women? Because of the seriousness of the case, use an extremely small significance level: 0.0001.

Exercise 5.22 A political scientist wishes to determine whether there is significant difference between the preferences of voters in 2 similar neighborhoods of a city with respect to an upcoming race for mayor. Samples of 30 voters are taken from each of the neighborhoods. In 1 neighborhood, 12 voters prefer the incumbent while in the other neighborhood 19 do so. Use a 10% significance level to decide whether there is a significant difference between the neighborhoods.

Exercise 5.23 A medical researcher samples 100 records of patients with diagnosed coronary heart disease from 1 city, taking care to ensure that the sample is random. The plasma LDL-C level for these individuals was found to be 226 mg/dl, and the sample standard deviation was 30 mg/dl. The researcher then took a random sample of 200 people from the same city who do not have diagnosed coronary heart disease. The mean cholesterol value for this sample was 152 mg/dl, and the sample standard deviation was 50 mg/dl. Do individuals without diagnosed heart disease have a significantly larger cholesterol value than those with diagnosed heart disease?

Exercise 5.24 A small college soccer team won its conference championship 9 times in the first 20 years of its existence. Then for the next 20 years it won only 3 times. Is this significant? Is it very significant?

Exercise 5.25 When the president of the company in Exercise 5.20 discovered that the questionnaire used in the postcampaign sample included the word "still," she was incensed: the question seemed biased in favor of a yes answer. Accordingly, she immediately proceeded to write her own questionnaire and take a new random sample. The only change was the omission of the word "still." In this new sample of 200 individuals, only 160 felt that the company was a major polluter. Did the alteration of the questionnaire have a significant effect? Did the public relations campaign have a significant effect?

Exercise 5.26 Using the information in Exercise 5.23, give a 90% confidence interval for the following:

1. the individual plasma LDL-C levels of individuals without heart disease;
2. the individual plasma LDL-C levels of individuals with heart disease;
3. the mean plasma LDL-C level of all individuals without heart disease;
4. the mean plasma LDL-C level of all individuals with heart disease.

Exercise 5.27 If a set of grades on a statistics examination are approximately normally distributed with mean 82 and standard deviation 6.9, find:

The lowest passing grade if the lowest 10% of the students are given F's.

The highest B if the top 5% of the students are given As.

Exercise 5.28 The average life of a certain type of engine is 10 years, with a standard deviation of 3.5 years. The manufacturer replaces for free all engines that fail while under guarantee. If the manufacturer is willing to replace only 2% of the engines that fail, how long a guarantee should be offered? Assume a normal distribution.

Exercise 5.29 The braking distances of 2 cars, F and C, from 50 k.p.h. are normally distributed, with mean 35 m and standard deviation 5 m. If they both approach each other on a narrow mountain road and first see each other when they are 100 m apart, what is the probability that they avoid a collision?

Exercise 5.30 If the probability of a male birth is 0.512, what is the probability that there will be fewer boys than girls in 1000 births?

Exercise 5.31 A multiple-choice quiz has 100 questions each with 4 possible answers of which only 1 is the correct answer. What is the probability that sheer guesswork yields from 10 to 30 correct answers for 40 of the 100 problems about which the student has no knowledge?

Exercise 5.32 A firm wishes to estimate (with a maximum error of 0.05 and a 98% confidence level) the proportion of consumers who use its product. How large a sample will be required in order to make such an estimate if the preliminary sales report indicates that about 25% of all consumers use the firm's product? How large a sample would be needed if no preliminary information were available?

Exercise 5.33 A sponsor of a weekly television program is interested in estimating the proportion of the city population that regularly watches its program. The sponsor wishes the estimate to be made with a 90% confidence level and an error of at most 4%. The sponsor has no information concerning the proportion of viewers that watches the program. How large a sample will be required to make the estimate?

Exercise 5.34 A person has just hired a building contractor to build a house. The house will be built in three stages. First, the contractor lays the foundation; second, the frame and exterior are built; and last a subcontractor puts in the wiring, plumbing and interior. Each stage must be completed before the next is started. In attempting to get an estimate of when the house will be totally completed, the purchaser is able to get the following information from those in charge of each stage.

Stage	Expected Time of Completion of the Stage (in Weeks)	Standard Deviation (in Weeks)
I	3	1
II	8	2
III	5	2

What are the expected value and standard deviation of completion time for the house, assuming the completion times of the stages are independent?

Exercise 5.35 The College Entrance Examination Board verbal and mathematical aptitude scores are approximately normally distributed with mean 500 and standard deviation 100 except that scores above 800 and below 200 are arbitrarily reported as 800 and 200, respectively. What percentage of the students taking the verbal exam score above 800 or below 200?

Exercise 5.36 You are the head of a polling company, and you have a contract to determine the percentage of the electorate in favor of a candidate. There are 1,000,000 members of the electorate, and each member chooses his/her opinion independently. Your contract specifies that you must determine the percentage to within 1% with 5% confidence or to within 5% with 1% confidence. Which is cheaper?

Exercise 5.37 A professor at a small college walks to school each day. On average, the trip takes 15 minutes with a standard deviation of 3 minutes. Assume a normal distribution. If her first class is at 10:30 AM, when must she leave home in order to be 95% certain of arriving on time? If the college serves coffee from 10:00 AM to 10:30 AM, how often would she have coffee before class if the professor leaves home at 10:10 AM every day? Assume it takes 3 minutes to have coffee.

Exercise 5.38 Suppose that resistors can be purchased each with a resistance that is uniformly distributed between 900 Ω and 1100 Ω. If 10 such resistors are connected in parallel, what is the probability that their total resistance will be within 5% of 10,000 Ω?

Exercise 5.39 When a thumbtack is tossed, it falls on its flat head with probability p. What must you do to find p to within $p/10$ at a significance level of 0.05? At a significance level of 0.01?

Exercise 5.40 You own a telephone company that services 2 cities A and B, each having 5000 customers. You would like to link your exchange with the more distant city, C. You estimate that during the busiest time each customer will require a line to C with probability .01. You want to be sure that there are enough lines to C so that there is only a 1% chance that at the busiest time some customer will be unable to get a line to C. Each landline to C will cost $10,000. You have 2 options. Either link A and B as if they were separate exchanges or link the entire exchange to C. In the second option, additional equipment costing $50,000 would be needed. Which option is cheaper?

Exercise 5.41 An experimental study is conducted using a rodent model to determine if a certain type of drug has an effect on the incidence of a particular disease. A sample of 100 rats was kept in a controlled environment and 50 of the rats were given the drug. Of the group not given the drug (the control group), there were 12 incidences of the disease, while 9 of the other group contracted it. Compute a 90% confidence interval for the difference in probability of contracting the disease between a rat given the drug and a control rat.

Exercise 5.42 A statistician named Burr relates the following story. "Having bought a bag of roasted chestnuts, the author walked home in the dark eating them with much gusto. After eating about 20, he arrived home, and, in opening the remaining 10 under the light, he found that 7 contained worms. What is the probability that none of the 20 contained worms? Or to phrase the problem better for statistical analysis: If there were only 7 wormy chestnuts among the original 30, what is the probability of drawing the first 20 all free from worms?"

Exercise 5.43 Since there is no reason to believe that the salaries of individuals will be normally distributed, only in very large samples can we expect the mean to be normally distributed. With this in mind reexamine Exercise 5.18. Regardless of the distribution of salaries of the female professionals, half the salaries will be above the median salary. Suppose that the salaries of the male professionals are known to have a median that is 97% of the mean. This situation would be typical since the presence of a few very high salaries can cause the mean to be somewhat unrepresentative. Does it now appear that the female professionals are gaining or losing relative to the males?

Exercise 5.44 (Certificate) In a survey of the work of chemical research workers, it was found, on the basis of extensive data, that on average each one required no fume hood 60% of the time, one 30% and two 10%; three or more were never required. If a group of four chemists worked independently of one another, how many fume hoods should be available in order to provide adequate facilities for at least 95% of the time?

Compute the probability distribution of the number of fume hoods needed by the four chemists. Then use this to answer the question.

Exercise 5.45 In Exercise 5.44, how many fume hoods would be required to satisfy a group of 50 chemists at least 95% of the time? Use a normal approximation.

Exercise 5.46 A governmental agency is responsible for protecting the fish populations of the lakes in a certain region. By means of many observations in the past, it sets lower bounds for populations, in each lake, of various species of fish. If it is later found that one species in a given lake has gone below the specified lower bound, the agency has the power to enforce limits on the pollutants which the factories bordering on the lake may discharge into the lake. How can the agency determine whether the lower limit has been reached? One way to do so is to employ the procedure described in Exercise 2.22. Let the lower limit be L. We first capture n fish, tag them and return them to the lake. Some time later, we drop a net, capture m fish and count how many are tagged. Describe how to find a number T so that there is only a 5% chance that T or more tagged fish will be found when there are L or more fish in the lake.

Exercise 5.47 (Silvey) An investigation was carried out on two suggested antidotes to the consequences of drinking, these being (a) 2 lb of mashed potatoes and (b) a pint of milk. Ten volunteers were used, five for each antidote, the allocation to antidote being random. One hour after each had drunk the same quantity of alcohol and swallowed the appropriate antidote, a blood test was carried out and the following levels (mg/ml) of alcohol in the blood were recorded:

(a)	76	52	92	80	70
(b)	110	96	74	105	125

By means of the runs test, decide whether there is sufficient evidence to conclude that one treatment is more effective than the other.

Exercise 5.48 (Guenther) Suppose that it is hypothesized that twice as many automobile accidents resulting in deaths occur on Saturday and Sunday as on other days of the week. That is, the probability that such accidents occur on Saturday is 2/9, on Sunday is 2/9, and on each other day of the week is 1/9. From the national record file, cards for 90 accidents are selected at random. These yield the following distribution of accidents according to the days of the week:

Sun	Mon	Tues	Wed	Thurs	Fri	Sat
30	6	8	11	7	10	18

Do these data tend to support or contradict the hypothesis? Use a 5% significance level.

Exercise 5.49 We wish to test whether or not the successive outcomes of a roulette wheel are random. For simplicity, we will only record whether the ball fell into a red or a black slot of the wheel. In 20 spins of the wheel, we observe the sequence: RRBRRBBBBBRBRRRRRBBBR. Applying the runs test and using a 5% significance level, are the successive outcomes random? What does this suggest about this roulette wheel?

Exercise 5.50 (Pazer & Swanson) A political scientist wishes to determine if the political preference of homeowners is independent of their immediately adjacent neighbors. A sequence of 16 homeowners, along the same side of a street, were interviewed, and based upon their responses were designated as either more conservative than their median C, or less conservative than their median L. Here is the resulting sequence:

$$L, C, L, C, C, C, C, L, L, C, L, L, L, L, C, C$$

Using a run test and a 5% level of significance, determine whether there is any evidence that political opinions are independent of one's neighbors (at least for this particular street).

Exercise 5.51 We all have taken laboratory courses at some time or other, and the temptation to "fudge" data on our report has certainly occurred to us. What we may not have realized is that one can devise a statistical test to determine whether such fudging took place. Suppose that a biologist wishes to prove that a certain genetic trait follows the classical Mendelian laws. In this theory a trait is determined by two genes, one acquired from each of the two parents. Let us say that there are two alleles (possibilities) for a given gene, one dominant A and one recessive a. Then there are three different genotypes: AA, Aa and aa. Let us suppose, as it often happens, that AA and Aa are indistinguishable. By successive inbreeding, the biologist has access to two individuals known to have genotypes AA and aa. When these are crossed, the offspring all have genotype Aa. But when two of the offspring are crossed, we find that the three genotypes AA, Aa and aa appear among their offspring with probabilities 1/4, 1/2, and 1/4, respectively. Of course, since we cannot actually distinguish AA from Aa, this means that on the average 3/4 of the offspring exhibit the dominant trait and 1/4 exhibit the recessive one. Let us suppose that the biologist produces 10,326 offspring from a pair of Aa parents. He observes that 7746 have the dominant trait and 2580 the recessive one. These are very close to the expected numbers 7744.5 and 2581.5 so he concludes that the experiment tends to support the hypothesis that this trait obeys the Mendelian laws.

Compute the probability that such an experiment would actually result in as close a fit with the theory as the biologist actually found. At the 5% level can one reject the hypothesis that the experiment was properly carried out?

Exercise 5.52 How many times must one toss a fair coin in order to have 95% confidence that it really is fair? Compare the number obtained by using the Bienaymé-Chebyshev inequality with what we get using the Central Limit Theorem.

Exercise 5.53 Let X be a standard random variable. Using the Bienaymé-Chebyshev inequality, compute the smallest α in each of the following:

1. $P(-\alpha \leq X \leq \alpha) \geq .95$

2. $P(-\alpha \leq X \leq \alpha) \geq .99$

3. $P(X \geq \alpha) \leq .05$

4. $P(X \geq \alpha) \leq .01$

Compare these values with the corresponding ones for case of X being $N(0,1)$.

Exercise 5.54 Let X be a nonnegative random variable. Prove that

$$P(X \geq \alpha) \leq E(X)/\alpha$$

for any $\alpha > 0$, whether X has a variance or not. This is known as Markov's inequality. Show that Markov's inequality implies the Bienaymé-Chebyshev inequality.

Exercise 5.55 The Weak Law of Large Numbers states that if $X_1, X_2, \ldots$ are equidistributed with finite mean m and a finite variance, then for every $\epsilon < 0$ there exists an N such that

$$P\left(\left|\frac{X_1 + X_2 + \cdots + X_n}{n} - m\right| < \epsilon\right) < 1 - \epsilon$$

for every $n < N$. Prove it using the Central Limit Theorem. Actually, the variance need not be finite, but that is much harder to prove.

Exercise 5.56 Show that $\sum_{n=1}^{\infty} X_n'/n$ converges with probability 1, where X_n''s are independent and are equally likely to be either +1 or -1.

5.9 Answers to Selected Exercises

Answer to 5.2.

$$
\begin{aligned}
Cov(X,Y) &= E(XY) - E(X)E(Y) \\
&= E(X^3) - E(X)E(X^2)
\end{aligned}
$$

We know that $E(X) = 0$ by symmetry, but X^3 is also symmetrical:

$$
\begin{aligned}
P(X^3 = n) &= P(-X^3 = -n) \\
&= P\left((-X)^3 = -n\right) \\
&= P(X^3 = -n) \quad \text{by the symmetry of } X
\end{aligned}
$$

Therefore, $E(X^3) = 0$. Hence, $Cov(X, Y) = 0$. But X and Y are dependent. Therefore, the converse to the Product Rule for Expectations in Section 5.1 is false.

Answer to 5.3. Let $X_1, \ldots, X_n$ be random variables which are not necessarily independent. Then, since the Addition Rule for Expectations does not require independence,

$$
\begin{aligned}
& Var(X_1 + \cdots + X_n) \\
&= E\left((X_1 + \cdots + X_n)^2\right) - E(X_1 + \cdots + X_n)^2 \\
&= E\left(\sum_{i=1}^{n} X_i^2 + 2\sum_{i<j} X_i X_j\right) - \left(\sum_{i=1}^{n} E(X_i)\right)^2 \\
&= \sum_{i=1}^{n} E(X_i^2) + 2\sum_{i<j} E(X_i X_j) - \sum_{i=1}^{n} E(X_i)^2 - 2\sum_{i<j} E(X_i) E(X_j) \\
&= \sum_{i=1}^{n} \left(E(X_i^2) - E(X_i)^2\right) + 2\sum_{i<j} \left(E(X_i X_j) - E(X_i) E(X_j)\right) \\
&= \sum_{i=1}^{n} Var(X_i) + 2\sum_{i<j} Cov(X_i, X_j)
\end{aligned}
$$

We now apply this formula to the gaps in the Uniform process. We know that $X_{(k)} = L_1 + \cdots + L_k$ and that the gaps are exchangeable (see Section 6.6 for the details). Therefore,

$$
\begin{aligned}
Var(X_{(k)}) &= \sum_{i=1}^{k} Var(L_i) + 2 \sum_{i<j\leq k} Cov(L_i, L_j) \\
&= kVar(L_1) + 2\binom{k}{2} Cov(L_1, L_2)
\end{aligned}
$$

Now the special case $k = n + 1$ is easy because $X_{(n+1)} = L_1 + \cdots + L_{n+1} = a$ is a constant, so $Var(X_{(n+1)}) = 0$. Therefore,

$$
0 = Var(X_{(n+1)}) = (n+1)Var(L_1) + 2\binom{n+1}{2} Cov(L_1, L_2)
$$

Thus, $Cov(L_1, L_2) = \frac{Var(L_1)}{n}$, and the computation of the variance of any order statistic reduces to the evaluation of one integral: $E(L_1^2)$. Moreover, by suitable manipulation this one can be reduced to an evaluation of $E(X_{(2)})$, but in another Uniform process as follows:

$$E(L_1^2) = \int_0^a x^2 \mathrm{dens}(L_1 = x)\, dx$$

$$= \int_0^a x^2 \frac{n}{a}\left(1 - \frac{x}{a}\right)^{n-1} dx$$

$$= a\int_0^a x\frac{n}{a}\frac{x}{a}\left(1 - \frac{x}{a}\right)^{n-1} dx$$

This integral looks very similar to the one for $E(X_{(2)})$ but for the Uniform process of sampling $n+1$ points from $[0, a]$:

$$E(X_{(2)}) = \int_0^a x\,\mathrm{dens}(X_{(2)} = x)\, dx$$

$$= \int_0^a x\binom{n}{1}\frac{n+1}{a}\frac{x}{a}\left(1 - \frac{x}{a}\right)^{n-1} dx$$

$$= (n+1)\int_0^a x\frac{n}{a}\frac{x}{a}\left(1 - \frac{x}{a}\right)^{n-1} dx$$

Thus,

$$E(L_1^2) = \frac{a}{n+1}E(X_{(2)}) = \frac{a}{n+1}\frac{2a}{n+2} = \frac{2a^2}{(n+1)(n+2)}$$

Therefore,

$$Var(L_1) = E(L_1^2) - E(L_1)^2$$

$$= \frac{2a^2}{(n+1)(n+2)} - \left(\frac{a}{n+1}\right)^2$$

$$= \frac{a^2 n}{(n+1)^2(n+2)}$$

Furthermore,

$$Cov(L_1, L_2) = \frac{Var(L_1)}{n} = \frac{a^2}{(n+1)^2(n+2)}$$

Thus,

$$Var(X_{(k)}) = \frac{ka^2 n}{(n+1)^2(n+2)} + \frac{k(k-1)a^2}{(n+1)^2(n+2)}$$

$$= \frac{k(n+1-k)a^2}{(n+1)^2(n+2)}$$

Answer to 5.6. Recall that the sample mean and sample variance are

$$\overline{m} = \frac{S_n}{n}$$

$$\overline{\sigma^2} = \frac{(X_1 - m)^2 + (X_2 - m)^2 + \cdots + (X_n - m)^2}{n}$$

$$\overline{\overline{\sigma^2}} \;=\; \frac{(X_1 - \overline{m})^2 + (X_2 - \overline{m})^2 + \cdots + (X_n - \overline{m})^2}{n - 1}$$

Each of these is a random variable. We will compute their expectations.

$$
\begin{aligned}
E(\overline{m}) \;&=\; E\!\left(\frac{S_n}{n}\right) \\
&=\; \frac{E(X_1 + \cdots + X_n)}{n} \\
&=\; \frac{E(X_1) + \cdots + E(X_n)}{n} \\
&=\; \frac{nm}{n} = m
\end{aligned}
$$

$$
\begin{aligned}
\overline{\sigma^2} \;&=\; E\!\left(\frac{(X_1 - m)^2 + (X_2 - m)^2 + \cdots + (X_n - m)^2}{n}\right) \\
&=\; \frac{E\!\left((X_1 - m)^2\right) + E\!\left((X_2 - m)^2\right) + \cdots + E\!\left((X_n - m)^2\right)}{n} \\
&=\; \frac{n\sigma^2}{n} = \sigma^2
\end{aligned}
$$

The computation of $E\!\left(\overline{\overline{\sigma^2}}\right)$ is more complicated. Recall that

$$E(X_i X_j) = E(X_i)E(X_j) = m^2$$

when $i \neq j$, because X_i and X_j are independent. Furthermore, $E(X_i^2) = \sigma^2 + m^2$, by definition of σ^2. Hence,

$$
\begin{aligned}
E\!\left(\overline{\sigma^2}\right) \;&=\; \sum_{i=1}^{n} E(X_i^2 - 2X_i \overline{m} + \overline{m}^2) \\
&=\; \sum_{i=1}^{n} E(X_i^2) - 2\sum_{i=1}^{n} E(X_i \overline{m}) + \sum_{i=1}^{n} E(\overline{m}^2) \\
&=\; n(\sigma^2 + m^2) - 2nE(X_1 \overline{m}) + nE(\overline{m}^2)
\end{aligned}
$$

because the X_i's are equidistributed. We now expand the second term above:

$$
\begin{aligned}
-2nE(X_1 \overline{m}) \;&=\; -2\sum_{j=1}^{n} E(X_1 X_j) \\
&=\; -2E(X_1^2) - 2\sum_{j=2}^{n} E(X_1)E(X_j) \\
&=\; -2\sigma^2 - 2m^2 - 2(n-1)m^2 \\
&=\; -2\sigma^2 - 2nm^2
\end{aligned}
$$

Finally, we expand the last term

$$
\begin{aligned}
nE(\overline{m}^2) &= \frac{1}{n}\sum_{j=1}^{n}\sum_{k=1}^{n}E(X_jX_k) \\
&= \frac{1}{n}\sum_{j=1}^{n}E(X_j^2) + \frac{1}{n}\sum_{j\neq k}^{n}E(X_jX_k) \\
&= \sigma^2 + m^2 + \frac{1}{n}\sum_{j\neq k}^{n}E(X_j)E(X_k) \\
&= \sigma^2 + m^2 + \frac{1}{n}n(n-1)m^2 \\
&= \sigma^2 + m^2(n-1)m^2 \\
&= \sigma^2 + nm^2
\end{aligned}
$$

Adding the three terms together gives

$$
\begin{aligned}
E\left(\sum_{i=1}^{n}(X_i-\overline{m})^2\right) &= n\sigma^2 + nm^2 - 2\sigma^2 - 2nm^2 + \sigma^2 + nm^2 \\
&= (n-1)\sigma^2
\end{aligned}
$$

Consequently, $E\left(\overline{\overline{\sigma^2}}\right) = \sigma^2$, but this works only if the denominator in the definition of $\overline{\overline{\sigma^2}}$ is $n-1$ *not* n. For n large, however, the distinction is small.

Answer to 5.11. Hearing about this new policy, 1000 conscientious experimenters formulate 1000 random scientific hypotheses. On average, 10 of them would find a significant result, and these 10 would then be entitled to publish their results. Let us say that these 10 articles constitute the first issue of the journal after the new policy is instituted. We would find that the journal policy allowed 100% of the published results to be spurious. Clearly, the journal policy is a result of a misunderstanding of the nature of statistical hypothesis testing. Significant at the .01 level does *not* mean that there is only a 1% chance that one is wrong.

Answer to 5.12. The misleading feature of the statement is the exclamation point! On average, one would expect that out of five experiments, each significant at the .05 level, one will be significant at the .01 level *by chance alone*. Furthermore, it would not be at all surprising if out of five experiments, each significant at the .05 level, *two* were significant at the .01 level. The chance of this happening being $\binom{5}{2}\left(\frac{1}{5}\right)^2\left(\frac{4}{5}\right)^3 \approx 0.2$.

For example, suppose that we wish to test whether a coin is fair, as in Section 5.4. This may seem like a trivial example, but most tests of significance of a hypothesis are logically equivalent to this test. In this test, we toss a coin 100 times. We get a significant result at the .05 level if 40 or fewer or 60 or more heads occur. At the .01 level we would find a significant result if 37 or fewer or 63 or more heads occur. Now, *on average*, if we test 100 fair coins we will find that 5 are significantly unfair at the .05 level and that 1 is significantly unfair at the .01 level. And we will come to this conclusion despite the fact that the coins are all fair.

As another example, we have all seen experimental results thrown out on flimsy excuses such as "contaminated reagents" or "interference." In most cases there is a good reason to discard the data. But there are cases, particularly those that rest heavily on statistics for their validity, where significant results occur only *because* the insignificant repetitions of the experiment were discarded.

The conclusion of this discussion is simply that an experimenter should be aware of how weak statistical reasoning and statistical measurements can be. Statistics never relieves one of the necessity of understanding the phenomenon being studied and can only make a contribution, albeit an important one, toward achieving this understanding.

Answer to 5.13.   Observing 40 or fewer in such a sample, where the model is a Bernoulli process with $p = 0.45$, is

```
> pbinom(40, 100, 0.45)
[1] 0.1830569
```

So the result is not significant.

Answer to 5.14.   Observing 40 in such a random sample is 11 fewer than the expected value. The probability of being this far from the expected value in a Bernoulli process with $p = 0.51$ is

```
> pbinom(40, 100, 0.51) + (1 - pbinom(61, 100, 0.51))
[1] 0.03516299
```

Observing 40 or fewer in a random sample from a Bernoulli process with $p = 0.51$ is

```
> pbinom(40, 100, 0.51)
[1] 0.01768271
```

Both of these are significant. Consequently, we can reject the hypothesis. Since the percentage of the population that is female is known very accurately, this means that the sample may not have been a random sample.

Answer to 5.15.   The problem requires finding a single-tail test of significance. We will introduce the use of 3 models for this phenomenon: Bernoulli, Poisson and Normal. Let D be the number of defective screws produced during 1 hour of operation. On average, during normal operation, we expect 100 defective screws to be produced, i.e., $E(D) = 100$. We will adjust the machine if there are significantly more than 100 defective screws. In probabilistic language this would be written as follows. Choose d so that:

$$P(D>d) = \beta$$

where β is the desired significance level. Then the machine is out of adjustment at significance level β if D is found to be greater than d.

Bernoulli model   A defective screw is produced with probability $p = 0.01$ and each screw is independent of the others. Then $D = S_{10000}$ and

$$P(D>d) = 1 - P(D{\leq}d) = 1 - \sum_{k=0}^{d} \binom{10000}{k} p^k q^{10000-k}$$

Poisson model   This process will be introduced in Chapter 7. It is included here for later reference. Assume the process is continuous with a density of defects $\alpha = 0.01$. This will not be as good as the Bernoulli model, but it is a good approximation and is much simpler computationally. For this model $D = N(10000)$ and

$$P(D>d) = 1 - P(N(10000){\leq}d) = \sum_{k=0}^{d} \frac{100^k}{k!} e^{-100}$$

Normal model   We use the Central Limit Theorem in the Bernoulli model. $D = S_{10000}$ has mean 100 and variance 99. Thus, $\frac{D-100}{\sqrt{99}}$ is approximately a standard normal distribution, and

$$P(D>d) = P\left(\frac{D-100}{\sqrt{99}} > \frac{d-100}{\sqrt{99}}\right)$$

It is easy to compute these models in R. In each case, we assume that a `support` has been defined:

Bernoulli model

```
> b <- function(k) exp(lchoose(10000,k)+k*log(.01)+
+       (10000-k)*log(.99))
> plot(support,
+       sapply(support, function(d) 1-sum(sapply(0:d, b))),
+       pch=1)
```

Poisson model

```
> p <- function(k)
+       exp(k*log(100)-lfactorial(k)-100)
> plot(support,
+       sapply(support, function(d) 1-sum(sapply(0:d, p))),
+       pch=2)
```

Normal model

```
> n <- function(d) 1-pnorm(d, mean=100, sd=sqrt(99))
> plot(support, sapply(s, n), pch=3)
```

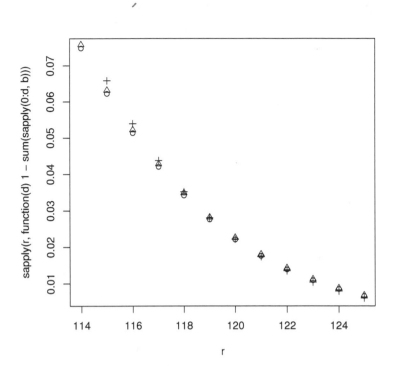

The **support** was determined with a little experimentation. All three models agree that $d = 116$ for the significance level .05 and $d = 123$ for the significance level .01.

Answer to 5.17. The question is whether the fraction of subscribers who use the product is at least 30%. We take this as the hypothesis, and perform a one-tailed test:

```
> pbinom(10,50,0.30)
[1] 0.07885062
```

The probability that 10 or fewer subscribers in the sample would use the product given that 30% of the subscribers use the product is less than 8%. So we can reject our hypothesis at the 10% significance level. In other words, the sample is not consistent with the hypothesis. By the stated requirement, the advertising campaign was not successful.

Answer to 5.19. The standard deviation is 2.09 ppm so the sample mean is well within one standard deviation of the required level. It should be clear that one cannot reject the hypothesis that the drug company is producing pills at an acceptable level of the impurity.

Answer to 5.20. Using the Bernoulli process with $p = 0.90$, the probability that 174 or fewer individuals in the sample would consider the firm to be a major polluter is:

```
> pbinom(174,200,0.9)
[1] 0.1004573
```

This is not significantly smaller than the previous percentage, suggesting that the campaign did not have a significant effect.

Answer to 5.25. To test whether the alteration was significant, we use a Bernoulli process with bias $p = 334/400 = 0.835$, which we approximate with a normal distribution with variance $npq = 400(0.835)(0.165) = 55.11$. The hypothesis is that the two samples have the same distribution. So their difference should be normally distributed with mean 0 and variance 55.11. The probability of observing a difference of -14 is

```
> pnorm(-14,0,sqrt(400*.835*.165))
[1] 0.02965592
```

which is significantly low. So the alteration did have a significant effect. Running the previous computation with the new sample gives:

```
> pbinom(160,200,0.9)
[1] 1.700733e-05
```

which shows that the public relations campaign had a significant effect as well.

Answer to 5.26. 296 ± 49.35; 10 ± 82.35; 296 ± 5; 310 ± 6

Answer to 5.29. .9999

Answer to 5.30. 0.215

Answer to 5.34. 16 weeks and 3 weeks

Answer to 5.35. about 0.3%

Answer to 5.36. The number of members of the electorate is an irrelevant datum in the problem. Assume that the percentage desired is p and that the electorate may be modeled by a Bernoulli process. We proceed as in Exercise 5.11. Note that there is an ambiguity in the problem. Do we want $p\pm.01$ or $p\pm.01p$, i.e., do we want an absolute or relative accuracy? We will assume that the desired accuracy is absolute, as would normally be the case in such a situation. We want to choose n so that

$$P(p - .01{\le}\bar{p}{\le}p + .01) \approx .95$$

or so that

$$P(p - .05{\le}\bar{p}{\le}p + .05) \approx .99$$

We rewrite these as

$$P\left(-.01\sqrt{\frac{n}{pq}}\le\frac{\bar{p} - p}{\sqrt{pq/n}}\le.01\sqrt{\frac{n}{pq}}\right) \approx .95$$

and

$$P\left(-.05\sqrt{\frac{n}{pq}}\le\frac{\bar{p} - p}{\sqrt{pq/n}}\le.05\sqrt{\frac{n}{pq}}\right) \approx .99$$

Since $\dfrac{\bar{p} - p}{\sigma}$ is close to being $N(0, 1)$, we conclude that n must be chosen either so that $.01\sqrt{\dfrac{n}{pq}} \approx 1.96$ or so that $.05\sqrt{\dfrac{n}{pq}} \approx 2.58$, i.e., either n must be 38,416 pq or n must be 2663 pq. Regardless of what p happens to be, the latter choice is better: it is cheaper to find p to within 5% with 99% confidence than to find p to within 1% with 95% confidence.

Answer to 5.37. Computing a one-sided confidence interval gives:

```
> qnorm(0.95,15,3)
[1] 19.93456
```

So she should leave 20 minutes early. Computing the one-sided confidence gives

```
> pnorm(17,15,3)
[1] 0.7475075
```

So she has time for coffee about 3 mornings out of 4.

Answer to 5.39. The only way we know of for computing p is to toss the thumbtack. How many times must we do it? The problem requires that n be large enough that $\bar{p}$ is within $p/10$ of p with probability 0.95, i.e., that

$$P\left(p - \frac{p}{10} \le \bar{p} \le p + \frac{p}{10}\right) = 0.95$$

Now for large n, $\bar{p} = \dfrac{S_n}{n}$ is close to being $N\left(p, \dfrac{pq}{n}\right)$. So rewrite the above condition as:

$$P\left(-\frac{p\sqrt{n}}{10\sqrt{pq}} \le \frac{\bar{p} - p}{\sqrt{pq/n}} \le \frac{p\sqrt{n}}{10\sqrt{pq}}\right) \approx 0.95$$

and $\dfrac{\bar{p} - p}{\sqrt{pq/n}}$ is close to being $N(0,1)$. Therefore, n should be chosen so that $\dfrac{p\sqrt{n}}{10\sqrt{pq}} \approx 1.96$, i.e., $n = \dfrac{384q}{p}$. The lack of symmetry between p and q arises from the fact that we want to find p to an accuracy of $p/10$. When p is small, this is very accurate indeed. As one tosses the tack, one will get a better idea of what p is and hence a better idea of what n ought to be.

Answer to 5.40. In both cases, we will use a Bernoulli model, but for computations we also use the normal distribution. Let Q_A, Q_B and Q be the number of requests for landlines at the busiest time from city A, city B and from the combined pair of cities, respectively. In the Bernoulli process with bias $p = .01$, Q_A and Q_B have the distribution of S_{5000}, while Q has the distribution of S_{10000}.

Case 1: Two cities serviced separately Each city requires l landlines, where

$$P(Q_A > l) = P(Q_B > l) = P(S_{5000} > l) = .01$$

Now S_{5000} is approximately $N(50, 49.5)$. We can compute the required level in R either with qbinom or qnorm:

```
> qbinom(.99, 5000, .01)
[1] 67
> qnorm(.99, 50, sqrt(49.5))
[1] 66.36731
```

Therefore, $l = 67$. This means that 134 landlines would be required in all, for a total cost of $1,340,000.

Case 2: Two cities jointly serviced   The cities require a total of l landlines, where

$$P(Q{>}l) = P(S_{10000}{>}l) = .01$$

Now S_{10000} is approximately $N(100, 99)$. Computing as before we get:

```
> qbinom(.99, 10000, .01)
[1] 124
> qnorm(.99, 100, sqrt(99))
[1] 123.1469
```

Therefore, $l = 124$. This means that only 124 landlines would be needed, but we also require $50,000 for additional equipment. The total cost is then $1,290,000. Despite the extra equipment, we find that this option is less expensive.

Answer to 5.43.   The median salary of the women is 83.5% while 88% of 97% is 85.4%. So it appears that they are losing ground, but the result is not significant. Using a Bernoulli process to model this, if the median salary of all women were 85.4%, then each woman's salary would have probability 0.5 of exceeding this figure. We observe that 3 out of the 8 salaries do so. The probability that 3 (or fewer) would do so by chance alone is:

```
> pbinom(3,8,0.5)
[1] 0.3632813
```

which is too large for us to reject that it happened by chance alone.

Answer to 5.51.   This is the reverse of what one normally uses for a statistical test. Instead of computing the probability of the tails of the distribution, one uses the central interval. The probability that one would observe 7746 or more with the dominant trait is

```
> 1-pbinom(7746,10326,0.75)
[1] 0.4826266
```

Using the same probability for the other tail gives a total probability of about 0.965 to be so far from the expected value. So the probability of being this close to the expected value is $1 - 0.965 = 0.035$, which is significant. One can reject the hypothesis that the experiment was properly carried out.

CHAPTER 6

Conditional Probability

Thomas Bayes was a British theologian and mathematician. As he did not conform to the religion that dominated England at the time, the universities at Oxford and Cambridge were closed to him. He published two papers in his lifetime, and several others were published posthumously, but overall he published very little. Nevertheless, his work had an impact. Probability at the time was concerned with determining the likelihood of certain events occurring, especially ones occurring in games of chance, in which the exact circumstances for the event are known. For example, one knows exactly how many cards are in a deck or how many white and black balls are in an urn, before one deals a hand or chooses a ball. These are sometimes called "forward probabilities." De Moivre posed the problem of whether one can "reverse" probabilities. For example, given the color of the selected ball, how many balls of each color were in the urn? Bayes solved the specific problem posed by de Moivre. His solution was later rediscovered independently by Laplace, who also generalized it. It is also possible that Bayes was not the first to solve the de Moivre problem. Nevertheless, the solution to this problem led to the notion that probability can be interpreted as a partial belief rather than just as frequency of occurrence. This allows one to apply probability to all sorts of propositions to which a frequency interpretation is unrealistic, such as whether there is life on Mars, because the events are not repeatable and so have no frequency interpretation. During the mid-twentieth century, statisticians were sharply divided about whether a probability could be given an interpretation in which such events made sense, and the term "Bayesian" was given to interpretations of this kind, mostly to contrast them with the more traditional interpretations that are now referred to as being "frequentist."

The theory of probability consists largely in making precise the probabilistic language that already forms part of our language. In effect the purpose of this course is to learn to "speak probability" properly. The lowest level of our probabilistic language is the event. This corresponds to simple phrases that are either true or false. For example, in the Bernoulli process H_i is the event

"The ith toss is heads." Random variables represent the next level: simple quantitative questions. For example, one might ask: "How long must one toss a coin until the first head appears?"

Conditional probability allows probabilistic reasoning. That is, we may now ask compound questions. For example, "If the first toss of a coin is tails, how long must one wait until the first head appears?" Moreover, we can split apart and combine such questions into new questions. The precise meaning of such expressions is not always obvious and is the source of many seeming paradoxes and fallacies. As a simple example, the question, "If the first toss is heads, is the second toss also heads?" is very different from the question, "Are the first two tosses heads?" The probability of the first is p while that of the second is p^2.

6.1 Discrete Conditional Probability

We begin with the definition and properties of the conditional probability of events.

Definition. Let A and S be events such that $P(S) > 0$. The **conditional probability** of A given S is

$$P(A \mid S) = \frac{P(A \cap S)}{P(S)}$$

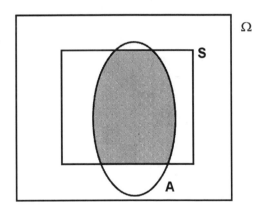

The event S is called the *condition.* The conditional probability $P(A \mid S)$ answers the question: "If S has occurred, how probable is A?" In effect, we have altered our sample space. Since we know that S has occurred, the sample space is now S. The event A given that S has occurred must now be interpreted as $A \cap S$, and the probability is $P(A \cap S)$ normalized by the probability of S so that the total probability is 1. Ordinary probabilities are the special case of conditional probabilities where the condition is the sample space Ω: $P(A) = P(A \mid \Omega)$.

Law of Alternatives

Suppose that instead of knowing that a certain event has occurred, we know that one of several possibilities has occurred, which are mutually exclusive. Call these alternatives $A_1, A_2, \ldots$ There may possibly be infinitely many alternatives. More precisely:

Definition. A set of events $A_1, A_2, \ldots$ form a **set of alternatives** if

1. The A_i's are pairwise disjoint.
2. $P(\bigcup_i A_i) = 1$.
3. $P(A_i) > 0$ for every i.

Usually, we will have the stronger condition that $\bigcup_i A_i$ is the entire sample space Ω. However, it is enough just for the union to be all of Ω except for an event of probability zero. We will see an example later where the union of all alternatives is not all of Ω.

Law of Alternatives. For any event B and set of alternatives $A_1, A_2, \ldots$

$$P(B) = P(B \mid A_1) P(A_1) + P(B \mid A_2) P(A_2) + \cdots$$

Proof. To verify this law we simply expand and cancel. The A_i's are disjoint so the events $B \cap A_i$ are also disjoint.

$$P(B \mid A_1) P(A_1) + P(B \mid A_2) P(A_2) + \cdots$$

$$= \frac{P(B \cap A_1)}{P(A_1)} P(A_1) + \frac{P(B \cap A_2)}{P(A_2)} P(A_2) + \cdots$$

$$= P(B \cap A_1) + P(B \cap A_2) + \cdots$$

$$= P(B \cap A_1 \cup B \cap A_2 \cup \cdots)$$

$$= P(B \cap (A_1 \cup A_2 \cup \cdots))$$

$$= P(B)$$

If the alternatives A_i are not exhaustive, we can still make sense of the Law of Alternatives provided all probabilities involved are conditioned by the event $A = \bigcup_i A_i$. More precisely:

Definition. A set of events $A_1, A_2, \ldots$ is a *set of alternatives* for A if

1. The A_i's are pairwise disjoint
2. $A = \bigcup_i A_i$
3. $P(A_i) > 0$ for every i.

Conditional Law of Alternatives. For any event B and a set of alternatives $A_1, A_2, \ldots$ for A:

$$P(B \mid A) = P(B \mid A_1) P(A_1 \mid A) + P(B \mid A_2) P(A_2 \mid A) + \cdots$$

Bayes' Law

One of the features of probability theory as we have developed it so far is that all events are treated alike. In principle, no events are singled out as "causes" while others become "effects." Bayes' law, however, is traditionally stated in terms of causes and effects. Although we will do so also, one should be careful not to ascribe metaphysical significance to these terms. Historically, this law has been misapplied in a great number of cases precisely because of such a misunderstanding.

We are concerned with the following situation. Suppose we have a set of alternatives $A_1, A_2, \ldots$ which we will refer to as "causes." Suppose we also have an event B which we will call the "effect." The idea is that we can observe whether the effect B has or has not occurred but not which of the causes A_i has occurred. The question is to determine the probability that a given cause occurred given that we have observed the effect. We assume that we know the probability for each of the causes to occur, $P(A_i)$, as well as the conditional probability for B to occur given each cause, $P(B \mid A_i)$. The probability $P(A_i)$ is called the **a priori** or **prior probability** of A_i, and we seek the probability $P(A_i \mid B)$, which we call the **a posteriori** or **posterior probability** of A_i. If the alternatives $A_1, A_2, \ldots$ represent various experimental hypotheses, and B is the result of some experiment, then Bayes' law allows us to compute how the observation of B changes the probabilities of these hypotheses.

Bayes' Law. For any event B and a set of alternatives $A_1, A_2, \ldots$ for A:

$$P(A_i \mid B) = \frac{P(B \mid A_i)\, P(A_i)}{\sum_j P(B \mid A_j)\, P(A_j)}$$

Proof. Let A and B be any two events having positive probabilities. By the definition of conditional probability,

$$P(B \mid A) = \frac{P(A \cap B)}{P(A)} \quad \text{and} \quad P(A \mid B) = \frac{P(A \cap B)}{P(B)}$$

As a result, we have two ways to express $P(A \cap B)$:

$$P(B \mid A)\, P(A) = P(A \cap B) = P(A \mid B)\, P(B)$$

Solving for $P(A \mid B)$ gives:

$$P(A \mid B) = \frac{P(B \mid A)\, P(A)}{P(B)}$$

Now apply this fact to the case for which A is A_i and use the Law of Alternatives to compute the denominator. The resulting expression is Bayes' law.

Law of Successive Conditioning

Suppose we have n events $B_1, B_2, \ldots, B_n$ such that $P(B_1 \cap B_2 \cap \cdots \cap B_n) > 0$. Then we can compute $P(B_1 \cap B_2 \cap \cdots \cap B_n)$ using a sequence of conditional probabilities.

Law of Successive Conditioning.

$$P(B_1 \cap B_2 \cap \cdots \cap B_n) =$$
$$P(B_1)\, P(B_2 \mid B_1)\, P(B_3 \mid B_1 \cap B_2) \cdots P(B_n \mid B_1 \cap B_2 \cap \cdots \cap B_{n-1})$$

Proof. The result follows almost immediately: just expand and cancel.

This law corresponds to the intuitive idea that the probability of several events occurring is the product of their individual probabilities. This idea is correct provided we interpret "individual probability" to mean the appropriate conditional probability.

By using the Law of Alternatives and the Law of Successive Conditioning we split the computation of an ordinary or a conditional probability into a succession of conditional probabilities. In effect, we form compound, nested, conditional questions out of simple questions.

Independence

Suppose that A and B are two events. If either A or B has zero probability of occurring, then A and B are trivially independent events. If $P(A) > 0$ and $P(B) > 0$, then the concept of the independence of the events A and B is best stated by using conditional probability. Namely, the following statements are equivalent:

- A and B are independent

- $P(A \mid B) = P(A)$

- $P(B \mid A) = P(B)$

Using this terminology we can see much more clearly that the independence of two events A and B means that knowing one has occurred does not alter the probability that the other will occur, or equivalently that the measurement of one does not affect the measurement of the other.

It can happen that two events that are dependent can become independent when they are conditioned on a third event. More precisely, two events A and B are said to be **conditionally independent** with respect to an event C when

$$P(A \cap B \mid C) = P(A \mid C)\, P(B \mid C)$$

The notion of conditional independence also applies to random variables. Two random variables X and Y are conditionally independent with respect to a random variable Z if every pair of events $P(X \leq x)$ and $P(Y \leq y)$ is conditionally independent with respect to every event $P(Z \leq z)$.

Conditional Random Variables

Suppose that we have a random variable X whose distribution is known, but that some event A occurs. Does this change the distribution? The answer is that the distribution of X must now be conditioned on the event A. In other words, every probability $P(X = x)$ for x in the support of X, must now be replaced by the conditional probability $P(X = x \mid A)$. The result is a new random variable called the **conditional random variable** which is written $X \mid A$. The distribution of $X \mid A$ is called the **conditional probability distribution** of X given A. A conditional random variable is just like any other random variable. The conditioning is just changing the sample space on which it is defined. Ordinary random variables can be regarded as being conditional random variables on the sample space Ω.

As we have seen, events are most commonly defined by random variables. So it is no surprise that we will be conditioning on events defined by one or more random variables. To keep the notation from becoming too cumbersome, it is a common practice to use abbreviations. The meaning of these abbreviations should be clear from the context. For example, if $X, Y_1, Y_2, \ldots, Y_n$ are random variables, then the conditional probability

$$P(X = x \mid (Y_1 = y_1) \cap (Y_2 = y_2) \cap \cdots \cap (Y_n = y_n))$$

is often abbreviated to:

$$P(X = x \mid Y_1 = y_1, Y_2 = y_2, \ldots, Y_n = y_n)$$

Think of the comma as an abbreviation for the intersection. Somewhat less commonly, this will be further abbreviated to

$$P(X \mid Y_1, Y_2, \ldots, Y_n)$$

leaving the values $x, y_1, y_2, \ldots, y_n$ implicit. When X is conditioned by one or more random variables, it is called the *conditional random variable X given* $Y_1, Y_2, \ldots Y^n$, and written $X \mid Y_1, Y_2, \ldots, Y^n$, again leaving the values of the random variables implicit.

6.2 Gaps and Runs in the Bernoulli Process

Recall that in the Bernoulli process, T_i is the gap between the $(i-1)^{st}$ and the ith head. We claimed that the T_i's are independent random variables, using an intuitive probabilistic argument. We now have the terminology for making this argument rigorous. The key notion is the Law of Alternatives.

Consider the probability distribution of the gap T_{i+1} when all of the preceding gaps are known. In other words, we want the conditional probability

$$P(T_{i+1} \mid T_1 = k_1, T_2 = k_2, \ldots, T_i = k_i)$$

Computing this conditional probability when $T_{i+1} = n$ is quite easy, for it corresponds to exactly two patterns of H's and T's. The denominator in the definition of conditional probability has this pattern:

$$\overbrace{\underbrace{TT\ldots TH}_{k_1}\underbrace{TT\ldots TH}_{k_2}\underbrace{TT\ldots TH}{}\underbrace{TT\ldots TH}_{k_i}}^{k=k_1+k_2+\cdots+k_i}$$

from which we see that:

$$P(T_1 = k_1, T_2 = k_2, \ldots, T_i = k_i) = q^{k-i}p^i$$

The numerator has this pattern:

$$\overbrace{\underbrace{TT\ldots TH}_{k_1}\underbrace{TT\ldots TH}_{k_2}\underbrace{TT\ldots TH}{}\underbrace{TT\ldots TH}_{k_i}\underbrace{TT\ldots TH}_{n}}^{k=k_1+k_2+\cdots+k_i}$$

from which we see that:

$$P(T_1 = k_1, T_2 = k_2, \ldots T_i = k_i, T_{i+1} = n) = q^{k+n-i-1}p^{i+1}$$

Therefore,

$$P(T_{i+1} = n \mid T_1 = k_1, T_2 = k_2, \ldots T_i = k_i) = \frac{q^{k+n-i-1}p^{i+1}}{q^{k-i}p^i} = q^{n-1}p$$

Although the above computation is not very difficult, there is an easier way to see it. Think of the condition:

$$(T_1 = k_1, T_2 = k_2, \ldots, T_i = k_i)$$

as *changing* our sample space. The new sample space consists of all infinite sequence of H's and T's, but renumbered starting with

$$k + 1 = k_1 + k_2 + \cdots + k_i + 1$$

This new sample space is *identical* to the Bernoulli sample space except for the renumbering and the fact that T_{i+1} is now the waiting time for the *first* head. Therefore,

$$P(T_{i+1} = n \mid T_1 = k_1, T_2 = k_2, \ldots, T_i = k_i) = q^{n-1}p$$

The key to the effective use of conditional probability is that it changes the sample space and hence the interpretation of the random variables defined on the old sample space.

We now apply the Law of Alternatives. The events

$$(T_1 = k_1, T_2 = k_2, \ldots, T_i = k_i)$$

as the k_j's take on all positive integer values, form a set of alternatives; for the set of sample points belonging to none of them is an event whose probability is zero.

$$P(T_{i+1} = n) = \sum_{k_1,\ldots k_i} P(T_{i+1} = n \mid T_1 = k_1, T_2 = k_2, \ldots, T_i = k_i)$$

$$P(T_1 = k_1, T_2 = k_2, \ldots, T_i = k_i)$$

$$= \sum_{k_1,\ldots k_i} q^{n-1} p P(T_1 = k_1, T_2 = k_2, \ldots T_i = k_i)$$

$$= q^{n-1} p \sum_{k_1,\ldots k_i} P(T_1 = k_1, T_2 = k_2, \ldots T_i = k_i)$$

$$= q^{n-1} p$$

Notice that the only fact we used about the events

$$(T_1 = k_1, T_2 = k_2, \ldots, T_i = k_i)$$

was that they form a set of alternatives. An immediate consequence is that the gaps are equidistributed. Furthermore, if we use the definition of conditional probability, we have that

$$P(T_{i+1} = n) = q^{n-1} p = P(T_{i+1} \mid T_1 = k_1, T_2 = k_2, \ldots, T_i = k_i)$$

for any choice of k_j's. In other words, T_{i+1} is independent of $T_1, \ldots, T_i$. By induction, all the T_i's are independent.

We can now see more clearly how the T_i's are related. They have the same distribution, but they are *not* the same. They have this property because the measurement of the ith gap "really" occurs in a different sample space than the first gap, but this new sample space is identical to the ordinary Bernoulli sample space, except for how we number the tosses. We went into detail for this argument to illustrate a nontrivial use of the Law of Alternatives. We will be more abbreviated in the future.

As an illustration of the Law of Alternatives, we consider a problem mentioned in Chapter 2. Namely, what is the probability that a run of h heads occurs before a run of t tails? Let A be this event. We solve this problem by using the following fact: when a run of less than h heads is "broken" by getting a tail, we must "start over," and similarly for runs of tails. Using the Law of Alternatives:

$$P(A) = P(A \mid X_1 = 1) P(X_1 = 1) + P(A \mid X_1 = 0) P(X_1 = 0)$$

Write $u = P(A \mid X_1 = 1)$ and $v = P(A \mid X_1 = 0)$ so that

$$P(A) = up + vq$$

Next, we use the Conditional Law of Alternatives for each of $P(A \mid X_1 = 1)$ and $P(A \mid X_1 = 0)$. Consider the first one. We know that we got a head on the first toss so the run has started. We then "wait" to see if the run will be broken. That is, let B_t be the waiting time for the first tail *starting* with the second toss. Either we get a tail and break the run or we get enough heads so that A occurs. More precisely,

$$P(A \mid X_1 = 1) \quad = \quad P(A \mid X_1 = 1, B_t < h) P(B_t < h)$$
$$+ P(A \mid X_1 = 1, B_t \geq h) P(B_t \geq h)$$

For the first alternative, the run of heads has been broken by a tail. Hence,

$P(A \mid X_1 = 1, B_t{<}h) = P(A \mid X_1 = 0)$, for the earlier heads have no effect on subsequent tosses. All that matters is that we "started" with a tail. On the other hand, $P(A \mid X_1 = 1, B_t{\geq}h) = 1$ because the condition implies that A has in fact occurred. Therefore,

$$
\begin{aligned}
u = P(A \mid X_1 = 1) &= v{\times}P(B_t{<}h) + 1{\times}P(B_t{\geq}h) \\
&= v(1 - p^{h-1}) + p^{h-1}
\end{aligned}
$$

Remember that for B_t, we start counting on the second toss, so that $P(B_t{<}h)$ is really the conditional probability $P(B_t{<}h \mid X_1 = 1)$.

The computation for $P(A \mid X_1 = 0)$ is analogous to that above. Let B_h be the waiting time for the first head starting with the second toss. Then

$$
\begin{aligned}
P(A \mid X_1 = 0) &= P(A \mid X_1 = 0, B_h{<}t)\, P(B_h{<}t) \\
&\quad + P(A \mid X_1 = 1, B_h{\geq}t)\, P(B_h{\geq}t)
\end{aligned}
$$

Here $P(A \mid X_1 = 0, B_h{\geq}t) = 0$ because the condition implies that A has *not* occurred. The probability $P(A \mid X_1 = 0, B_h{<}t)$ is u because the run of tails has been broken. Therefore,

$$
\begin{aligned}
u = P(A \mid X_1 = 0) &= u{\times}P(B_h{<}t) + 0{\times}P(B_h{\geq}t) \\
&= u(1 - q^{t-1})
\end{aligned}
$$

Combining the two equations above gives us the system of equations,

$$
\begin{aligned}
u &= v(1 - p^{h-1}) + p^{h-1} \\
v &= u(1 - q^{t-1})
\end{aligned}
$$

Solve for u and v and substitute to obtain:

$$
P(A) = up + vq = \frac{p^{h-1}(1 - q^t)}{p^{h-1} + q^{t-1} - p^{h-1}q^{t-1}}
$$

We check this by considering the special case $h = t$ and $p = q = 1/2$. As we expect by symmetry, $P(A) = 1/2$ in this case.

6.3 Sequential Sampling

In most sampling situations, for example sampling people in a population, we generally sample the population without replacement, i.e., the same individual cannot be chosen more than once in one sample. For such a sampling procedure, the successive choices are *not* independent of one another; for with each choice, the population (and hence the sample space) gets smaller.

For very large populations, this would seem to be a small effect. But in smaller populations it can be pronounced. For example, suppose we play a card guessing game. We draw a card at random from a deck, try to guess the suit, look to see if our guess was correct and then place the card aside. If we continue to sample the cards this way, the probabilities for getting a card of a given suit change constantly. Indeed, we will always know for certain what the suit of the last card drawn will be.

The problem of sequential sampling is to describe the dependence of each of the choices on the other choices. The idealized model is the following. We have an urn containing r red balls and b black balls for a total of N balls in all. We select a ball at random from the run, note its color and then place it aside. This procedure is repeated until n balls have been chosen from the urn. Define the random variables X_i by:

$$X_i = \begin{cases} 1 & \text{if the } i^{th} \text{ ball is red} \\ 0 & \text{if the } i^{th} \text{ ball is black} \end{cases}$$

The problem is to find the distributions and the correlations of the X_i's.

Notice that we have switched the roles of the objects and the containers in the occupancy model. In the occupancy model, we place objects into containers. In the sampling model, the objects become the positions in the sample and the containers become individuals in a population. In the sequential sampling model, it is traditional to view the population more picturesquely as a collection of colored balls in an urn.

Consider the first choice X_1. The probability distribution of X_1 is

$$p_0 = P(X_1 = 0) \quad = \quad \frac{b}{N}$$

$$p_1 = P(X_1 = 1) \quad = \quad \frac{r}{N}$$

Next, consider the second choice X_2. To compute its distribution we must use the Law of Alternatives. For example, $P(X_2 = 0 \mid X_1 = 0)$ is $\dfrac{b-1}{r+b-1}$ because there is one fewer black ball in the urn. Similarly for $P(X_2 = 0 \mid X_1 = 1)$. Now put these together:

$$\begin{aligned} P(X_2) = 0 \quad &= \quad P(X_2 = 0 \mid X_1 = 0)\, P(X_1 = 0) \\ &\quad + \ P(X_2 = 0 \mid X_1 = 1)\, P(X_1 = 1) \\ &= \quad \frac{b-1}{N-1} \times \frac{b}{N} + \frac{b}{N-1} \times \frac{r}{N} \\ &= \quad \frac{(b-1)b + br}{(N-1)N} \\ &= \quad \frac{b(N-1)}{(N-1)N} \\ &= \quad \frac{b}{N} \end{aligned}$$

Similarly,

$$P(X_2 = 1) = \frac{r}{N}$$

The random variables X_1 and X_2 are equidistributed! This is quite unexpected. One wonders whether this is an accident of algebra or if there is some deep principle here. If the latter, we would expect that all the X_i's have the same distribution. As we shall see, this is indeed the case. The seeming paradox arises from the fact that we are *not* considering the random variables conditionally. For example, in the card guessing game above, if we chose not to look at the first 51 cards sampled, we would have no reason to suppose that the last card sampled has any special properties: it does not "know" that the other cards have been sampled.

Exchangeability

As often happens in mathematics, the situation only becomes clear when we consider it from a broader perspective. Consider the joint distribution of *all* the X_i's:

$$P(X_1 = i_1, X_2 = i_2, \ldots, X_n = i_n)$$

where the $i_1, \ldots, i_n$ take on the values 0 and 1 arbitrarily. We compute this by using the Law of Successive Conditioning:

$$P(X_1 = i_1, X_2 = i_2, \ldots, X_n = i_n) =$$
$$P(X_1 = i_1)\, P(X_2 = i_2 \mid X_1 = i_1)\, P(X_3 = i_3 \mid X_1 = i_1, X_2 = i_2) \cdots$$

For example, if $n = 5$ and $(i_1, \ldots, i_5) = (0, 1, 1, 0, 1)$ then

$$
\begin{aligned}
P(X_1 = 0, X_2 = 1, \ldots, X_5 = 1) &= \frac{b}{N} \times \frac{r}{N-1} \times \frac{r-1}{N-2} \times \frac{b-1}{N-3} \times \frac{r-2}{N-4} \\
&= \frac{b!}{(b-2)!}\frac{r!}{(r-3)!}\frac{(N-5)!}{N!} = \frac{\binom{N-5}{r-3}}{\binom{N}{r}}
\end{aligned}
$$

Each factor is the number of balls of the appropriate color at the time divided by the number of balls in the urn at the time. More generally, if we have drawn a sequence of k reds and l blacks, then the probability is

$$P(X_1 = i_1, X_2 = i_2, \ldots, X_n = i_n) = \frac{\binom{N-n}{r-k}}{\binom{N}{r}}$$

The probability of drawing a given sequence of reds and blacks depends only on the number of reds and blacks drawn. In other words,

$$P(X_1 = i_1, X_2 = i_2, \ldots, X_n = i_n)$$

is the same if we permute the $i_1, \ldots, i_n$ leaving the X_j's alone (or equivalently if we permute the X_j's leaving the i_j's alone).

Since we can compute the individual distributions of the X_j's from the joint distribution by taking marginals, we immediately get that the X_j's are equidistributed. Moreover, the joint distribution of any pair of X_j's is the same as any other. For example, the joint distribution of X_5 and X_7 is the same as that of X_1 and X_2:

$$P(X_5 = m, X_7 = n) = P(X_1 = m, X_2 = n)$$

and the latter is easy to compute. In general, the joint distribution of any n of the X_j's is the same as that of the first n of them.

This allows us to compute the probability that there will be k reds and l blacks, in *any* order whatsoever. Since the probability will be the same no matter what order the balls are chosen, we just multiply the probability of one of the sequences by the number of sequences. There are $\binom{n}{k}$ sequences $i_1, \ldots, i_n$ such that there are k reds and l blacks, so the probability is

$$\frac{\binom{n}{k}\binom{N-n}{r-k}}{\binom{N}{r}} = \frac{\frac{n!}{k!l!}\frac{(N-n)!}{(r-k)!(b-l)!}}{\frac{N!}{r!b!}} = \frac{\frac{r!}{k!(r-k)!}\frac{b!}{l!(b-l)!}}{\frac{N!}{n!(N-n)!}} = \frac{\binom{r}{k}\binom{b}{l}}{\binom{N}{n}}$$

which is exactly as if the set of k reds was chosen from all r reds, and then one independently chooses the set of l blacks from all b blacks. This is surprising since successive selections without replacement are not independent, but because of exchangeability, it is as if they were. This distribution is known as the **hypergeometric distribution**, and it has applications in statistics.

All these facts follow from the fact that the joint distribution of the X_j's is unchanged when we permute the X_j's. This is the real reason that the choices in sequential sampling are equidistributed.

Definition. Random variables (either integer or continuous) $X_1, X_2, \ldots, X_n$ are said to be **exchangeable** when their joint distribution (or density) is a symmetric function (i.e., unchanged when the arguments of the function are permuted).

An example of a set of exchangeable random variables we have already seen is a set of independent, equidistributed random variables. If X_1, X_2 and X_3 are independent, equidistributed integer random variables, then

$$\begin{aligned} P(X_1 = i_1, X_2 = i_2, X_3 = i_3) \quad &= \quad P(X_1 = i_1)\, P(X_2 = i_2)\, P(X_3 = i_3) \\ &= \quad p_{i_1} p_{i_2} p_{i_3} \end{aligned}$$

But as we have just seen, being exchangeable is not as strong a condition as being independent and equidistributed.

We mention in passing that being exchangeable is not really that much more general than being independent and equidistributed. There is a deep theorem of probability theory which, roughly speaking, says that every set of exchangeable random variables can be "synthesized" from independent, equidistributed random variables by suitable conditioning.

The Pólya Urn Process

A slightly more general sampling model than sampling either with replacement or without replacement is called the **Pólya Urn Process**. In this process, we begin with an urn containing r red balls and b black balls. We draw a ball at random. If it is red, we put the drawn ball plus c more red balls into the urn. If it is black, we put the drawn balls plus d more black balls into the urn. We then repeat this. Sampling with replacement is the case $c = d = 0$, and sampling without replacement is the case $c = d = -1$.

This process was originally introduced as a model of epidemics. If we think of the red balls as diseased individuals, then each discovery of a red ball increases the likelihood other balls will be red ($c>0$). There are obvious defects in such a model which we will not pursue. We will just think of this process as a general form of sampling.

As usual, let $X_1, \ldots, X_n$ be the successive results of drawing n balls in the Pólya Urn Process. The computation of the joint distribution of the X_j's is much the same as before. For example,

$$P(X_1 = 1, X_2 = 1, X_3 = 0, X_4 = 0, X_5 = 1) =$$
$$\frac{r}{r+b} \times \frac{r+c}{r+b+c} \times \frac{b}{r+b+2c} \times \frac{r+d}{r+b+2c+d} \times \frac{r+2c}{r+b+2c+2d}$$

In general, the X_j's will not be exchangeable, but if $c = d$, they are.

6.4 Continuous Conditional Probability

Consider the Uniform process of sampling n points from the interval $[0, a]$. Suppose we know that the kth point in order, $X_{(k)}$, was t. Given this information, what is the smallest point, $X_{(1)}$? It seems reasonable to answer this question with the conditional probability distribution

$$F(x) = P\big(X_{(1)} \leq x \mid X_{(k)} = t\big)$$

of $X_{(1)}$ given that $\big(X_{(k)} = t\big)$. Unfortunately, we know that $P\big(X_{(k)} = t\big) = 0$; so that, technically speaking, the above conditional probability does not make sense!

On the other hand, it is easy to determine what $P\big(X_{(1)} \leq x \mid X_{(k)} = t\big)$ ought to mean. The event $\big(X_{(k)} = t\big)$ means that exactly $k - 1$ points have fallen in the interval $[0, t]$, and the random variable $X_{(1)}$ should then be reinterpreted as the first order statistic of the Uniform process of dropping $k - 1$ points in the interval $[0, t]$. Therefore,

$$P\big(X_{(1)} \leq x \mid X_{(k)} = t\big) = 1 - \left(\frac{t-x}{x}\right)^{k-1}$$

Notice that we do not have to "choose" the $k - 1$ points which are to fall in $[0, 1]$. This choice is already implicit in the fact that we have conditioned by $\big(X_{(k)} = t\big)$.

Although we cannot make sense, in general, of a conditional probability $P(A \mid B)$ when $P(B) = 0$, we can do so when B is the event $(X = t)$ for a continuous random variable X. We will call this the **continuous conditional probability** (although we shall often drop the adjective "continuous"). The following is the formal definition of this concept. But one rarely uses the definition directly. As with the ordinary conditional probability, the best way to compute a continuous conditional probability is to regard the condition as defining a new sample space and to reinterpret the events and random variables of the old sample space in this new sample space.

Definition. For an event A and a continuous random variable X, the *continuous conditional probability of A given that $X = t$* is

$$P(A \mid X = t) = \lim_{\epsilon \to 0} P(A \mid t < X \le t + \epsilon)$$

$$= \lim_{\epsilon \to 0} \frac{P(A \cap (t < X \le t + \epsilon))}{P(t < X \le t + \epsilon)}$$

provided that this limit exists.

Notice that we do *not* divide by ϵ. The reason is that ϵ appears in both the numerator and the denominator. If you wish, $P(A \mid X = t)$ is the limit:

$$\lim_{\epsilon \to 0} \frac{P(A \cap (t < X \le t + \epsilon)) / \epsilon}{P(t < X \le t + \epsilon) / \epsilon}$$

Both the numerator and the denominator in this limit have "densities" as their limits.

Just to make sure, we will compute $P(X_{(1)} \le x \mid X_{(k)} = t)$ directly from the definition to see that we get what we computed earlier. We know from our computation in Section 4.1 that

$$P(t < X_{(k)} \le t + \epsilon) = n\binom{n-1}{k-1} \frac{t^{k-1} \epsilon (a - t - \epsilon)^{n-k}}{a^n} + \frac{\epsilon^2}{a^2} (\text{CE})$$

Next, we compute $P(X_{(1)} > x, t < X_{(k)} \le t + \epsilon)$. Except for a term having a factor of ϵ^2, this event corresponds to having $k - 1$ points in the interval $[x, t]$, one point in $[t, t + \epsilon]$ and the rest in $[t + \epsilon, a]$. Therefore,

$$P(X_{(1)} > x, t < X_{(k)} \le t + \epsilon)$$

$$= n\binom{n-1}{k-1} \frac{(t - x)^{k-1} \epsilon (a - t - \epsilon)^{n-k}}{a^n} + \frac{\epsilon^2}{a^2} (\text{CE})$$

We now combine the above two computations.

$$\frac{P(X_{(1)} > x, t < X_{(k)} \le t + \epsilon)}{P(t < X_{(k)} \le t + \epsilon)}$$

$$= \frac{n\binom{n-1}{k-1}(t - x)^{k-1} \epsilon (a - t - \epsilon)^{n-k} + \epsilon^2 (\text{CE})}{n\binom{n-1}{k-1} t^{k-1} \epsilon (a - t - \epsilon)^{n-k} + \epsilon^2 (\text{CE})}$$

$$\to \left(\frac{t - x}{t} \right)^{k-1}$$

Which is the same result as before. Needless to say, this is the hard way to compute the probability.

Consider one more example. Suppose we know that the first point, X_1, is t. Given this, what is the smallest point, $X_{(1)}$? Again, the answer is a probability distribution:

$$F(x) = P(X_{(1)} \leq x \mid X_1 = t)$$

We split this into two cases.

Case 1: $x < t$. By the independence of the X_i's in the Uniform process, $P(X_{(1)} \leq x \mid X_1)$ should be interpreted as $P(X_{(1)} \leq x)$ but in the Uniform process of sampling $n - 1$ points from $[0, a]$. That is, knowing that X_1 is t does not influence whether any other points are smaller than x. Therefore,

$$P(X_{(1)} \leq x \mid X_1 = t) = \left(\frac{a - x}{a} \right)^{n-1}$$

Case2: $x \geq t$. Here the fact that $X_1 = t$ means that $(X_{(1)} \leq x)$ has occurred. Therefore, $P(X_{(1)} \leq x \mid X_1 = t) = 1$.

Combining these two cases, we find that $P(X_{(1)} \leq x \mid X_1 = t)$ is not a continuous function. When we condition by $(X_1 = t)$, the random variable $X_{(1)}$ becomes discontinuous. This will often be the case for conditional distributions.

The Continuous Law of Alternatives

One of the most important facts about continuous conditioning is that the Law of Alternatives has a continuous version. Indeed, continuous conditional probabilities are important primarily because of this, and the continuous form of this law is by far one of the most useful tools of probability.

Recall that if a set of events $A_1, A_2, \ldots$ is a set of alternatives, then the probability of any event B is

$$P(B) = \sum_i P(B \mid A_i) P(A_i)$$

For continuous conditional probabilities, we replace the alternative events A_i by the "alternatives" $(X = t)$, where t takes on all real values, and we replace the sum by an integral.

Continuous Law of Alternatives. For any continuous random variable X and event A for which the continuous conditional probabilities $P(A \mid X = t)$ exist,

$$P(A) = \int_{-\infty}^{\infty} P(A \mid X = t) \, \text{dens}(X = t) \, dt$$

Furthermore, if Y is another continuous random variable then

$$\text{dens}(Y = s) = \int_{-\infty}^{\infty} \text{dens}(Y = s \mid X = t) \, \text{dens}(X = t) \, dt$$

Proof. We now give a rigorous proof of this law. The key fact we need is the Mean Value Theorem of Calculus. Recall what this says. If $f(x)$ is a continuous function on the interval $[a, b]$, then for some point ξ between a and b,

$$f(\xi) = \frac{1}{b-a} \int_a^b f(x)dx$$

Now let $\epsilon > 0$ be a small number. Divide up the real line into intervals of length ϵ by the points $t_n = n\epsilon$. This partitions the real line into intervals $[t_i, t_{i+1}]$ of length $\Delta t_i = \epsilon$. Take B_i to be the event $(t_i < X \leq t_{i+1}) = (t_i < X \leq t_i + \epsilon)$. Then the B_i's form a set of alternatives. By the (ordinary) Law of Alternatives,

$$
\begin{aligned}
P(A) &= \sum_i P(A \mid B_i) \, P(B_i) \\
&= \sum_i P(A \mid t_i < X \leq t_i + \epsilon) \, P(t_i < X \leq t_i + \epsilon)
\end{aligned}
$$

By the Mean Value Theorem applied to $f(t) = \mathrm{dens}(X = t)$, there is some ξ_i in the interval $[t_i, t_{i+1}]$ such that

$$f(\xi_i) = \frac{1}{\Delta t_i} \int_{t_i}^{t_{i+1}} f(t)dt = \frac{1}{\Delta t_i} P(t_i < X \leq t_i + \epsilon)$$

or

$$P(t_i < X \leq t_i + \epsilon) = f(\xi_i)\Delta t_i$$

Therefore,

$$P(A) = \sum_i P(A \mid t_i < X \leq t_i + \epsilon) \, f(\xi_i)\Delta t_i$$

Now as $\Delta_i = \epsilon \to 0$ this last sum approaches

$$\int_{-\infty}^{\infty} P(A \mid X = t) \, \mathrm{dens}(X = t) \, dt$$

by the definition of the integral. The case of another continuous random variable Y conditioned by densities is easily established by simply applying the first case to the probability distribution function of Y, and then differentiating under the integral sign. We have therefore proved the Continuous Law of Alternatives.

Notice that we used the fact that the density of X is continuous. However, in practice it will only be piecewise continuous. This technical detail will never bother us. The Continuous Law of Alternatives holds in this case also.

6.5 Conditional Densities

In most computations concerning continuous random variables, the densities are much easier to handle than the distribution function. We can give density versions of conditional probability, continuous conditional probability and the Continuous Law of Alternatives.

Let us begin with the simplest case: conditional density. Suppose we have an event B such that $P(B) > 0$, and a random variable Y. The distribution of Y given that B has occurred is

$$F(s) = P(Y \le s \mid B)$$

In general, it is possible that a continuous random variable can fail to be continuous after conditioning, as we saw in Section 6.4. But if it is still continuous, we may speak of the **conditional density** of Y given B:

$$\text{dens}(Y = s \mid B) = F'(s) = \frac{d}{ds}F(s) = P(Y \le s \mid B)$$

The conditional density has a Law of Alternatives that corresponds to the ordinary Law of Alternatives. See Section 6.6. If the event B is of the form $(X = t)$, then the conditional density can be defined by a limiting process just as we did in Section 6.4. More precisely, the **continuous conditional density** of Y given $X = t$ is

$$\text{dens}(Y = s \mid X = t) = \lim_{\epsilon \to 0} \text{dens}(Y = s \mid t < X \le t + \epsilon)$$

if this limit exists. If $\text{dens}(X = t) \ne 0$ then

$$\text{dens}(Y = s \mid X = t) = \frac{\text{dens}(Y = s, X = t)}{\text{dens}(X = t)}$$

exactly as one would expect.

Consider again the example of Section 6.4. The conditional density

$$\text{dens}\big(X_{(1)} = x \mid X_{(k)} = t\big)$$

when $x < t$, is the same as the density $\text{dens}\big(X_{(1)} = x\big)$ but for the process of dropping $k - 1$ points on $[0, t]$, i.e.,

$$\text{dens}\big(X_{(1)} = x \mid X_{(k)} = t\big) = \frac{(k-1)(t-x)^{k-2}}{t^{k-1}}$$

We could also compute this using a ratio of densities:

$$\text{dens}\big(X_{(1)} = x \mid X_{(k)} = t\big) = \frac{\text{dens}\big(X_{(1)} = x, X_{(k)} = t\big)}{\text{dens}\big(X_{(k)} = t\big)}$$

The two densities on the right were computed in Section 4.5:

$$\text{dens}\big(X_{(k)} = t\big) = n\binom{n-1}{k-1}\frac{t^{k-1}(a-t)^{n-k}}{a^n}$$

$$\text{dens}\big(X_{(1)} = x, X_{(k)} = t\big) = \binom{n}{0,1,k-2,1,n-k}\frac{(t-x)^{k-2}(a-t)^{n-k}}{a^n}$$

Therefore,

$$\text{dens}\big(X_{(1)} = x \mid X_{(k)} = t\big) = \frac{\binom{n}{0,1,k-2,1,n-k}(t-x)^{k-2}(a-t)^{n-k}}{n\binom{n-1}{k-1}t^{k-1}(a-t)^{n-k}}$$

$$= \frac{(k-1)(t-x)^{k-2}}{t^{k-1}}$$

The Continuous Bayes' Law

By using conditional densities one can formulate a continuous version of Bayes' law. Suppose we have two random variables X and Y. We call X the "cause" and Y the "effect." For example, X may represent a parameter in an experiment, which we cannot measure directly, while Y is some directly measurable quantity. We want to determine the effect on the distribution of X given a particular observation of Y. As in the discrete Bayes' law, we assume that we know the *a priori* distribution of X, $\text{dens}(X = x)$, as well as the conditional densities of Y given a value of X, $\text{dens}(Y = y \mid X = x)$. By a calculation almost identical to the one for the discrete Bayes' law, we have this formula:

$$\text{dens}(X = x \mid Y = y) = \frac{\text{dens}(Y = y \mid X = x)\,\text{dens}(X = x)}{\int_{-\infty}^{\infty}\text{dens}(Y = y \mid X = t)\,\text{dens}(X = t)\,dt}$$

Continuous Law of Successive Conditioning

In a similar manner as that above, one can state a continuous analog of the Law of Successive Conditioning as follows:

If $X_1.X_2\ldots, X_n$ are continuous random variables, then their joint density is given by:

$$\text{dens}(X_1 = t_1, X_2 = t_2, \ldots, X_n = t_n) = \text{dens}(X_1 = t_1)$$
$$\text{dens}(X_2 = t_2 \mid X_1 = t_1)\cdots\text{dens}(X_n = t_n \mid X_1 = t_1, \ldots, X_{n-1} = t_{n-1})$$

6.6 Gaps in the Uniform Process

As an application and illustration of the conditioning techniques just introduced, we give a detailed and *rigorous* treatment of the gaps in the Uniform process. We begin with a problem called "needles on a stick." If we drop a set of needles, each of length h, on a stick of length b, what is the probability that none of the needles overlap? We then show that the gaps in the Uniform process are exchangeable.

Needles on a Stick

We first restate the problem in terms of the Uniform process. The position of a given needle is completely determined by its left endpoint. The process of dropping n needles of length h on a stick of length b is then the same as dropping n points on the interval $[0, b - h]$. Write $a = b - h$. Here are five needles of length h on a stick of length b.

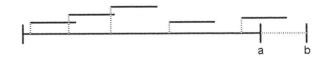

Now two needles are nonoverlapping if and only if their left endpoints are at least distance h apart. Let $A_{a,n}$ be the event "When n needles are dropped on a stick of length $b = a + h$ they do not overlap." Then

$$
\begin{aligned}
P(A_{a,n}) &= P(L_2 \geq h, L_3 \geq h, \ldots, L_n \geq h) \\
&= P((\min{}_{2 \leq i \leq n} L_i) \geq h)
\end{aligned}
$$

Unfortunately, the gaps are not independent, so we cannot simplify this formula as it is. So we first compute the probability of a slightly different event, then apply the Continuous Law of Alternatives to compute the probability of the original event. Accordingly, let $B_{a,n}$ be the event "When n points are dropped on $[0, a]$, they are all at least distance h from each other *and* from the right endpoint." This is exactly the same as $A_{a,n}$ but we have added the condition that the last gap, L_{n+1}, is also larger than h, i.e.,

$$
\begin{aligned}
A_{a,n} &= (L_2 \geq h, L_3 \geq h, \ldots, L_n \geq h) \\
B_{a,n} &= (L_2 \geq h, L_3 \geq h, \ldots, L_n \geq h, L_{n+1} \geq h)
\end{aligned}
$$

To compute $B_{a,n}$ we condition on the position of the largest point:

$$
\begin{aligned}
P(B_{a,n}) &= \int_{-\infty}^{\infty} P(B_{a,n} \mid X_{(n)} = t) \operatorname{dens}(X_{(n)} = t)\, dt \\
&= \int_{(n-1)h}^{a-h} P(B_{a,n} \mid X_{(n)} = t) \frac{n}{a^n} t^{n-1}\, dt
\end{aligned}
$$

Here the limits of integration stem from the fact that $B_{a,n}$ can only occur if the largest point falls so that the rightmost gap, L_{n+1}, is larger than h and so that there is enough "room" for $n-1$ gaps of size h to the left of $X_{(n)}$. Now $P(B_{a,n})$ is the same as the probability of dropping $n-1$ points on $[0, t]$ so that all gaps are at least h and also so that the largest point, $X_{(n-1)}$ falls at least distance h from t. Thus

$$
P(B_{a,n} \mid X_{(n)} = t) = P(B_{t,n-1})
$$

We may therefore use mathematical induction. If we write $p_n = P(B_{a,n})$, then

$$
p_n(a) = \int_{(n-1)h}^{a-h} p_{n-1}(t) \frac{n}{a^n} t^{n-1}\, dt
$$

Consider $p_1(a)$. This is the probability that a point dropped on $[0, a]$ falls farther than distance h from a. So $p_1(a) = \dfrac{a - h}{a}$. More generally, the above inductive formula can be used to deduce that $p_n = \left(\dfrac{a - hn}{a} \right)^n$.

To compute $P(A_{a,n})$ from what we now know about $P(B_{a,n})$, we condition on the largest point $X_{(n)}$. It is easy to see that $P(A_{a,n} \mid X_{(n)} = t) = P(B_{a,n})$.

Therefore,

$$
\begin{aligned}
P(A_{a,n}) &= \int_{-\infty}^{\infty} P\big(A_{a,n} \mid X_{(n)} = t\big)\, \mathrm{dens}\big(X_{(n)} = t\big)\, dt \\[2mm]
&= \int_{(n-1)h}^{a} P(B_{t,n-1})\, \frac{n}{a^n} t^{n-1}\, dt \\[2mm]
&= \frac{n}{a^n} \int_{(n-1)h}^{a} \left(\frac{t - (n-1)h}{t} \right)^{n-1} t^{n-1}\, dt \\[2mm]
&= \frac{n}{a^n} \int_{(n-1)h}^{a} (t - (n-1)h)^{n-1}\, dt \\[2mm]
&= \frac{n}{a^n} \left[\frac{(t - (n-1)h)^n}{n} \right]_{(n-1)h}^{a} \\[2mm]
&= \frac{(a - (n-1)h)^n}{a^n} \\[2mm]
&= \left(\frac{(b - nh)}{b - h} \right)^n \qquad \text{provided } b \geq nh
\end{aligned}
$$

Exchangeability of the Gaps

Recall from Section 4.5 that the gaps L_i in the Uniform process are equidistributed but that we gave only an intuitive justification. We can now give a rigorous proof using conditional density.

Consider the first two gaps. The density of L_1 is $\mathrm{dens}(L_1 = t_1) = \frac{n(a-t_1)^{n-1}}{a^n}$. Therefore, by the Law of Alternatives,

$$
\mathrm{dens}(L_2 = t_2) = \int_{-\infty}^{\infty} \mathrm{dens}(L_2 = t_2 \mid L_1 = t_1)\, \frac{n(a - t_1)^{n-1}}{a^n}\, dt_1
$$

Now the conditional density $\mathrm{dens}(L_2 = t_2 \mid L_1 = t_1)$ is the same as that of the *first* gap in the Uniform process of dropping $n-1$ points on the interval $[t_1, a]$. Therefore,

$$
\mathrm{dens}(L_2 = t_2 \mid L_1 = t_1) = \begin{cases} \dfrac{(n-1)(a-t_1-t_2)^{n-2}}{\left(a - t_1^{n-1}\right)^{n-1}} & \text{if } 0 \leq t_2 \leq a - t_1 \\ 0 & \text{otherwise} \end{cases}
$$

Hence:

$$
\begin{aligned}
\mathrm{dens}(L_2 = t_2) &= \int_0^{a-t_2} \frac{(n-1)(a - t_1 - t_2)^{n-2}}{(a - t_1)^{n-1}}\, \frac{n(a - t_1)^{n-1}}{a^n}\, dt_1 \\[2mm]
&= \int_0^{a-t_2} \frac{(n-1)n}{a^n}(a - t_1 - t_2)^{n-2}\, dt_1 \\[2mm]
&= \frac{(n-1)n}{a^n} \left[\frac{-(a - t_1 - t_2)^{n-1}}{n-1} \right]_0^{a-t_2}
\end{aligned}
$$

$$= \frac{n}{a^n}(a - t_2)^{n-1}$$

Therefore, L_1 and L_2 are equidistributed.

We now consider all of the gaps. To show that they are exchangeable we must compute their joint density. We do this by using the Law of Successive Conditioning. Writing this out for the gaps it is:

$$\text{dens}(L_1 = t_1, L_2 = t_2, \ldots, L_{n+1} = t_{n+1}) = \text{dens}(L_1 = t_1)$$
$$\text{dens}(L_2 = t_2 \mid L_1 = t_1) \, \text{dens}(L_3 = t_3 \mid L_1 = t_1, L_2 = t_2) \cdots$$

Each factor in this product is a conditional density

$$\text{dens}(L_j = t_j \mid L_1 = t_1, L_2 = t_2, \ldots, L_{j-1} = t_{j-1})$$

This conditional density is the same as the density of the first gap in the Uniform process of dropping $n - j + 1$ points on $[t_1 + \cdots + t_{j-1}, a]$; namely,

$$\frac{(n - j + 1)(a - t_1 - t_2 - \cdots - t_j)^{n-j}}{(a - t_1 - t_2 - \cdots - t_{j-1})^{n-j+1}}$$

Therefore,

$$\text{dens}(L_1 = t_1, L_2 = t_2, \ldots, L_{n+1} = t_{n+1}) = \begin{cases} \frac{n!}{a^n} & \text{if } t_1 + \cdots + t_{n+1} = a \\ 0 & \text{otherwise} \end{cases}$$

by cancellation. Since the joint density does not depend on any of the t_i's, it is symmetric. Therefore, the gaps are exchangeable. At first this does not seem correct, since the various gaps seem to be different from one another. In particular, the first and last gaps seem like they should be different from the others since they are at the ends of the interval. You need to "adjust your intuition" to deal with this. It helps to use the "points on a circle" interpretation of the Uniform process as in Section 4.5.

An immediate, and by no means obvious, consequence of the exchangeability of the gaps is that the covariance of any pair of them is the same as that of the first two. This implies, paradoxically, that the correlation between the first two gaps is the same as that between L_1 and any L_i! You really need to do some serious adjusting of your intuition to believe this property of the gaps.

Table of Conditioning Laws

Probabilities	Densities
$P(A\mid B) = \frac{P(A \cap B)}{P(B)}$	$dens(X = s\mid B) = \frac{d}{ds}P(X \leq s\mid B)$
$P(A\mid Y = s) =$	$dens(Y = s\mid X = t) = \frac{dens(Y=s, X=t)}{dens(X=t)}$
$\lim_{\epsilon \to 0} P(A\mid s < Y \leq s + \epsilon)$	$= \lim_{\epsilon \to 0} \frac{d}{ds}P(Y \leq s\mid t < X \leq t + \epsilon)$

Discrete and Continuous Conditioning

	Probabilities	**Densities**
Discrete	$P(B) = \sum_i P(B\vert A_i)P(A_i)$	$\text{dens}(Y = s) = \\ \sum_i \text{dens}(Y = s\vert A_i)P(A_i)$
Continuous	$P(B) = \int_{-\infty}^{\infty} P(B\vert X = t)\text{dens}(X = t)dt$	$\text{dens}(Y = s) = \\ \int_{-\infty}^{\infty} \text{dens}(Y = s\vert X = t)\text{dens}(X = t)dt$

Laws of Alternatives

$$P(A_i\vert B) = \frac{P(B\vert A_i)P(A_i)}{P(B)} = \frac{P(B\vert A_i)P(A_i)}{\sum_j P(B\vert A_j)P(A_j)}$$

$$\text{dens}(X = x\vert Y = y) = \frac{\text{dens}(Y = y\vert X = x)\text{dens}(X = x)}{\int_{-\infty}^{\infty}\text{dens}(Y = y\vert X = t)\text{dens}(X = t)dt}$$

Bayes' Law

$$P(B_1 \cap B_2 \cap \cdots \cap B_n) = P(B_1)P(B_2\vert B_1)P(B_3\vert B_1 \cap B_2)\cdots \\ P(B_n\vert B_1 \cap B_2 \cap \cdots \cap B_{n-1})$$

$$\text{dens}(X_1 = t_1, X_2 = t_2, \ldots, X_n = t_n) = \\ \text{dens}(X_1 = t_1)\text{dens}(X_2 = t_2\vert X_1 = t_1)\cdots \\ \text{dens}(X_n = t_n\vert X_1 = t_1, \ldots, X_{n-1} = t_{n-1})$$

Laws of Successive Conditioning

6.7 The Algebra of Probability Distributions

Very early in our study of random variables we noted that we can perform
algebraic operations on them to get new random variables. We did not, however, make any systematic study of what effect an algebraic operation has
on the distribution of the random variables involved. For example, if X is a
continuous random variable with density $f(x)$, what is the density of $2X$?
The answer is most assuredly *not* $2f(x)$. In fact, the correct answer is $\frac{1}{2}f\left(\frac{x}{2}\right)$.
This illustrates a basic fact about algebraic operations on random variables:
the effect of a simple operation on a random variable is seldom reflected in a
simple way on its density. In this section, we consider two kinds of operations
on random variables: "change of variables" on a single random variable and
the sum of two independent random variables.

Change of Variables

Let X be a continuous random variable, whose density is $f(x)$. Let $g(x)$ be an increasing function. We wish to determine the distribution of the random variable $g(X)$. To do so, we must consider the distribution function of X, not just its density. Accordingly, let $F(x)$ be $P(X \leq x)$, so that $f(x) = F'(x)$. The distribution function of $g(X)$ is given by $P(g(X) \leq x)$. Now we assumed that $g(x)$ was an increasing function, so it has an inverse function $G(y)$ which is also increasing. Therefore,

$$P(g(X \leq x)) = P(G(g(x)) \leq G(x)) = P(X \leq G(x) = F(G(x))$$

To get the density of $g(X)$ we differentiate by using the chain rule:

$$\text{dens}(g(X) = x) = F'(G(x))G'(x) = f(G(x))G'(x)$$

By the inverse function principle of Calculus, $G'(x) = 1/g'(G(x))$. Therefore, we have shown:

Change of Variables Formula. For a continuous random variable X and an increasing, differentiable function $g(x)$,

$$\text{dens}(g(X) = x) = \frac{f(G(x))}{g'(G(x))}$$

where $G(y)$ is the inverse function of $g(x)$.

An immediate consequence of this formula is that for any continuous random variable X whose distribution function is $F(x)$, the random variable $F(X)$ is uniformly distributed on $[0, 1]$. Thus every continuous random variable is, by a change of variables, expressible in terms of any other. If we wish to simulate a random variable X whose distribution function is $F(x)$, we just use G(unif_rand), where $G(y)$ is the inverse function of $F(x)$. This technique for simulating a random variable is called the **inverse transformation method**. This method is the basis for most random number generators, such as the ones supported by R. Indeed, if one looks at the source code for R, one finds that the random number generators ultimately depend on the random number generator for the Uniform process on $[0, 1]$. This function is called unif_rand. Note that this function is an internal function in the C language that is not available in the R language. However, the runif function can be used to obtain nearly direct access to unif_rand.

Of course, no computer can generate an actual uniformly distributed random variable. That would require infinitely many bits of precision. In practice, a computer will only generate a number with some maximum precision. On most computers today, the maximum precision for a real number is 53 bits or 16 decimal digits. So a uniformly distributed random number is equivalent to the first 53 tosses of a randomly generated Bernoulli sample point.

The inverse transformation method is very commonly used for random number generation, but it requires that we compute the inverse of a function. This can be difficult when the distribution to be simulated is very complicated. The

rejection method (also called rejection sampling) is a simulation technique that can be used in many cases where the inverse method is infeasible. To use this method we need a simulation for another random variable Y that is "similar" to X. More precisely, if $h(x)$ is the density function of Y, we require that there be a constant c such that

$$\frac{f(x)}{h(x)} \leq c \quad \text{for all x}$$

We then use the following two-step algorithm to generate a random number according to the density function $f(x)$:

1. Generate a random number x with the density $h(x)$.

2. Generate a uniformly distributed random number u from the interval $[0, 1]$. If $u \leq \dfrac{f(x)}{ch(x)}$, then output x. Otherwise *reject* the random number x, and return to step 1.

We leave the proof as Exercise 6.52. Here is a simple example. Suppose we want to randomly generate numbers according to the density

$$\frac{x(1 - x)e^x}{3 - e}$$

on $[0, 1]$. It is easy to plot this function and verify that it is always less than 2. The inverse to the distribution function is not easy to find, but the density function is bounded by 2, so we use rejection sampling to generate the random number from the uniform distribution:

```
> rdist <- function(n) {
+    for (i in 1:n) {
+       x <- runif(1,0,1)
+       u <- runif(1,0,1)
+       ifelse(2*u<=x*(1-x)*exp(x)/(3-exp(1)),
+              return(x), -1)
+    }
+    return(-1)
+ }
> rdist(1000)
[1] 0.5103522
```

The parameter to the function is the maximum number of iterations. The probability of acceptance on each iteration can be shown to be $1/2$ in this case (and $1/c$ in general), so 1000 should be enough. The `for` statement performs the iteration. It is also called a `for` loop. The name for the statement is analogous to the mathematical concept of *for all* or *for every*. The `for` loop runs a statement for each element of a vector. The `for` loop variable (in this case i) is assigned successively to each element of the vector, starting with the first one, and ending with the last one, unless the loop is terminated in some way. The `ifelse` function computes its first parameter. If it is `TRUE`, then the value of the function is the second parameter. Otherwise, the value is the third parameter. The `return` function is used for returning a value from a function. In this case, if the first `ifelse` parameter is `TRUE`, then the sample is accepted, and the randomly chosen value is returned. If the `ifelse` parameter is `FALSE`, then the sample is rejected, and a new one is computed. The value computed by `ifelse` in this case is not used, and it does not matter what value we put in that position. If the number of iterations is not sufficient, then the iteration completes, and the function failed to compute an acceptable random number so it returns -1.

The change of variables formula we have given applies only to an increasing function $g(x)$. For a decreasing function, the only change is that the sign of the right-hand side must be reversed. For more complicated functions $g(x)$, one must partition the domain of $g(x)$ into intervals on which it is increasing or decreasing and apply the change of variables formula to each such interval. The results must then be combined to get the density of $g(X)$. Needless to say, this can get quite intricate.

Sums of Independent Random Variables

Suppose that X and Y are two random variables. Their sum is again a random variable, $X + Y$. For example, in the Uniform process, $X_{(2)} = L_1 + L_2$. Now if we know the distributions of X and Y, can we compute the distribution of the sum $X + Y$? In general, the answer is no, for we need the joint distribution in order to compute the distribution of the sum. In the above example, we cannot compute the distribution of $X_{(2)}$ from the distributions of L_1 and L_2 alone, we must know also the joint distribution of L_1 and L_2.

On the other hand, if X and Y are independent, we can compute their joint distribution from their marginal distributions. As a result, we expect that the distribution of $X + Y$ has some reasonable expression in terms of the distributions of X and Y. Suppose for the moment that X and Y are independent, integer random variables with distributions

$$p_n = P(X = n)$$

$$q_n = P(Y = n)$$

Then by the Law of Alternatives,

$$P(X + Y = n) = \sum_k P(X + Y = n \mid Y = k) P(Y = k)$$

$$= \sum_k P(X = n - k \mid Y = k) P(Y = k)$$

$$= \sum_k P(X = n - k) P(Y = k)$$

since X and Y are independent

$$= \sum_k p_{n-k} q_k$$

The distribution $r_n = \sum_k p_{n-k} q_k$ is called the **discrete convolution** of the distributions p_n and q_n. In the case of integer random variables, we can see clearly what the convolution means: $P(X + Y = n)$ is the sum of all possible "ways" that $X + Y$ can equal n: $X = n - k$ and $Y = k$, for all possible k.

In the continuous case, the sum is replaced by an integral, but the idea is the same. Suppose that X and Y are continuous random variables having densities:

$$f(x) \quad = \quad \text{dens}(X = x)$$

$$g(x) \quad = \quad \text{dens}(Y = x)$$

The density of their sum is:

$$\text{dens}(X + Y = t) \quad = \quad \int_{-\infty}^{\infty} \text{dens}(X + Y = t \mid Y = s)\, \text{dens}(Y = s)\, ds$$

$$= \quad \int_{-\infty}^{\infty} \text{dens}(X = t - s \mid Y = s)\, \text{dens}(Y = s)\, ds$$

$$= \quad \int_{-\infty}^{\infty} \text{dens}(X = t - s)\, \text{dens}(Y = s)\, ds$$

because X and Y are independent

$$= \quad \int_{-\infty}^{\infty} f(t - s) g(s)\, ds$$

The function

$$h(t) = \int_{-\infty}^{\infty} f(t - s) g(s)\, ds$$

is called the **convolution** of f and g, which we shall write

$$h = f * g$$

The convolution of two functions is an important operation which appears in numerous contexts, for example, dynamical systems in engineering and optics in physics, to name just two. Its appearance in probability theory is perhaps the most easily understood context in which the convolution arises. Actually, there are many operations similar to the one above that go by the name "convolution." For example, if we consider the special case of two continuous, *positive*, independent random variables X and Y, the density of their sum is

$$h(t) = \int_0^t f(t - s) g(s)\, ds$$

because $f(t - s) = 0$ if $s > t$ and $g(s) = 0$ if $s < 0$. This is the form of the definition one sees most commonly.

Although it is not obvious from the definition, convolution is a commutative, associative operation. That is, for densities f, g and h:

$$f * g = g * f \qquad \text{and} \qquad (f * g) * h = f * (g * h)$$

These follow from the fact that addition of random variables is commutative and associative, respectively.

As an example of a convolution, we show a result which is implicit in many of the calculations in Chapter 5: the sum of normally distributed random variables is again normal. We will just take the case of two standard normal random variables, and leave the general case as an exercise. By definition, the convolution of the standard normal density with itself is given by:

$$\int_{-\infty}^{\infty} \frac{1}{\sqrt{2\pi}} \exp\left(\frac{-(y-x)^2}{2}\right) \frac{1}{\sqrt{2\pi}} \exp\left(\frac{-x^2}{2}\right) dx$$

$$= \frac{1}{2\pi} \int_{-\infty}^{\infty} \exp\left(\frac{-y^2 + 2xy - x^2 - x^2}{2}\right) dx$$

$$= \frac{1}{2\pi} \exp\left(-y^2/2\right) \int_{-\infty}^{\infty} \exp\left(\frac{+2xy - 2x^2}{2}\right) dx$$

$$= \frac{1}{2\pi} \exp\left(-y^2/2\right) \int_{-\infty}^{\infty} \exp\left(x(y-x)\right) dx$$

Using the change of variables $u = x - \frac{y}{2}$, this becomes:

$$\frac{1}{2\pi} \exp\left(-y^2/2\right) \int_{-\infty}^{\infty} \exp\left(-\left(u+\frac{y}{2}\right)\left(u-\frac{y}{2}\right)\right) du$$

$$= \frac{1}{2\pi} \exp\left(-y^2/2\right) \int_{-\infty}^{\infty} \exp\left(-u^2 + \frac{y^2}{4}\right) du$$

$$= \frac{1}{2\pi} \exp\left(-y^2/2 + y^2/4\right) \int_{-\infty}^{\infty} e^{-u^2} du$$

$$= \frac{1}{2\pi} \exp\left(-y^2/4\right) \sqrt{\pi}$$

$$= \frac{1}{\sigma\sqrt{2\pi}} \exp\left(\frac{-y^2}{2\sigma^2}\right),$$

where $\sigma = \sqrt{2}$.

6.8 Exercises

Exercise 6.1 A game is played with six double-sided cards. One card has "1" on one side and "2" on the other. Two cards have "2" and "3" on the two sides. And the last three have "3" and "4" on them. The six cards are shuffled by one person. A random card is then drawn and held in a random orientation between two other persons, each of whom sees only one side of the card. The winner is the one seeing the smaller number. Suppose that the first person chooses the "2/3" card. Compute the probabilities each of the two persons thinks he/she has for winning.

Exercise 6.2 A person is given an urn and is told it contains two red balls and two black balls. He draws two of the balls at random without replacing them, and both turn out to be red. He puts these aside. What is the probability that the next ball drawn is black? Another person in the room has been blindfolded during all of the preceding. After taking off her blindfold, she takes a ball out of the urn at random. She knows which balls were originally in the urn and that two have been drawn so far but does not know their colors. What does she think the probability of drawing a black ball is? How could the fact that she was blindfolded have any effect on the probability of the next drawing of a ball? Explain.

Exercise 6.3 Place k objects into n containers at random. If the first container is empty, what is the probability that the second is also?

Exercise 6.4 During a poker game, a kibitzer manages to get a brief glimpse of one of the hands (and no other hands). In this glimpse he sees only that one card in the hand is an Ace. He did not notice which Ace it was. What is the probability that the hand has at least two Aces? If the kibitzer noticed that one card is a black Ace, what is the probability that the hand has at least two Aces? Finally, suppose the kibitzer saw that the hand had the Ace of spades. Discuss whether such glimpses are really possible. The "moral" of this example is that (conditional) probabilities of events change considerably when one learns information that seems to have no obvious relevance.

Exercise 6.5 Three prisoners are informed by their jailer that one of them has been chosen at random to be executed and that the other two are to be freed. They are told they will learn their fate in one week's time. Prisoner A asks the jailer to tell him privately the name of a fellow prisoner who will be set free, claiming that there would be no harm in divulging this information, since he already knows that at least one will go free, and he cannot inform the prisoner in question about his good fortune. The jailer refuses to tell prisoner A, pointing out that if A knew the name of one of his fellows to be set free, then his own probability of being executed would rise from $1/3$ to $1/2$, since he would then be one of two prisoners; and this would be cruel. What do you think of the jailer's reasoning? *Be precise.*

Exercise 6.6 You are playing bridge. Assume that the deck is thoroughly shuffled. If you receive four hearts, how many did your partner receive? Generalize the exercise to the case of receiving N hearts.

Exercise 6.7 (Neyman-Pearson errors) A commuter has the choice of taking the train to work or of driving to work. If she takes the train she will get to work on time about one time in four. If she takes her car, she is almost certain of getting to work on time, but at considerable inconvenience. Although she calls the transit company every morning, their information is wrong a third of the time. So she adopts the following strategy: if the transit company says the train will be late, she always takes her car, and if not she takes her car a third of the time at random. Compute how probable it is that she will be late. What is the probability that she takes her car even though she would have been on time if she had taken the train? This kind of "mistake" is known as a **Neyman-Pearson Error of Type I**. She makes a **Neyman-Pearson Error of Type II** if the train is late when she takes it. Compute the probability that she makes an error of type II. Note that the probability of either kind of error is a conditional probability. Furthermore, in order to make the above computations, one must assume a number of independence properties of the various events. State explicitly any assumptions you must make in this problem. The probability of an error of type I is called the **significance level** of the decision, and 1 minus the probability of an error of type II is called the *power* of the decision. One should reexamine statistical hypothesis testing using this terminology. When several tests are available, one clearly would like the one with the largest power for a given significance level. Unfortunately, however, in practice one rarely knows precisely what model will be implied by the *rejection* of the hypothesis being tested, so computing the power of a test is not as easy as it seems.

Exercise 6.8 Suppose that the commuter in Exercise 6.7 scores the two inconveniences of being late and of driving the car at 1 and 2, respectively. What is her optimal strategy? Do the same with 1 and 2 interchanged.

Exercise 6.9 An event A of positive probability is said to be *favorable* to an event B if $P(B \mid A) \geq P(B)$; in other words, if we know that A has occurred then the probability that B has occurred also is the same as it was or is greater. Note that if A is independent of B, then it is favorable to B. Suppose we have a family having two children. Let A be the event "the first child is a girl," let B be the event "the second child is a girl," and let C be the event "the two children have different gender." Show that A and B are both favorable to C but that $A \cap B$ is not. Similarly show that C is favorable to both A and B but not to $A \cap B$. Give examples to show that an event can be favorable to two others without favoring their union and conversely that two events can favor a third without their union doing so.

Exercise 6.10 (Simpson's paradox) A new treatment for a disease has just become available but is still experimental and is very expensive. In a teaching hospital with a large budget, a random sample of 100 patients with the disease is randomly divided into 2 groups, one having 90 patients the other 10. The larger group is treated, and 30 of the patients show clear improvement. Only 1 of the members of the control does. In a city hospital with a smaller budget, a similar test is made but now the smaller group gets the treatment. In this group 9 patients show improvement, while in the untreated group half improve and half do not. In either case, the treatment seems to be effective. However, if we view this as 1 sample of 200, from which 100 are treated and the other 100 are not, then a different picture emerges. Among the untreated patients, 46 improve, while only 39 treated patients improve. This seems to suggest that the treatment actually *decreases* one's chance for improvement. Explain the apparent paradox here.

Exercise 6.11 There are three children in a family. A friend is told that at least two of them are boys. What is the probability that all three are boys? The friend is then told that the two are the oldest two children. Now what is the probability that all three are boys? Use Bayes' Law to explain this. Assume throughout that boys are as likely as girls and that each child is independently either a boy or a girl.

Exercise 6.12 A student is about to take a quiz. If he studies, he will pass with probability 0.99, but if he goes to the dorm party his chances of passing decline to 1/2. The next day he passes the exam. Did he go to the party?

Exercise 6.13 Use Bayes' law to compute the probability in Exercise 6.7 that the commuter took the train given that she was on time.

Exercise 6.14 Three boxes each contain two coins. One has two silver coins, one has two gold coins, and one has one of each. A box is chosen completely at random and a coin is chosen at random from that box. It is gold. Is the other coin in the box gold also?

Exercise 6.15 The manufacturer of screws in Exercise 5.15 is producing good screws 99% of the time but now the machine that detects the flawed screws is itself out of adjustment, producing an incorrect decision 10% of the time. What is the probability that a discarded screw is really flawed?

Exercise 6.16 A lie detector test is known to be 80% reliable when the person is guilty and 95% reliable when the person is innocent. If a suspect was chosen from a group of suspects of which only 1% have ever committed a crime, and the test indicates that the person is guilty, what is the probability of innocence?

Exercise 6.17 In the optimal choice problem (Exercise 4.30), the best strategy is to make no decision for a certain length of time (say k days) and then to choose the best candidate of all those seen up to that point. If the monarch chooses the jth candidate, what is the probability that she is the best candidate?

Exercise 6.18 Using the strategy in Exercise 6.17, compute the probability that the monarch will choose the jth candidate. Use this to find the probability that this strategy succeeds. For which k will this be maximized? Generalize the exercise to N candidates.

Exercise 6.19 Use Bayes' law to compute the probability of each of the kinds of hand in Exercise 2.20, given that one has at least one pair.

Exercise 6.20 A target is a disk of radius 1 m. A bullet is fired at the disk and hits it. Assume the bullet's mark has a uniform distribution, i.e., the probability that it hits a region A is proportional to the area of A. How far from the center does the bullet hit?

Exercise 6.21 A *scimitar* is a sword shaped like a circular arc (at least for this problem). Suppose that during a Turkish festival, a group of n Turks throw their scimitars independently and at random along a circle of circumference a. Suppose that each scimitar has arc length h along this circle. What is the probability that none of the scimitars overlap?

Exercise 6.22 In the Uniform process of sampling $n>2$ points from $[0, a]$, what is the probability that the first three gaps are all less than b?

Exercise 6.23 For $1 \leq i < j < k \leq n$, compute the joint density of $X_{(i)}$ and $X_{(k)}$ given that $\left(X_{(j)} = t\right)$.

Exercise 6.24 Let $L_1, L_2, \ldots, L_{n+1}$ be the gaps in the Uniform process of sampling n points from $[0, a]$. Find the distributions of the order statistics of the gaps, i.e., put the gaps in order, getting the random variables

$$L_{(1)}, L_{(2)}, \ldots, L_{(n+1)}$$

Then compute their expectations. Compare this with the scimitar problem (Exercise 6.21) and with the broken DNA problem (Exercise 4.20).

Exercise 6.25 Compute the distribution of the median gap in the Uniform process of sampling n points from $[0, u]$.

Exercise 6.26 In the Uniform process of sampling n points from $[0, a]$, what is the probability that the largest gap is at least twice as large as the smallest gap?

Exercise 6.27 Given positive numbers $t_1, t_2, \ldots, t_{n+1}$, what is the probability that in the Uniform process, for all $i = 1, 2, \ldots, n+1$, the ith gap is greater than t_i?

Exercise 6.28 Give a rigorous statement and proof of the identity

$$\text{dens}(Y = s \mid X = t) = \frac{\text{dens}(Y = s, X = t)}{\text{dens}(X = t)}$$

Exercise 6.29 Give a rigorous (ϵ, δ) proof of the Continuous Law of Alternatives.

Exercise 6.30 Define a *cluster* of size k and width ϵ to be a sequence of k points contained in an interval of length ϵ. In the Uniform process how many clusters of size k and width ϵ are there?

Exercise 6.31 Drop r red points and b black points $(r + b = n)$ at random uniformly on $[0, a]$. What is the probability that a run of h red points precedes a run of t black points?

Exercise 6.32 Compute the probability that at least one of the four players in a bridge game is dealt a yarborough. Note that the four hands are not independent but are exchangeable. Compare this answer with what you would get if you dealt the four hands independently from four different decks. Use the result of Exercise 2.11.

Exercise 6.33 Give a much easier proof of the result in Exercise 4.6 that the spread in the Uniform process has the same distribution as the $(n-2)^{nd}$ order statistic. Hint: Use the exchangeability of the gaps.

Exercise 6.34 (Discrete Needles on a Stick Problem) Choose k numbers from the set $\{1, 2, \ldots, n\}$ at random. What is the probability that no two are closer than l units apart?

Exercise 6.35 (Discrete Scimitars on a Circle Problem) Choose k numbers from the set of integers modulo n. What is the probability that no two have a difference congruent modulo n to one of the elements of the R vector (-a):(a-1)?

Exercise 6.36 Find the distributions of the following random variables in terms of the random variable X:

1. $Y = X + c$, where c is a constant,
2. $Y = aX + b$, where a and b are constants,
3. $Y = |X|$,
4. $Y = \sqrt{X}$, where X is a positive random variable, and
5. $Y = \log(X)$, where X is a positive random variable.

Exercise 6.37 A point is dropped at random (uniformly) on a square of side a. What is the distance of this point from the center of the square?

Exercise 6.38 Let S be the speed of a molecule in a uniform gas at equilibrium. Then S is a positive random variable whose density is given by $\text{dens}(S = s) = 4\sqrt{b^3/\pi}s^2 e^{-bs^2}$ for $s>0$, where b is a constant which depends on the absolute temperature and mass of the molecule. Find the density of the kinetic energy E of the molecule, where E is defined by $\frac{1}{2}mS^2$.

Exercise 6.39 Suppose that a long DNA molecule of length a is broken at random into two pieces. Compute the distribution of the ratio of the length of the longer piece by that of the shorter piece.

Exercise 6.40 For the situation in Exercise 6.39 compute the expectation of the ratio of the lengths of the longer and shorter pieces. Do the same for a molecule broken into three pieces.

Exercise 6.41 For the situation in Exercise 6.39 and Exercise 6.40, perform the same analyses for a molecule broken into three pieces.

Exercise 6.42 (Student's t-distribution) We remarked in Chapter 5 that if we do not know the variance of a random variable, then we can approximate it using the sample variance. Unfortunately, if we only have a small sample, the fact that the sample variance is being used instead of the actual variance can result in the sample mean having a distribution considerably different from a normal distribution, even if the original sequence of random variables were all normally distributed. This fact was first pointed out by William Gosset, who wrote under the pseudonym of "Student." We will now perform his computation.

Let $X_1, \ldots, X_n$ be a sequence of independent, normally distributed random variables with mean m and variance σ^2. The sample mean is $\overline{m} = \dfrac{1}{n}\sum_i X_i$ and the sample variance is $\overline{\overline{\sigma^2}} = \dfrac{1}{n-1}\sum_i (X_i - \overline{m})^2$. We wish to compute the distribution of the random variable $t = \dfrac{(\overline{m} - m)\sqrt{n}}{\overline{\overline{\sigma}}}$. Note that this random variable is not defined for $n = 1$. For the purposes of computing the distribution of t, we may assume that $X_1, \ldots, X_n$ follow the same distribution $N(0,1)$. For $n = 2$, t is the random variable $\dfrac{X_1 + X_2}{|X_1 - X_2|}$, and it is easy to check that $X_1 + X_2$ and $X_1 - X_2$ are independent and both have distribution $N(0,2)$. More generally, show that t has the same distribution as the ratio $\dfrac{X}{\sqrt{Y_1^2 + \cdots + Y_{n-1}^2}}$ where $X, Y_1, \ldots, Y_{n-1}$ are independent and have the standard normal distribution. The distribution of t is called the **Student's t-distribution** with $n-1$ degrees of freedom. When $n = 2$, the distribution of t is the same (up to a scale change) as the Cauchy distribution. See Section 5.7. Prove this.

Exercise 6.43 Let X and Y be independent, uniformly distributed random variables on $[0,1]$. Prove that $cos(2\pi X)\sqrt{-2\log(Y)}$ has the distribution $N(0,1)$. This fact is useful for generating a sequence of independent, normally distributed pseudo-random numbers by computer.

Exercise 6.44 Let $X_1, \ldots, X_n$ be a sequence of independent, standard normal random variables. Compute the distributions of the order statistics

$$X_{(1)}, \ldots, X_{(n)}$$

of these random variables. Write a program that uses a numerical integration to find an approximation for $E(X_{(j)})$ accurate to 3 decimal places. Then make a table of $E(X_{(1)})$ for n between 1 and 20.

Exercise 6.45 In a college cafeteria, ice cream is available for the evening meal in servings that vary in weight according to a normal distribution with a standard deviation of 100 gm. The cafeteria workers maintain about 15 servings for students to choose from. Every day student A chooses the smallest serving available, while student D chooses the largest. Over the school year (200 meals), how much more ice cream does student D eat than student A?

Exercise 6.46 Drop n points at random independently and uniformly on a square of a side a. How close is the point closest to the center of the square?

Exercise 6.47 (Rayleigh distribution) Let X and Y be independent random variables having the same distribution $N(0, \sigma^2)$. Find the distribution of $\sqrt{X^2 + Y^2}$. We can interpret this as the distribution of the deviation of an object from a target point when the object is dropped onto the target from above. X and Y are the deviations in the x and y directions with respect to the Cartesian coordinate system whose origin is at the target point.

Exercise 6.48 For the situation described in Exercise 6.47, consider a circle and a square of the same area, both centered at the target point. Which is more likely to contain the point where the object lands? Hint: use probabilistic reasoning.

Exercise 6.49 (χ^2-distribution) Return to Exercise 5.19. The FDA should be just as concerned with the variance as with the mean quantity of impurity. For example, if a company produces pills with an average of 4 ppm impurities but a variance of 4 ppm^2, 31% of the pills it is producing are below standard. One can test a sample variance just as one can test a sample mean. The distribution of the sample variance of a random sample of size n from a normally distributed population is called the **chi-square distribution** with $n - 1$ degrees of freedom. If the mean is known and not simply computed from the data of the random sample, then the distribution is chi-square with n degrees of freedom. More precisely, the chi-square distribution is defined as the distribution of the sum of the squares of n independent standard normal random variables. The distribution of the sample variance is obtained from the chi-square by multiplying by a constant. Compute the variance of the chi-square distribution. Use this to compute the variance of the sample variance of a random sample of size n. Do this both when mean is known and when the mean is not known. For large samples, the chi-square is approximately normally distributed, because of the Central Limit Theorem. Now suppose that the FDA determines that 9 ppm impurity is possibly hazardous. It would seem reasonable to require that no more than 1 pill per thousand can have this much of the impurity. Does the drug company conform to this requirement? Use 95% one-sided confidence intervals both for the mean and for the variance.

Exercise 6.50 Return to Exercise 5.38. Suppose that the resistors are in parallel rather than in series. Use a suitable normal approximation to find the 95% confidence interval for the resistance of this circuit.

Exercise 6.51 Compute the density of the sum of n independent, uniformly distributed random variables on $[0, a]$.

Exercise 6.52 Prove that the rejection sampling method generates a random number according to the density $h(x)$.

Exercise 6.53 Choose four points at random on a circle. Call them X_1, X_2, X_3 and X_4. What is the probability that the chords $\overline{X_1 X_2}$ and $\overline{X_3 X_4}$ intersect? Hint: use a symmetry argument.

Exercise 6.54 What is the probability that a random quadratic polynomial, ax^2+bx+c, has real roots? Here the coefficients are independent and uniformly distributed on $[0, 1]$.

Exercise 6.55 (Epstein) A gambling house offers the following game. After paying an entrance free E, a coin is tossed until the number of heads exceeds the number of tails. The player is then paid the number of dollars equal to the number of times the coin was tossed. What is the average amount of money that the player expects to receive?

Exercise 6.56 A random walk in two or more dimensions is simply two or more independent one-dimensional random walks acting simultaneously. What is the probability that a two-dimensional random walk, starting from the origin, eventually returns to the origin? If it returns, how long, on average, does it take to do so? Do the same for a three-dimensional random walk.

Exercise 6.57 If the gambling house in Exercise 6.55 has only N dollars available for winners, what is a fair entrance fee E for the game described there?

6.9 Answers to Selected Exercises

Answer to 6.1.   This problem is rather ambiguously stated, but probability problems in the "real world" generally are. Exercise 6.5 is an even more complicated example of this. Here it should be clear that the probability required is the *conditional* probability that each person has for winning. It is important that each person *knows* that the cards are shuffled at random.

The sample space then consists of 12 sample points (including two orientations for each card), each one equally probable. Let L be the event that the person on the left wins, and R be the event that the person on the right wins. Write A_i for the event that the person on the left sees the number i on the card, and write B_i for the the event that the person on the right sees the number i on the card. The conditional probabilities L given each of the A_i's are easy to compute:

$$P(L \mid A_1) = \frac{P(L \cap A_1)}{P(A_1)} = \frac{1/12}{1/12} = 1$$

$$P(L \mid A_2) = \frac{P(L \cap A_2)}{P(A_2)} = \frac{2/12}{3/12} = \frac{2}{3}$$

$$P(L \mid A_3) = \frac{P(L \cap A_3)}{P(A_3)} = \frac{3/12}{5/12} = \frac{3}{5}$$

$$P(L \mid A_4) = \frac{P(L \cap A_4)}{P(A_4)} = \frac{0/12}{3/12} = 0$$

The probabilities with L replaced by R and A by B are the same.

Thus if the "2/3" card is held up, the person on the left will compute a conditional probability of 2/3 of winning and the person on the right will compute a probability of 3/5 of winning. The paradox is that $\frac{2}{3} + \frac{3}{5} > 1$; *both* persons think their chances of winning are over 50%! The paradox arises from the mistaken belief that conditional probabilities can be added when the events in question are complementary. However, this is only true when the events are in the *same* sample space. In this case, the events $(L \mid A_2)$ and $(R \mid B_3)$ are events in *different* samples spaces, and so in general have no relationship to one another.

Roughly speaking, *all* probabilities are conditional probabilities, the condition being all of your *a priori* assumptions about the mathematical model and all of the information you have up to the time when the probability is computed. As a result, contradictions of the kind above often occur when one discusses probabilistic questions without using precise language.

Answer to 6.2.   This is an example of the use of one phrase, "probability of drawing a black ball," to describe a variety of different mathematical concepts. The man has more information than the woman, so he will naturally compute a different probability than the woman.

Answer to 6.3.   One must be careful in this problem: the occupancy numbers are not independent. We want $P(\Theta_2 = 0 \mid \Theta_1 = 0)$ in the sampling process of placing k objects in n containers. By a direct computation:

$$
\begin{aligned}
P(\Theta_2 = 0 \mid \Theta_1 = 0) &= \frac{P(\Theta_2 = 0, \Theta_1 = 0)}{P(\Theta_1 = 0)} \\
&= \left(\frac{(n-2)^k}{n^k}\right)\left(\frac{(n-1)^k}{n^k}\right)^{-1} \\
&= \left(\frac{n-2}{n-1}\right)^k \\
&= \left(1 - \frac{1}{n-1}\right)^k
\end{aligned}
$$

Answer to 6.4.   Are such glimpses really possible? Anyone who has seen a deck of cards would doubt this: the Ace of spades is markedly different in design from any other Ace. So the statement that the glimpsed card is a black Ace would mean that the glimpsed card is the Ace of clubs. A similar difficulty occurs in Exercise 6.5. Your knowledge of the eyesight, honesty, motivation,... of the kibitzer all strongly influence your computation.

In any case, let us consider the problem as it is stated. Define these events:

$$
\begin{aligned}
A_i &= \text{"The hand has at least } i \text{ Aces"} \\
B &= \text{"The hand has a black Ace"} \\
S &= \text{"The hand has the Ace of spades"}
\end{aligned}
$$

We can then compute using R:

```
> A2 <- choose(4,2)*choose(48,3) + choose(4,3)*choose(48,2) +
```

```
+ choose(4,4)*choose(48,1)
> A1 <- A2 + choose(4,1)*choose(48,4)
> B <- choose(2,1)*choose(50,4) + choose(2,2)*choose(50,3)
> S <- choose(51,4)
> BA2 <- choose(2,1)*choose(2,1)*choose(48,3) +
+ choose(2,1)*choose(2,2)*choose(48,2) +
+ choose(2,2)*choose(50,3)
> SA2 <- choose(3,1)*choose(48,3) +
+ choose(3,2)*choose(48,2) + choose(48,1)
> A2/A1
[1] 0.1221849
> BA2/B
[1] 0.1895877
> SA2/S
[1] 0.2213685
```

The first conditional probability is $P(A_2|A_1)$; the second is $P(A_2|B)$; and the third is $P(A_2|S)$. Your initial reaction to this problem might be that information about an Ace known to be in the player's hand should have very little, if any, effect on what *other* cards are in the hand. Yet we find that the information has quite a substantial effect.

Answer to 6.5. This problem has generated quite a few debates, some very heated. Here is some flavor of some of the arguments. The problem is important because it is one of the simplest examples where the Bayesian point of view occurs in a nontrivial way. As a result, we will return to it again later.

The computations in this exercise are very easy ones, and complicated derivations are not what is required. The real problem lies in translating the vague language of the problem into the language of probability theory. Let A, B and C be the events that prisoners A, B and C are set free. These events are *not* independent. For example, $A \cap B = \overline{C}$. There are four points of view in this problem corresponding to the following probabilities:

$P(A) = \frac{2}{3}$ Prisoner A before being told anything

$P(A \mid B) = \frac{1}{2}$ Prisoner A after being told that B will be set free

$P(A \mid C) = \frac{1}{2}$ Prisoner A after being told that C will be set free

$P(A) = 0 \; or \; 1$ Jailer's point of view

Now the prisoner is right that his probability of being executed will not change no matter what the jailer says, because $P(A) = \frac{2}{3}$. However, it seems clear that the jailer is referring not to $P(A)$ but to $P(A \mid B)$ or $P(A \mid C)$, since the jailer refers to the respective probabilities. The prisoner, on the other hand, is referring only to $P(A)$ for himself, although he does seem to recognize that if he could somehow inform the other prisoner about his fate, then that would affect the other prisoner. The fact that he cannot do this is part of prisoner A's argument.

This takes care of the problem as it was stated, vague though it is. Let us now try to look a little more deeply at the problem. Suppose that the jailer agrees to the prisoner's request. If A is going to be executed, the jailer has a choice: either tell A that B is going to be set free or tell A that C is going to be set free. One naturally assumes that the jailer will make one of these choices at random with probability $\frac{1}{2}$. Is such a model reasonable? In practice, one would expect some bias (the jailer being human after all). So assume that the choice is made to say B with probability p and C with probability $q = 1-p$. If we make this assumption, our whole model changes. We now have four events to consider:

$$\overline{B} \;\;=\;\; \text{``B is executed''}$$

$$\overline{C} \;\;=\;\; \text{``C is executed''}$$

$$D_1 \;\;=\;\; \text{``A is executed and the jailer says that B will go free''}$$

$$D_2 \;\;=\;\; \text{``A is executed and the jailer says that C will go free''}$$

By assumption:

$$P(\overline{B}) = P(\overline{C}) \;\;=\;\; P(D_1 \cup D_2) = \frac{1}{3}$$
$$P(D_1 \mid \overline{A}) \;\;=\;\; p$$
$$P(D_2 \mid \overline{A}) \;\;=\;\; q$$

Hence,

$$P(D_1) = P(D_1 \mid \overline{A}) \, P(\overline{A}) = \frac{p}{3}$$

and similarly,

$$P(D_2) = \frac{q}{3}$$

The event that the jailer tells the prisoner that B will be set free is now $D_1 \cup \overline{C}$. The probability that A is set free given that the jailer tells the prisoner that B is set free is now:

$$P(A \mid D_1 \cup \overline{C}) \;\;=\;\; \frac{P((A \cap D_1) \cup (A \cap \overline{C}))}{P(D_1 \cup \overline{C})}$$
$$=\;\; \frac{P(A \cap D_1) + P(A \cap \overline{C})}{P(D_1) + P(\overline{C})}$$
$$=\;\; \frac{0 + \frac{1}{3}}{\frac{p}{3} + \frac{1}{3}}$$
$$=\;\; \frac{1}{p+1}$$

Similarly, the probability that A is set free given that the jailer tells the prisoner that C is set free is

$$P(A \mid D_2 \cup \overline{B}) = \frac{1}{q+1}$$

Now consider some cases:

Unbiased jailer This is the case in which $p = q = \frac{1}{2}$. For this case the two conditional probabilities are:

$$P\left(A \mid D_1 \cup \overline{C}\right) \;\; = \;\; \frac{1}{\frac{1}{2} + 1} = \frac{2}{3}$$

$$P\left(A \mid D_2 \cup \overline{C}\right) \;\; = \;\; \frac{1}{\frac{1}{2} + 1} = \frac{2}{3}$$

So if the jailer agrees to be fair, the probability does not change!

Totally biased jailer For example, assume that $p = 1$ and $q = 0$. Now the conditional probabilities are:

$$P\left(A \mid D_1 \cup \overline{C}\right) \;\; = \;\; \frac{1}{1 + 1} = \frac{1}{2}$$

$$P\left(A \mid D_2 \cup \overline{C}\right) \;\; = \;\; \frac{1}{0 + 1} = 1$$

This looks paradoxical until you realize that the only way for the jailer to tell A that C is going to be set free is if B was going to be executed. So in this case the probabilities do change, possibly very dramatically.

Some argue that the first case above provides the correct model for this situation. In the absence of any other information, this is not an unreasonable assumption. Indeed, in Chapter 9 we will give a formal reason why: the case $p = q = \frac{1}{2}$ represents "maximum entropy." But one must make it clear that this *is* an assumption. To argue that the jailer is wrong and that the probabilities "don't change" without explicitly mentioning that this assumption is being made is a mistake.

The assessment of the value of p is one of the inputs into this problem. How does one compute or estimate it? Here is where the Bayesian point of view enters. Although we presume that p is a constant, we regard it as a random variable! To compute probabilities, we use the Law of Alternatives. We will return to this discussion when we discuss randomization in Chapter 8.

Answer to 6.11. The paradox in this problem is that it seems at first that being told "two children (at least) are boys" is the same as being told "the two oldest are boys." In either case, one would argue that by independence, the probability that all three children are boys is $1/2$. However, if we use the Bernoulli process: coin tossing model, to model this phenomenon (where boy=heads, girl=tails), then the two events quoted above are not the same. The first is:

$$\left(H_1 \cap H_2\right) \cup \left(H_1 \cap H_3\right) \cup \left(H_2 \cap H_3\right) = \left(S_3 \geq 2\right)$$

and the second event is

$$\left(H_1 \cap H_2\right) = \left(S_2 = 2\right)$$

By an easy computation

$$P\left(S_3 = 3 \mid S_3 \geq 2\right) \;\; = \;\; 1/4$$

$$P(S_3 = 3 \mid S_3 = 2) \quad = \quad 1/2$$

The explanation for the disparity between the two conditional probabilities above is similar to the explanation of the "reference point paradox" in Section 4.5. If we choose families having three children, subject to the criterion that at least two of the children be boys, we will be choosing among half the families. If we use the second criterion, we will be choosing among a more restricted set of choices. Both criteria are satisfied, however, by families having three boys. Thus, these families will seem more probable, not because there are more such families, but rather because there are *relatively* more of them.

We can use Bayes' Law in this problem as follows. The alternatives are $(S_3 = 0), \ldots, (S_3 = 3)$; i.e., the number of boys in the family. The event B in the statement of Bayes' Law is $(S_3 \geq 2)$ in the first case and $(S_3 = 2)$ in the second. Applying Bayes' Law in each of the two cases we have:

$$P(S_3 = 3 \mid S_2 \geq 2) \quad = \quad \frac{P(S_2 \geq 2 \mid S_3 = 3)\,P(S_3 = 3)}{\sum_{i=0}^{3} P(S_2 \geq 2 \mid S_3 = i)\,P(S_3 = i)}$$

$$= \quad \frac{1 \cdot \frac{1}{8}}{1 \cdot \frac{3}{8} + 1 \cdot \frac{1}{8}}$$

$$= \quad 1/4$$

$$P(S_3 = 3 \mid S_2 = 2) \quad = \quad \frac{P(S_2 = 2 \mid S_3 = 3)\,P(S_3 = 3)}{\sum_{i=0}^{3} P(S_2 = 2 \mid S_3 = i)\,P(S_3 = i)}$$

$$= \quad \frac{1 \cdot \frac{1}{8}}{\frac{1}{3} \cdot \frac{3}{8} + 1 \cdot \frac{1}{8}}$$

$$= \quad 1/2$$

The two computations differ in just one term. In the first case,

$$P(S_2 \geq 2 \mid S_3 = 3) = 1$$

while in the second case the corresponding term is

$$P(S_2 = 2 \mid S_3 = 3) = \frac{1}{3}$$

i.e., in families having exactly two boys, only a third have two oldest boys.

Answer to 6.19. The alternative events A_l are the different kinds of hand (l ranges from a to j). The event B is "the hand contains at least one pair." For the 10 different kinds of hand in Exercise 2.20, it is easy to compute

$P(B \mid A_l)$:

	$P(B \mid A_l)$	$P(B \mid A_l)\,P(A_l)$	$P(A_l \mid B)$
(a)	0	0	0
(b)	1	.4226	.8573
(c)	1	.0475	.0964
(d)	1	.0211	.0428
(e)	0	0	0
(f)	1	.0014	.0028
(g)	0	0	0
(h)	0	0	0
(i)	1	.002	.004
(j)	0	0	0

Therefore, $P(B)$ is 0.4929.

Answer to 6.20. Let D be the distance of the bullet from the center of the target. We want to compute the distribution of D, i.e., $P(D \leq d)$. We know that $P(D \leq d) = C\pi d^2$, for some constant C, because the event $(D \leq d)$ means that the bullet hit the disk of radius d centered at the center of the target. We also know that $P(D \leq 1) = 1$ because the bullet was assumed to have hit the target. Hence, $C\pi = 1$ or $C = \frac{1}{\pi}$. Thus $P(D \leq d) = d^2$. The density of D is then

$$\mathrm{dens}(D = d) = \begin{cases} 2d & \text{if } 0 \leq d \leq 1 \\ 0 & \text{otherwise} \end{cases}$$

Answer to 6.21. Let the position of a scimitar be determined by its tip, i.e., the first point of the scimitar encountered when going *saatin tersi yönde* (counterclockwise) around the circle. This situation is then equivalent to dropping n points on a circle of circumference a, and we want the probability that no two are closer than h apart. If we use one point to "cut" the circle, then this process is the same as dropping $n - 1$ points on $[0, a]$, and we want the probability that the smallest gap is greater than h. In other words,

$$p_{n,a} = P(L_{(1)} > h) = P(L_1 > h, L_2 > h, \ldots, L_n > h)$$

Conditioning on the position of the first (smallest) point,

$$\begin{aligned} p_{n,a} &= \int_h^{a-(n-1)h} P(L_1 > h, \ldots, L_n > h \mid L_1 = x)\,\mathrm{dens}(L_1 = x)\,dx \\ &= \int_h^{a-(n-1)h} (p_{n-1,a-x}) \frac{n(a-x)^{n-1}}{a^n}\,dx \\ &= \frac{n}{a^n} \int_h^{a-(n-1)h} (p_{n-1,a-x})(a-x)^{n-1}\,dx \end{aligned}$$

Now the special case $n = 2$ is easy to compute:

$$p_{2,a} = \begin{cases} 0 & \text{if } a \leq 2h \\ \frac{a-2h}{a} & \text{if } a > 2h \end{cases}$$

Thus, by induction, we find that

$$p_{n,a} = \begin{cases} \left(\frac{a-nh}{a}\right)^{n-1} & \text{if } a > nh \\ 0 & \text{if } a \leq nh \end{cases}$$

Answer to 6.33. Using the notation in Exercise 4.6, we can compute the density of S in the following much easier manner:

$$\begin{aligned}
\text{dens}(S = s) &= \text{dens}\big(X_{(n)} - X_{(1)} = s\big) \\
&= \text{dens}\big(a - (L_{(1)} + L_{(n+1)}) = s\big) \\
&= \text{dens}\big(a - (L_{(n)} + L_{(n+1)}) = s\big) \quad \text{by exchangeability} \\
&= \text{dens}\big(X_{(n-2)} = s\big)
\end{aligned}$$

Answer to 6.39. Because the molecule is long, we may model the breakage with a Uniform process of dropping one point on $[0, a]$. The shorter piece has length $L_{(1)}$ and the longer piece has length $L_{(2)}$. However, because we are dropping only one point, $X_{(1)} = X_1$, the random variable $L_{(1)}$ is the minimum of X_1 and $a - X_1$. Therefore, the distribution is

$$\begin{aligned}
P\big(L_{(1)} \leq x\big) &= P(X_1 \leq x, a - X_1 \leq x) \\
&= P(X_1 \leq x, X_1 \geq a - x) \\
&= \frac{2x}{a} \quad \text{when } x \leq \frac{a}{2}
\end{aligned}$$

So $L_{(1)}$ is uniformly distributed on $[0, a/2]$. Its expectation is therefore equal to $a/4$. Since $L_{(1)} + L_{(2)} = a$, the expectation of $L_{(2)}$ is $3a/4$. Accordingly, the ratio of the expected length of the longer piece to the expected length of the shorter piece is 3.

Answer to 6.40. Using the notation of the answer to Exercise 6.39, we consider the ratio $R = \dfrac{L_{(2)}}{L_{(1)}}$. The distribution of R can be computed using the Law of Alternatives:

$$\begin{aligned}
P(R \leq r) &= P\big(L_{(2)} \leq r L_{(1)}\big) \\
&= P\Big(L_{(2)} \leq r L_{(1)} \mid X_1 \leq \frac{a}{2}\Big) P\Big(X_1 \leq \frac{a}{2}\Big) \\
&\quad + P\Big(L_{(2)} \leq r L_{(1)} \mid X_1 > \frac{a}{2}\Big) P\Big(X_1 > \frac{a}{2}\Big) \\
&= P\Big(L_{(2)} \leq r L_{(1)} \mid X_1 \leq \frac{a}{2}\Big) P\Big(X_1 \leq \frac{a}{2}\Big) \\
&\quad + P\Big(L_{(2)} \leq r L_{(1)} \mid X_1 > \frac{a}{2}\Big) P\Big(X_1 > \frac{a}{2}\Big) \\
&= P\Big(a - X_1 \leq r X_1 \mid X_1 \leq \frac{a}{2}\Big) \frac{1}{2} \\
&\quad + P\Big(X_1 \leq r(a - X_1) \mid X_1 > \frac{a}{2}\Big) \frac{1}{2}
\end{aligned}$$

The two alternatives are symmetric so they will give the same probability. Therefore,

$$
\begin{aligned}
P(R \leq r) &= P(a - X_1 \leq rX_1) \\
&= P\left(X_1 \geq \frac{a}{r+1}\right) \\
&= \frac{\frac{a}{2} - \frac{a}{r+1}}{\frac{a}{2}} \\
&= 1 - \frac{2}{r+1} = \frac{r-1}{r+1}
\end{aligned}
$$

The support of R is $[1, \infty)$. Using the formula in Section 4.4, the expectation of R is:

$$
E(R) = 1 + \int_1^\infty \frac{2}{r+1} dr
$$

which is easily seen to diverge to ∞. This may seem surprising considering that the ratio of the expectations is only 3.

Answer to 6.49. The variance for chi-square with 1 degree of freedom is 2. From this it is easy to determine that the variance of the chi-square distribution for n degrees of freedom is $2n$. The variance of the sample variance for a random sample of size n is $2\sigma^2/n$ when the mean is known and $2\sigma^2/(n-1)$ when the mean is unknown. For the FDA sample, the number of degrees of freedom is 49. In this case, the drug company does not conform.

Answer to 6.53. If we choose 4 points completely at random on a circle, then there are 4!=24 permutations of these points that will differ from one another only by the order of dropping of the points. It is possible that 2 or more points will coincide, but it should be clear that such an event has probability zero. These 24 permutations divide up into 3 groups of 8 with respect to the interior chords. Thus for every set of 4 points on the circle, there are 3 *equally likely* ways to draw in the interior chords:

Only 1 of the 3 has the chords intersecting. Therefore, the probability of intersection is $1/3$.

Answer to 6.56. It returns with probability 1, but the expected time to return is infinity. For a three-dimensional random walk, the probability of return is about 0.239.

CHAPTER 7

The Poisson Process

Siméon-Denis Poisson was a French mathematician, geometer and physicist. His teachers were Laplace and Lagrange, and they were to remain friends for life. His most important works during his lifetime were a series of papers on definite integrals and his advances in Fourier series. This work was the foundation of later work in this area by Dirichlet and Riemann. However, he is today best known for the Poisson distribution that first appeared in an article he wrote in 1837. In addition to the Poisson distribution, many concepts have been named in his honor. Curiously, he was not highly regarded by other French mathematicians, either during his lifetime or after his death. His reputation was guaranteed by the esteem that he was held in by foreign mathematicians who seemed more able than his own colleagues to recognize the importance of his ideas. Poisson himself was completely dedicated to mathematics. It has been reported that Poisson frequently said, "Life is good for only two things, discovering mathematics and teaching mathematics."

The **Poisson process** is the third basic stochastic process, the first two being the Bernoulli and Uniform processes. It can be defined in many ways, and we will examine several of them. Like other stochastic processes it has many interpretations, and each one has a large collection of applications to specific models. We also compare the Uniform and Bernoulli processes with the Poisson process. We find that the Poisson process can be regarded as a continuous analog of the Bernoulli process and that the Poisson process can be obtained in the limit from the Uniform process as the length of the interval increases. We then turn things around by showing that the Uniform process can be obtained by conditioning the Poisson process!

7.1 Continuous Waiting Times

Suppose we toss a coin k times and that we get k tails. It is intuitively obvious that on the next toss there is the same probability for heads as ever: the coin does not remember what took place in the past. We can express the fact that a coin has no memory in terms of the single waiting time W_1 as follows:

$$P(W_1 > k + n \mid W_1 > k) = P(W_1 > n)$$

The probability that one will get a run of $k + n$ tails *given* that one has just gotten a run of k tails is simply the probability that one will get the additional n tails: the preceding tails neither help nor hurt. How long one must wait does not depend on how long one has already waited.

In real life, if one is waiting for an incident to take place there is no abstract entity flipping an abstract coin during small discrete time intervals determining when the incident is to occur. For example, one might be standing next to a Geiger counter waiting for a click. The waiting time in this case is *continuous,* but like the Bernoulli waiting time, the waiting time has no memory. We express this using conditional probability.

Definition. A positive continuous random variable W is said to have the **exponential distribution** when

$$P(W > t + s \mid W > s) = P(W > t)$$

for all positive t, s.

We will also call W a **continuous memoryless waiting time**, although we will see that the value of W need not represent time. The exponential distribution is an ubiquitous distribution appearing in the most unexpected places.

What is surprising about random variables having the exponential distribution is that the seemingly innocuous assumption we have made in defining this concept determines the probability distribution of W but for a single parameter. In Appendix B, we show that the probability distribution of a continuous waiting time W is

$$F(t) = P(W \le t) = 1 - e^{-\alpha t}$$

for some positive constant α; and its density is

$$f(t) = \alpha e^{-\alpha t}$$

We now see why we say that W is exponentially distributed. Here are the distributions of some continuous waiting times:

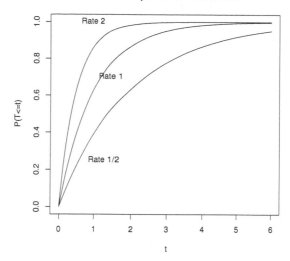

and here are their densities:

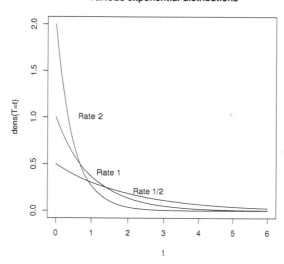

The parameter α may be interpreted as the frequency of the incidents in time: roughly speaking, α incidents occur per unit time "on average." We will call it the **intensity**, but it is usually called the *rate parameter*. Since W is a waiting time, its median is called the *half-life*. It is easy to determine the half-life of W by solving for $1/2$ in the distribution function. The half-life of an exponential random variable is $\log(2)/\alpha$. It is very common to specify an exponential distribution using its half-life rather than its intensity.

The power of probabilistic reasoning (made rigorous by conditional probability) is that we may compute the distribution of a random variable without referring to a sample space or to events of it. The distribution is defined purely in phenomenological terms, i.e., in terms of the observed phenomena only.

Consider, for example, that we have a collection of points dropped independently and uniformly throughout the entire infinite plane. By "uniformly" we mean that the probability of finding a point in a region of finite area t depends only on the area t, not on its shape or location. By "independently" we mean that for two disjoint regions, the probability of finding a point in one region is independent of finding a point in the other. Write $P(T>t)$ for the probability of finding no point in a region of area t. Then

$$P(T>t+s) = P(T>t)\,P(T>s)$$

Therefore, as above,

$$P(T>t) = e^{-\alpha t}$$

where α may be interpreted as the density of the points dropped on the plane. It is reasonable to regard T as a "waiting area," i.e., "how large an area must a region be in order to find a point in the region?"

Consider next a collection of stars distributed at random in a large region of space. How far away is the nearest neighbor to a star in this region? This is quite similar to the above problem, but we now have three dimensions. Instead of a region of some area t, we use a spherical volume of radius r whose center is the given star. If the average density of the stars is α, then

$$P(\text{Nearest neighbor is more than } r \text{ units away}) = \exp\left(-4\pi r^3/3\right)$$

Suppose we are in a forest with randomly located trees. How far can one see if one looks in one particular direction? By symmetry, one may assume that one looks along the positive x-axis from the origin. Assume also, for simplicity, that the trees are all ρ units in radius. Let T be the random variable "how far can one see along the x-axis?" If T is larger than t, then there are no centers of trees in the region shown in this diagram:

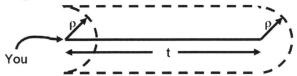

The indentation on the left side of the diagram is a consequence of the fact that one happens to be standing at that point. The area of the dotted region is the same as that of a rectangle of sides t and 2ρ. Therefore,

$$P(T>t) = e^{-2\alpha\rho t}$$

Needless to say, this is an idealized model (trees do not all have the same radius), but it illustrates the basic idea.

One gets a very similar model when one studies the effect of a beam of high-energy protons entering a detector consisting of a tank of liquid hydrogen. Here the "trees" are the nuclei of the hydrogen atoms, although we should build the model in three dimensions instead of two.

To summarize, every "waiting time" for which the future does not depend on the past exhibits exponential decay.

The Gamma Distribution

We just saw that the analog for continuous random variables of the random variable W_1 in the Bernoulli process is an exponential random variable. We now ask for the analog of the kth waiting time W_k in the Bernoulli process. Now the kth waiting time is the sum of the first k gaps in the Bernoulli process, $W_k = T_1 + T_2 + \cdots + T_k$, and the gaps are independent. Therefore, we could have computed the distribution of W_k by convolving the distributions of the gaps, all of which are geometric random variables. As an example, the distribution of W_2 is the convolution of the distribution of T_1, $q^{n-1}p$, with itself, i.e.,

$$P(W_2 = n) = \sum_{k=1}^{n-1} q^{(n-k)-1} p q^{k-1} p = \sum_{k=1}^{n-1} q^{n-2} p^2 = (n-1) q^{n-2} p^2$$

Consider now the continuous analog of a waiting time: the exponential distribution. The sum of two independent exponentially equidistributed random variables T_1 and T_2 may be regarded as the waiting time for the second occurrence, $W_2 = T_1 + T_2$, just as in the Bernoulli process. In Section 7.3, we shall build a more concrete model on which to define this random variable. Although we have not yet defined the T_i's on a specific sample space, we can nevertheless compute the distribution of the continuous waiting time W_2. It is the convolution of $\alpha e^{-\alpha t}$ with itself:

$$
\begin{aligned}
\text{dens}(W_2 = t) &= \int_0^t \alpha e^{-\alpha(t-s)} \alpha e^{-\alpha s} ds \\
&= \alpha^2 \int_0^t e^{-\alpha t} ds \\
&= \alpha^2 e^{-\alpha t} \int_0^t ds \\
&= \alpha^2 t e^{-\alpha t}
\end{aligned}
$$

More generally, the density of the kth waiting time W_k is the convolution

$$\underbrace{\alpha e^{-\alpha t} * \alpha e^{-\alpha t} * \cdots * \alpha e^{-\alpha t}}_{k \text{ times}}$$

W_k is the sum $T_1 + T_2 + \cdots + T_k$ of k independent exponentially distributed random variables, each of which has rate parameter α. This convolution is easily computed:

$$\text{dens}(W_k = t) = \frac{\alpha^k t^{k-1}}{(k-1)!} e^{-\alpha t}$$

This is called the **gamma distribution**. It is also known as the **Erlang distribution**. Notice that it has two parameters: α and k, where k is a positive integer.

In the formula for the gamma density function, every term is meaningful for any positive real number, except for the factorial. The factorial term serves to normalize the density so that the total density is 1. This gives us a motivation for defining the factorial on all real numbers so that it normalizes this density in general. More precisely, we wish to have a function Γ for which

$$\Gamma(k, \alpha) = \int_0^\infty \alpha^k t^{k-1} e^{-\alpha t} dt$$

By making the change of variables $u = \alpha t$, we find that

$$\Gamma(k, \alpha) = \int_0^\infty u^{k-1} e^{-u} du$$

and it follows that Γ does not depend on α. By construction, we know that

$$\Gamma(k) = (k - 1)!$$

for all $k \geq 1$. Other values of the gamma function have been computed, such as $\Gamma(1/2) = \sqrt{\pi}$. In R, the gamma function is called **gamma**, and its logarithm is **lgamma**.

We have already seen an example of a gamma distribution in which the parameter k is not an integer. The chi-square distribution with n degrees of freedom, introduced in Exercise 6.49, is an example of a gamma distribution for which the intensity α is $1/2$, and the parameter k is $n/2$.

Means and Variances of Continuous Waiting Times

We end this section by computing the means and variances of the continuous waiting times. It is an easy exercise to verify that if T is exponentially distributed, then $E(T) = 1/\alpha$, where α is the rate parameter of T. This coincides with our intuitive feeling that α is an "intensity."

The variance of T is easily computed by using integration by parts twice.

$$
\begin{aligned}
Var(T) &= E(T^2) - E(T)^2 \\
&= \int_0^\infty t^2 \alpha e^{-\alpha t} dt - \left(\frac{1}{\alpha}\right)^2 \\
&= \alpha \left[\frac{-t^2}{\alpha} e^{-\alpha t} - \frac{t}{\alpha^2} e^{-t} + \frac{2}{\alpha^3} e^{-t}\right]_0^\infty - \frac{1}{\alpha^2} \\
&= \frac{2\alpha}{\alpha^3} - \frac{1}{\alpha^2} \\
&= 1/\alpha^2
\end{aligned}
$$

So the standard deviation $\sigma(T)$ is the same as the mean $E(T)$: both are $1/\alpha$.

Because the kth waiting time W_k of the Poisson process is the sum of k independent, equidistributed exponential random variables, the variance of W_k is

$$Var(W_k) = \frac{k}{\alpha^2}$$

7.2 Comparing Bernoulli with Uniform

We could at this point simply *define* the Poisson process to be a sequence of independent, exponentially distributed random variables, having the same intensity α. But we prefer to take a different approach, which builds on the two processes we have already thoroughly studied. We therefore now give a detailed comparison of the Bernoulli and Uniform processes emphasizing their similarities and their differences. The Poisson process will be a "limit" of both processes so that, in a sense, it furnishes a formal link between them. In so doing, we will discover some new aspects of both these processes.

Process Parameters

The Bernoulli process depends on a single parameter: the bias p of the coin. The Uniform process depends on two parameters: the length a of the interval and the number n of points sampled. There is already a certain asymmetry here. The number of points per unit interval, $\alpha = n/a$ is called the *intensity* of the Uniform process. Different Uniform processes having the same intensity are quite similar, especially when n is large.

Sample Spaces

The sample space Ω of the Bernoulli process is the set of all sequences of 0's and 1's. To every such sequence, we can associate a set of natural numbers: the set of positions having 1's. The positions are the waiting times W_k for the sample point. For example,

$$(1, 1, 0, 1, 1, 0, 1, 1, 1, 0, \ldots) \text{ corresponds to } \{1, 2, 4, 5, 7, 8, 9, \ldots\}$$

This gives us a new way of looking at Ω. It is the set of all subsets of the natural numbers. We already saw how to extract this subset from a Bernoulli sample point in Section 3.1 by using the indexing feature of R. For example,

```
> range <- 1:10
> sample <- rbinom(length(range), 1, 0.5)
> range[sample==1]
[1] 1 2 4 5 7 8 9
```

By contrast, the sample space Ω of the Uniform process is the set of all sequences $(x_1, x_2, \ldots, x_n)$ of real numbers such that $0 \leq x_i \leq a$. There seems to be little similarity between this sample space and the Bernoulli sample space.

Events

The elementary events of the Bernoulli process are the subsets

$$H_n = (X_n = 1) = \text{``The } n^{th} \text{ toss is heads.''}$$

The elementary events of the Uniform process are the subsets

$$(s \leq X_i < t) = \text{``The } i^{th} \text{ point falls in } [s, t].\text{''}$$

In both cases, the events in general are obtained by intersections, complements and unions from the elementary events.

Random Variables

Up to now, we have viewed the random variables X_n of the Bernoulli process and X_i of the Uniform process as being fundamental. But there is an alternative point of view. We could equally well define the Bernoulli process by the random variables S_n, the number of heads in the first n tosses. We should write this $S_n^{(p)}$ to denote the fact that it depends on the parameter p. We know that $S_n^{(p)}$ has a binomial distribution with bias p:

$$P\left(S_n^{(p)} = k\right) = \binom{n}{k} p^k q^{n-k}$$

This was the point of view taken in R. The **rbinom** function generates the random variable $S_n^{(p)}$, and there is no separate random number generator for the Bernoulli process.

Similarly, we could define the Uniform process by the random variables $U_{n,a}(t)$, the number of points falling in the interval $[0, t]$. These are random variables we have not yet seen. For each t in $[0, the random variable a]$, $U_{n,a}(t)$ is an *integer* random variable. In fact, one can easily see that $U_{n,a}(t)$ has the binomial distribution, for points fall in $[0, t]$ or in $[t, a]$ with the same probability as a tossed coin with bias $p = \dfrac{t}{a}$ lands heads or tails, respectively. Therefore,

$$P(U_{n,a}(t) = k) = \binom{n}{k} \left(\frac{t}{a}\right)^k \left(1 - \frac{t}{a}\right)^{n-k}$$

We shall abbreviate $U_{n,a}(t)$ to simply $U(t)$. For each t, $U(t)$ is a new random variable.

Next, we compare the waiting time W_k for the kth head in the Bernoulli process with the kth order statistic $X_{(k)}$. If we think of $[0, a]$ as part of a time axis, there is clearly an analogy between these two random variables. Compare their distribution and density:

$$P(W_k = n) = \binom{n-1}{k-1} q^{n-k} p^k$$

$$\text{dens}\left(X_{(k)} = t\right) = \binom{n-1}{k-1} \frac{n}{a} \left(\frac{t}{a}\right)^{k-1} \left(1 - \frac{t}{a}\right)^{n-k}$$

These are quite similar indeed, except that in the latter case a factor of $\dfrac{t}{a}$ has become $\dfrac{n}{a}$ as a result of differentiation.

We finally come to the gaps in these two processes. In the Bernoulli process, the gaps T_i are equidistributed with the geometric distribution:

$$P(T_i = k) = q^{k-1}p$$

The gaps L_i of the Uniform process are also equidistributed, having the beta distribution:

$$\text{dens}(L_i = t) = \frac{n}{a}\left(1 - \frac{t}{a}\right)^{n-1}$$

The analogy between these two cases is quite striking.

However, the analogy breaks down because the gaps T_i are independent, whereas the gaps L_i are not. To be sure, the L_i's try as hard as they can to be independent — they are exchangeable — but this is not enough. Another way to see the difference between the two processes is to return to the "fundamental" random variables S_n and $U(t)$. The difference $S_n - S_m$ is the number of heads between m and n. Similarly, $U(t) - U(s)$ is the number of points falling between s and t. Now if $(m_1, n_1]$ and $(m_2, n_2]$ are disjoint intervals of integers, then the random variables $S_{n_1} - S_{m_1}$ and $S_{n_2} - S_{m_2}$ are independent. But even if $[s_1, t_1)$ and $[s_2, t_2)$ are disjoint subintervals of $[0, a]$, $U(t_1) - U(s_1)$ and $U(t_2) - U(s_2)$ are *not* independent.

The difficulty stems from the fact that the Uniform process is on a *finite* interval and a *finite* number of points whereas the Bernoulli process is not limited in this way. Whether we *choose* to limit the Bernoulli process to a given finite number of tosses is irrelevant, for we can always continue it if we wish. The Uniform process has no such option. We can always drop more points, but we cannot extend the interval to a longer one without totally altering our process.

Reconciling Bernoulli and Uniform

These considerations suggest that there is a third process that makes the analogy perfect. Just letting the length go to infinity does not work because we cannot make sense of sampling a single point or a finite number of points uniformly from an infinite interval. We would like to have a process that is both uniform on an infinite interval *and* samples an infinite number of points. This works provided we let a and n go to infinity simultaneously, while keeping the intensity $\alpha = \dfrac{n}{a}$ fixed. Intuitively, because the number of points per unit interval remains the same, in the limit one will have the same intensity, and the Uniform processes will converge to a new process.

For example, consider the density of a gap as a and n become large.

$$\text{dens}(L_i = t) \quad = \quad \frac{n}{a}\left(1 - \frac{t}{a}\right)^{t-1}$$

$$= \frac{\alpha\left(1 - \frac{t\alpha}{n}\right)^n}{\left(1 - \frac{t\alpha}{n}\right)}$$

We know from calculus that $\lim_{n \to \infty} \left(1 + \frac{x}{n}\right)^n = e^x$. Therefore, if we let n tend to ∞, the above expression becomes

$$\frac{\alpha e^{-t\alpha}}{1 - 0} = \alpha e^{-\alpha t}$$

That is, in the limit, the gaps become exponentially distributed. This is exactly what we would expect, for the gaps of the Bernoulli process are waiting times, and we would hope that the gaps of the new process will be continuous waiting times.

Consider another example: the joint density of two gaps. In the uniform process this density is not the product of the individual densities. In the limit, however, the joint density of two gaps is the product:

$$\text{dens}(L_1 = t_1, L_2 = t_2) \quad = \quad \text{dens}\left(X_{(1)} = t_1, X_{(2)} = t_1 + t_2\right)$$

$$= \quad \frac{n(n-1)}{a^2}\left(1 - \frac{t_1 + t_2}{a}\right)^{n-2}$$

$$= \quad \frac{\alpha\left(\alpha - \frac{1}{a}\right)\left(1 - \frac{\alpha(t_1 + t_2)}{n}\right)^n}{\left(1 - \frac{\alpha(t_1 + t_2)}{n}\right)^2}$$

$$\rightarrow \quad \frac{\alpha^2 \exp\left(-\alpha(t_1 + t_2)\right)}{(1 - 0)^2}$$

$$= \quad (\alpha \exp\left(-\alpha t_1\right))(\alpha \exp\left(-\alpha t_2\right))$$

of the densities. So in the new process, the gaps are independent and equidistributed, just as in the Bernoulli process. This helps to confirm our feeling that this is the correct approach.

As a final example, we consider the limit of the random function $U_{n,a}(t)$ as $a, n \to \infty$. Recall that

$$P(U_{n,a}(t) = k) \quad = \quad \binom{n}{k}\left(\frac{t}{a}\right)^k\left(1 - \frac{t}{a}\right)^{n-k}$$

$$= \quad \frac{n!}{k!(n-k)!}\left(\frac{\alpha t}{n}\right)^k\left(1 - \frac{\alpha t}{n}\right)^{n-k}$$

$$= \quad \frac{n!}{(n-k)!n^k}\frac{(\alpha t)^k}{k!}\left(1 - \frac{\alpha t}{n}\right)^{n-k}$$

Consider the last factor above. As in the last two examples, we can see that $\left(1 - \frac{\alpha t}{n}\right)^{n-k} \to e^{-\alpha t}$ because k is a *fixed* integer. The rearrangement in the last step above was done so that the second factor does not depend on either

n or a. This leaves the first factor. Multiplying out, it looks like this:

$$\frac{n!}{(n-k)!n^k} = \frac{n}{n}\frac{n-1}{n}\frac{n-2}{n}\cdots\frac{n-k+1}{n}$$

$$= 1\left(1-\frac{1}{n}\right)\left(1-\frac{2}{n}\right)\cdots\left(1-\frac{k-1}{n}\right)$$

Each of these factors approaches 1 as $n\to\infty$. Since there are a *fixed* number of them, their product also approaches 1. Therefore,

$$\lim_{\substack{n,a\to\infty \\ \frac{n}{a}=\alpha}} P(U_{n,a}=k) = \frac{(at)^k}{k!}e^{-at}$$

Since $U_{n,a}(t)$ is the fundamental random function of the uniform process, its limit will be the fundamental random function of the new process. We shall write $N_\alpha(t)$ or $N(t)$ for this limit:

$$P(N_\alpha(t)=k) = \frac{(at)^k}{k!}e^{-at}$$

Notice that the distribution of $N_\alpha(t)$ depends only on the product at. We call this distribution the **Poisson distribution**. More precisely:

Definition. An integer random variable X is said to have the *Poisson distribution with parameter* λ if

$$P(X=k) = \begin{cases} \frac{\lambda^k}{k!}e^{-\lambda} & \text{if } k\geq 0 \\ 0 & \text{if } k<0 \end{cases}$$

The expectation of a Poisson random variable is

$$\begin{aligned} E(X) &= \sum_k kP(X=k) \\ &= \sum_{k=1}^{\infty} k\frac{(\lambda)^k}{k!}e^{-\lambda} \\ &= e^{-\lambda}\sum_{k=1}^{\infty}\frac{(\lambda)^k}{(k-1)!} \\ &= \lambda e^{-\lambda}\sum_{k=1}^{\infty}\frac{(\lambda)^{k-1}}{(k-1)!} \\ &= \lambda e^{-\lambda}e^{\lambda} \\ &= \lambda \end{aligned}$$

Therefore, $E(N_\alpha(t)) = at$. By a similar calculation, we find that $E(N_\alpha(t)^2) = (at)^2 + at$. So the variance is $Var(N_\alpha(t)) = at$.

If we imagine that an infinite number of points are spread on the interval $[0,\infty)$ with density α, then $N(t)$ is the number of points that have fallen in the interval $[0, t]$. The average number of points that fall in $[0, t]$ is $E(N(t)) = \alpha t$, and the average number of points per unit interval is $\dfrac{E(N(t))}{t} = \alpha$. This justifies calling α the density or intensity of the process.

Notice that we may no longer speak of which point has fallen first, which is second, and so on. If we return to the Uniform process for a moment, we can see why this would be so. Originally, we used $X_1, X_2, \ldots, X_n$ as the defining random variables of the Uniform process. If we use the random function $U_{n,a}(t)$, we can no longer distinguish which point is which. All we know is which point is $X_{(1)}$, which is $X_{(2)}$, and so on. For example, $X_{(1)} = t$ if $U_{n,a}(t) = 0$ and $U_{n,a}(s) = 1$ for $s > t$. We can recover the entire Uniform process if we know all the random functions $U_{1,a}(t), U_{2,a}(t), \ldots, U_{n,a}(t)$. But when we let $a, n \to \infty$, we only use the random function $U_{n,a}(t)$. As a result, the order statistics make sense in the Poisson process, but there is no analog of the random variables X_i of the Uniform process.

7.3 The Poisson Sample Space

So far we have discussed the Poisson process from two points of view. We first considered it purely phenomenologically via the random variables T_i and W_k. Next, we considered it as the limit of uniform processes as the length of the interval increases. We must reconcile these two approaches, and we do so by an explicit construction of a model.

The **Poisson sample space** is $\Omega = \{$all rare sequences$\}$, where a **rare sequence** is a set of points in $[0, \infty)$ such that every finite interval has at most finitely many points of the rare sequence, i.e., the sequence does not cluster. To avoid confusion, the points of a rare sequence are called *incidents* or *blips*. Do not confuse this notion with the concept of a sample point. The sample points of Ω in this case are the rare sequences (*not* the blips).

Defining Ω this way seems very natural because the Poisson process is the limit of the Uniform process (of intensity α) on $[0, a]$ as $a \to \infty$. Unfortunately, we have allowed n to approach infinity as well. So the intensity is not an intrinsic part of the Poisson sample space as it was for the uniform sample space, which depends both on the number of points n that are dropped on the interval and the length a of the interval. The same situation occurs for the Bernoulli process. There, the sample space is the same whatever the bias of our coin. It is only through the definition of the probability P that we can say that "the average number of heads in n tosses is np." In a similar way, we shall define a probability P on the Poisson sample space so that the average number of points falling on an interval of length t is αt.

We define the probability P on Ω by means of the random function $N(t)$. We already saw in Section 7.2 what the distributions of the random variables $N(t)$ ought to be. We will see in a moment how this point of view leads immediately to probabilities on the elementary Poisson events. All the distributions of the other random variables on Ω will be derived from the distributions of the $N(t)$. We make three fundamental assumptions:

1. For every nonnegative real number t, $N(t)$ is a nonnegative *integer* random variable whose value is the number of blips in the interval $[0, t)$. More generally, for $s \leq t$, $N(t) - N(s)$ is also an integer random variable whose value is the number of blips in the interval $[s, t)$.

2. The number of blips on any subinterval $[s, t)$ of $[0, \infty)$ has a Poisson distribution:

$$P(N(t) - N(s) = k) = \frac{(\alpha(t - s))^k}{k!} e^{-\alpha(t-s)}, \quad \text{if } 0 \leq s \leq t.$$

Notice that we assume the density of the blips is independent of the location of the subinterval. In particular, when $s = 0$, we have that

$$P(N(t) = k) = \frac{(\alpha t)^k}{k!} e^{-\alpha t}$$

3. If $[s_1, t_1)$ and $[s_2, t_2)$ are disjoint subintervals of $[0, \infty)$, then $N(t_1) - N(s_1)$ and $N(t_2) - N(s_2)$ are independent random variables. In other words, what happens on disjoint subintervals are independent of one another. This property is called *stationary independent increments*.

The above three fundamental assumptions implicitly define the events of Ω. Assumption (1) implies that $(N(t) - N(s) = k)$ is an event of Ω for all s, t and k. Because of the importance of this event we introduce a notation for it:

$$\begin{bmatrix} s, t \\ k \end{bmatrix} = (N(t) - N(s) = k)$$

$$= \text{“All rare sequences having } k \text{ blips in } [s, t)\text{”}$$

The above fundamental assumptions may be rewritten in terms of events as follows.

1. The subsets $\begin{bmatrix} s, t \\ k \end{bmatrix}$ of Ω are the *elementary Poisson events*. An arbitrary event of Ω is obtained from elementary events by intersections, complements and unions.

2. The probability of an elementary Poisson event is:

$$P\left(\begin{bmatrix} s, t \\ k \end{bmatrix}\right) = \frac{(\alpha(t - s))^k}{k!} e^{-\alpha(t-s)}$$

3. If $[s_1, t_1)$ and $[s_2, t_2)$ are disjoint intervals, then $\begin{bmatrix} s_1, t_1 \\ k \end{bmatrix}$ and $\begin{bmatrix} s_2, t_2 \\ l \end{bmatrix}$ are independent events.

Unfortunately, there is a problem with the definition of P above. How do we know that conditions (1), (2) and (3) do not imply some subtle contradiction? For the Bernoulli and the Uniform processes it was quite obvious that our definition gives a unique value for $P(A)$ no matter how the event A is written in terms of elementary events. This approach does not apply to the Poisson process, because there are many relationships among the elementary events, and it is not obvious that they are compatible with our definition of the probability measure P. We sketch the proof that the Poisson process is consistently defined in Section 7.4.

Sums of Independent Poisson Random Variables

The Poisson process has an important property we now discuss. Imagine that we have two independent Poisson processes of intensities α and β. To distinguish them, we color the blips of the first process red and the blips of the second process blue. Now suppose we are color-blind. What we will see is a Poisson process with intensity $\alpha + \beta$. The resulting Poisson process is called the *mixture* of two independent Poisson processes. Let $N_{red}(t)$ and $N_{blue}(t)$ be the random functions of the red and blue processes, respectively. We claim that $N_{red}(t) + N_{blue}(t)$ has a Poisson distribution with the parameter $(\alpha + \beta)t$.

Let us prove this. Since we assumed $N_{red}(t)$ and $N_{blue}(t)$ are independent, the distribution of the function $N_{red}(t) + N_{blue}(t)$ is the convolution of their individual distributions.

$$
\begin{aligned}
P(N_{red}(t) + N_{blue}(t) = n) &= \sum_{k=0}^{n} P(N_{red}(t) = k)\, P(N_{blue}(t) = n - k) \\
&= \sum_{k=0}^{n} \frac{(\alpha t)^k}{k!} e^{-\alpha t} \frac{(\beta t)^{n-k}}{(n-k)!} e^{-\beta t} \\
&= \frac{t^n}{n!} e^{-(\alpha+\beta)t} \sum_{k=0}^{n} \frac{n!}{k!(n-k)!} \alpha^k \beta^{n-k} \\
&= \frac{t^n}{n!} e^{-(\alpha+\beta)t} (\alpha + \beta)^n \\
&= \frac{((\alpha + \beta)t)^n}{n!} e^{-(\alpha+\beta)t}
\end{aligned}
$$

The second last expression follows from binomial theorem, and the last expression is the Poisson distribution with parameter $(\alpha + \beta)t$. Thus, *the sum of independent Poisson random variables is again Poisson*. This is an important property of the Poisson process, and we will find some important applications of it in later sections.

Physical Systems and the Poisson Process

We have already mentioned several examples of exponentially distributed random variables in Section 7.1. The Poisson process is a sequence of independent exponentially distributed random variables so we should not be surprised at its ubiquity.

Geiger Counters

The first example that comes to mind immediately is the sequence of clicks of a Geiger counter. If we are measuring the radiation of a radioactive sample, the clicks are almost the blips of a Poisson process. Of course, we know that the intensity α will gradually decrease as the sample decays. However, if we choose to measure time so that α becomes constant, the Poisson process is an almost perfect model of the physical system. Even if we measure time in the usual units, the model is very close.

Quality Control

Suppose we have a continuous assembly process (say of a rope or wire) and that the process occasionally produces tiny defects randomly on the rope. If we know the length of the rope and the number of defects, then the Uniform process is a good model of this system. On the other hand, if we know only the average number of defects per unit length (from prior experience), then the Poisson process is a better model. Even if we also know the length of the rope to be produced, the Poisson process is the better model.

One can use such models for Quality Control. If a long length of rope is being produced, one can sample portions of the rope to determine if the number of defects per unit length is exceeding a prespecified level of acceptance, as might happen if an assembly machine is out of adjustment.

For example, if the average density of the defects on the rope is $\frac{1}{10}$ defect/foot, then the probability of *no* defects on a rope of length 10 feet is $\exp\left(-\left(10\frac{1}{10}\right)\right) = e^{-1} \approx 0.368$. The probability of *exactly* two defects is $\frac{\left(\frac{1}{10}10\right)^2}{2} \exp\left(-\frac{1}{10}10\right) = \frac{1}{2}e^{-1} \approx 0.184$.

Blips from Space

Suppose one is aiming a radio telescope toward one direction in the sky. The signal one is receiving is a sequence of irregularly spaced radio bursts or "blips." Is the signal simply noise or is it a broadcast from some station? By comparing statistical properties of the blips with the known properties of the Poisson process one can distinguish the random noise from a broadcast signal, with a certain probability of error.

Seeds on a Cornfield

There is a two-dimensional model analogous to the Poisson process. For a region A in the plane, let $\mu(A)$ be the area of A. We replace the random function $N(t)$ by a random function $N(A) =$ number of blips in the region A. The fundamental assumptions on $N(A)$ are:

1. For every region A, $N(A)$ is an *integer* random variable, the number of blips occurring in the region.

2. $P(N(A) = k) = \dfrac{(\alpha\mu(A))^k}{k!} e^{-\alpha\mu(A)}$, i.e., $N(A)$ has the Poisson distribution with parameter $\alpha\mu(A)$.

3. If A and B are disjoint regions, then $N(A)$ and $N(B)$ are independent random variables.

Of course, we could produce a model of this kind in any number of dimensions.

An example of such a process is the process of sprinkling seeds from an airplane randomly onto a field with some intensity α, the average number of seeds per unit area. As another example, we might have stars spread randomly throughout a large volume of space with some average density α of stars per unit volume.

These are just a small selection of an enormous range of examples of the Poisson process occurring in nature. In fact, it is the most common of the three basic stochastic processes. We shall now see how the Poisson process can enrich our understanding of the other two stochastic processes; moreover, it will give us a powerful tool for computing probability distributions in these processes.

Gaps and Waiting Times

We must now check that in our model the distributions of the gaps and the waiting times correspond to our earlier computations. Consider first the waiting time W_k for the kth blip. The event $(W_k \geq t)$ means that the kth blip has not yet occurred by time t, or equivalently, that $k - 1$ or fewer blips occurred in the interval $[0, t]$. In terms of $N(t)$:

$$(W_k \geq t) = (N(t) = 0) \cup (N(t) = 1) \cup \cdots \cup (N(t) = k - 1)$$

The events of the right-hand side being disjoint, we may compute:

$$
\begin{aligned}
P(W_k \geq t) &= P(N(t) = 0) + P(N(t) = 1) + \cdots + P(N(t) = k - 1). \\
&= e^{-\alpha t} + \alpha t e^{-\alpha t} + \frac{(\alpha t)^2}{2} e^{-\alpha t} + \cdots + \frac{(\alpha t)^{k-1}}{(k-1)!} e^{-\alpha t}
\end{aligned}
$$

The density of W_k is the derivative:

$$\frac{d}{dt} P(W_k \leq t) = \frac{d}{dt}(1 - P(W_k > t)) = -\frac{d}{dt}(P(W_k > t))$$

Since the distribution function for W_k is continuous, we can ignore the distinction between $\geq$ and $>$ to get:

$$
\begin{aligned}
\mathrm{dens}(W_k = t) &= -\frac{d}{dt}(P(W_k \geq t)) \\
&= -\frac{d}{dt}\left(e^{-\alpha t} + \alpha t e^{-\alpha t} + \frac{(\alpha t)^2}{2} e^{-\alpha t} + \cdots + \frac{(\alpha t)^{k-1}}{(k-1)!} e^{-\alpha t} \right)
\end{aligned}
$$

When we differentiate this expression, each term except for the first gives rise to two terms:

$$
\begin{aligned}
-\mathrm{dens}(W_k = t) \ =\ & -\alpha e^{-\alpha t} \\[4pt]
& + \alpha e^{-\alpha t} - \alpha^2 t e^{-\alpha t} \\[4pt]
& + \alpha^2 t e^{-\alpha t} - \frac{\alpha^3 t^2}{2} e^{-\alpha t} \\[4pt]
& + \cdots \\[4pt]
& + \frac{\alpha^{k-1} t^{k-2}}{(k-2)!} e^{-\alpha t} - \frac{\alpha^k t^{k-1}}{(k-1)!} e^{-\alpha t}
\end{aligned}
$$

All the terms then cancel, except for the last one so that

$$
\mathrm{dens}(W_k = t) = \frac{\alpha^k t^{k-1}}{(k-1)!} e^{-\alpha t}
$$

which is the gamma distribution. In particular, the first waiting time (the first gap) is exponentially distributed.

To show that the gaps are independent and exponentially distributed with intensity α, we use the Continuous Law of Successive Conditioning and the Continuous Law of Alternatives in exactly the same way that we used the ordinary Law of Successive Conditioning and the ordinary Law of Alternatives to compute the distributions of the gaps of the Bernoulli process in Section 6.2. The verification is left as an exercise.

The Uniform Process from the Poisson Process

We built the model of the Poisson process by thinking of what happens to the Uniform process as the length of the interval gets larger. We can turn this around; conditioning the Poisson process to have exactly n blips in the interval $[0, a]$, produces the Uniform process. To see this, we compute the conditional probability:

$$
P(N(t) = k \mid N(a) = n) = \frac{P(N(t) = k, N(a) = n)}{P(N(a) = n)}
$$

Now $(N(t) = k) \cap (N(a) = n)$ says that n blips occur in $[0, a]$ and that k of them occur in $[0, t]$. In other words, k blips occur in $[0, t]$ and $n - k$ occur in $[t, a]$. These are disjoint intervals; therefore

$$
\begin{aligned}
P(N(t) = k \mid N(a) = n) \ &=\ \frac{P(N(t) = k, N(a) - N(t) = n - k)}{P(N(a) = n)} \\[8pt]
&=\ \frac{P(N(t) = k)\, P(N(a) - N(t) = n - k)}{P(N(a) = n)}
\end{aligned}
$$

$$= \frac{\frac{(\alpha t)^k}{k!} e^{-\alpha t} \frac{(\alpha(s-t))^{n-k}}{(n-k)!} e^{-\alpha(a-t)}}{\frac{(\alpha a)^n}{n!} e^{-\alpha a}}$$

$$= \frac{n!}{k!(n-k)!} \frac{t^k (a-t)^{n-k}}{a^n}$$

$$= \binom{n}{k} \left(\frac{t}{a}\right)^k \left(1 - \frac{t}{a}\right)^{n-k}$$

We recognize this as the distribution of $U_{n,a}(t)$. Since we can express all computations about order statistics in terms of the random function $U_{n,a}(t)$, we can, in principle, compute anything about order statistics by conditioning the Poisson process.

For example, we can compute the densities of the order statistics without using a limit argument as we did in Section 4.5.

$$\mathrm{dens}\big(X_{(k)} = t\big) \quad = \quad \mathrm{dens}(W_k = t \mid N(a) = n)$$

$$= \quad \frac{\mathrm{dens}(W_k = t, N(a) = n)}{P(N(a) = n)}$$

If you feel uneasy about the use of a mixture of a density and a probability in the above computation, just rewrite it as:

$$\mathrm{dens}(W_k = t, N(a) = n) = \frac{d}{dt} P(W_k \leq t, N(a) = n)$$

The event $(W_k < t)$ is the same as saying that at least k blips occur in $[0, t]$, i.e.,

$$(W_k < t) = (N(t) = k) \cup (N(t) = k + 1) \cup \cdots$$

On the other hand, if we know that the kth blip has occurred at time t, then $(N(a) = n)$ and $(N(a) - N(t) = n - k)$ are the same event. Therefore,

$$\mathrm{dens}\big(X_{(k)} = t\big) \quad = \quad \frac{\mathrm{dens}(W_k = t, N(a) - N(t) = n - k)}{P(N(a) = n)}$$

$$= \quad \frac{\mathrm{dens}(W_k = t)\, P(N(a) - N(t) = n - k)}{P(N(a) = n)}$$

$$= \quad \frac{\alpha^k t^{k-1}}{(k-1)!} e^{-\alpha t} \frac{\alpha^{n-k}(a-t)^{n-k}}{(n-k)!} e^{-\alpha(a-t)} \left(\frac{\alpha^n a^n}{n!} e^{-\alpha a}\right)^{-1}$$

$$= \quad \frac{n!}{(k-1)!(n-k)!} \frac{t^{k-1}(a-t)^{n-k}}{a^n}$$

$$= \quad \binom{n-1}{k-1} \frac{n}{a} \left(\frac{t}{a}\right)^{k-1} \left(1 - \frac{t}{a}\right)^{n-k}$$

Notice that when a Poisson process is conditioned by the event $(N(a) = n)$, the result is always the Uniform process of sampling n points from $[0, a]$, no matter what the intensity was in the original Poisson process. This is further confirmation that the Poisson process is the process of sprinkling points at random on $[0, \infty)$.

Bernoulli	Poisson	Uniform
p = bias	α = intensity or average number of blips per unit interval	n = number of points a = length of interval $\frac{n}{a}$ = intensity
$X_i = i^{th}$ toss independent equidistributed Bernoulli distribution		$X_i = i^{th}$ point independent equidistributed Uniform distribution
S_n = number of heads in first n tosses Binomial distribution	$N(t)$ = number of blips in interval $[0, t)$ Poisson distribution	$U(t)$ = number of points in interval $[0, t)$ Binomial distribution
$W_k = k^{th}$ waiting time Negative binomial distribution	$W_k = k^{th}$ waiting time Gamma distribution	$X_{(k)} =$ k^{th} order statistic Beta distribution
$T_i = i^{th}$ gap independent equidistributed Geometric distribution	$T_i = i^{th}$ gap independent equidistributed Exponential distribution	$L_i = i^{th}$ gap not independent exchangeable Beta distribution

Table of Analogies: Bernoulli, Poisson and Uniform Processes

Distribution	Type	Parameter(s)	Model(s)
Exponential	continuous	α	W_1 or any T_k
Gamma	continuous	α, k	W_k
Poisson	integer	λ	$N(t)$ where $\lambda = \alpha t$

Distribution	Mean	Variance	Density or Probability
Exponential	$1/\alpha$	$1/\alpha^2$	$f(t) = \alpha e^{-\alpha t}$
Gamma	k/α	k/α^2	$f(t) = \dfrac{\alpha^k t^{k-1}}{(k-1)!} e^{-\alpha t}$
Poisson	λ	λ	$p_k = \dfrac{\lambda^k}{k!} e^{-\lambda}$

Table of Poisson Distributions

7.4 Consistency of the Poisson Process

Unlike the other basic processes, it is not obvious that the Poisson process is consistently defined. The difficulty is that there are many relationships among the elementary events. In this section, we examine some of these relationships, and show that the definition of the Poisson probability measure is consistent with these relationships.

For example, consider the event $\overline{\begin{bmatrix} s,t \\ 0 \end{bmatrix}}$, i.e., the event that one or more blips occur in the interval $[s,t)$. We could also write this as:

$$\overline{\begin{bmatrix} s,t \\ 0 \end{bmatrix}} = \begin{bmatrix} s,t \\ 1 \end{bmatrix} \cup \begin{bmatrix} s,t \\ 2 \end{bmatrix} \cup \cdots$$

The probability of the event on the left-hand side above is:

$$P\left(\overline{\begin{bmatrix} s,t \\ 0 \end{bmatrix}}\right) = 1 - P\left(\begin{bmatrix} s,t \\ 0 \end{bmatrix}\right)$$

$$= 1 - e^{-\alpha(t-s)}$$

On the other hand, because the events $\begin{bmatrix} s,t \\ 1 \end{bmatrix}$, $\begin{bmatrix} s,t \\ 2 \end{bmatrix}$, $\ldots$ are disjoint, the probability of the right-hand side is the following:

$$P\left(\begin{bmatrix} s,t \\ 1 \end{bmatrix}\right) + P\left(\begin{bmatrix} s,t \\ 2 \end{bmatrix}\right) + \cdots = e^{-\alpha(t-s)}\left(\alpha(t-s) + \frac{(\alpha(t-s))^2}{2} + \cdots\right)$$

Now the second factor above is the same as the Taylor series expansion of $e^{\alpha(t-s)}$, except for the first term so the sum reduces to:

$$e^{-\alpha(t-s)}\left(e^{\alpha(t-s)} - 1\right) = 1 - e^{-\alpha(t-s)}$$

So we get the same answer either way, although it was not entirely obvious.

As another example, if $r \leq s \leq t$, then k blips occur in $[r,t)$ if and only if some occur in $[r,s)$ and the rest occur in $[s,t)$. In symbols:

$$\begin{bmatrix} r,t \\ k \end{bmatrix} = \bigcup_{l=0}^{k}\left(\begin{bmatrix} r,s \\ l \end{bmatrix} \cap \begin{bmatrix} s,t \\ k-l \end{bmatrix}\right)$$

We can therefore compute $\begin{bmatrix} r,t \\ k \end{bmatrix}$ in two ways. The first way is:

$$P\left(\begin{bmatrix} r,t \\ k \end{bmatrix}\right) = \frac{(\alpha(t-r))^k}{k!}e^{-\alpha(t-r)}$$

The second way is the following. It is quite complicated and requires all of our assumptions on P.

$$P\left(\begin{bmatrix} r,t \\ k \end{bmatrix}\right) = P\left(\bigcup_{l=0}^{k}\left(\begin{bmatrix} r,s \\ l \end{bmatrix} \cap \begin{bmatrix} s,t \\ k-l \end{bmatrix}\right)\right)$$

$$= \sum_{l=0}^{k} P\left(\begin{bmatrix} r,s \\ l \end{bmatrix}\right) P\left(\begin{bmatrix} s,t \\ k-l \end{bmatrix}\right)$$

$$= \sum_{l=0}^{k} \frac{(\alpha(s-r))^l}{l!} e^{-\alpha(s-r)} \frac{(\alpha(t-s))^{k-l}}{(k-l)!} e^{-\alpha(t-s)}$$

$$= \sum_{l=0}^{k} \frac{\alpha^l((s-r))^l}{l!} \frac{\alpha^{k-l}((t-s))^{k-l}}{(k-l)!} e^{-\alpha(t-r)}$$

$$= e^{-\alpha(t-r)} \sum_{l=0}^{k} \frac{\alpha^k}{l!(k-l)!} ((s-r))^l((t-s))^{k-l}$$

$$= \alpha^k e^{-\alpha(t-r)} \sum_{l=0}^{k} \frac{1}{l!(k-l)!} ((s-r))^l((t-s))^{k-l}$$

$$= \frac{\alpha^k}{k!} e^{-\alpha(t-r)} \sum_{l=0}^{k} \frac{k!}{l!(k-l)!} ((s-r))^l((t-s))^{k-l}$$

$$= \frac{\alpha^k}{k!} e^{-\alpha(t-r)} \sum_{l=0}^{k} \binom{k}{l} ((s-r))^l((t-s))^{k-l}$$

$$= \frac{\alpha^k}{k!} e^{-\alpha(t-r)} ((s-r)+(t-s))^k \quad \text{by the binomial theorem}$$

$$= \frac{\alpha^k(t-r)^k}{k!} e^{-\alpha(t-r)}$$

Once again, we get the same answer either way, although it was even more complicated this time.

Are there other relations among the events $\begin{bmatrix} s,t \\ k \end{bmatrix}$ not obtainable from the above two examples? The answer is no, but this is not easy to prove, and we leave this as an exercise. In any case, we have now shown that P is consistently defined. One can, in fact, show that in some sense every possible probability on the Poisson sample space is a perturbation of the probability P we have just defined. The simplest example of such a perturbation is a process in which α varies in time.

7.5 Exercises

Exercise 7.1 You are the captain of the *Bicentennial Eagle*, a spaceship that has just returned from hyperspace to ordinary space, only to encounter the debris of a recently destroyed planet. The debris consists of approximately spherical rocks 20 m in radius. The destroyed planet was originally the same size as the earth, and its debris is now uniformly scattered throughout a region 10^6 km in radius. Your maneuvering jets are temporarily out of order. If you are headed directly toward the center of the debris, what are your chances of getting all the way through the debris without a collision? Assume that your ship has a circular cross-section of radius 10 m. Explain any assumptions you may be making.

Exercise 7.2 At 5×10^4 km/hour, how long would you have in Exercise 7.1 to repair your maneuvering jets before your chances of collision reach 10%? Explain precisely what you are computing in this problem.

Exercise 7.3 A beam of protons is accelerated to a high-energy level and is deflected so that it encounters a pool of liquid hydrogen. The tracks of the protons in the beam are visible in this detector, and one can easily see where a proton in the beam collides with a proton in the pool of liquid hydrogen. Describe how far a given proton travels before it collides with a proton in the pool.

Exercise 7.4 There exist enzymes that attack only a certain nucleotide sequence in a chromosome. Describe a means of testing whether or not a given nucleotide sequence appears randomly in a given chromosome.

Exercise 7.5 In the birthday coincidence problem (Exercise 2.8), the paradox comes from thinking that one is looking for another person with the same birthday as *your* birthday. Compute the distribution of the number of persons chosen at random you must ask until you find one with the same birthday as yours. What kind of distribution is it? Find an exponential distribution that approximates it. Compute the average value of this random variable as well as the number of persons one must ask in order to have a 50% chance of finding one with your birthday.

Exercise 7.6 How does the answer to Exercise 7.5 change if we include February 29th as a possible birthdate?

Exercise 7.7 Roughly speaking, the relationship between the birthday problem in Exercise 7.5 and the birthday coincidence problem in Exercise 2.8 is that in either case we have a certain number of *pairs* of persons from which we look for a birthday coincidence, but that in the former problem we consider a collection of pairs all of which have one given person in common, whereas in the latter we consider all pairs from a set of persons. For example, we saw that in a class of 23 students there is about a 50% chance of a birthday coincidence. Such a class has $\binom{23}{2} = \frac{23\cdot22}{2} = 253$ pairs of students. Compare this with the last part of your answer to Exercise 7.5.

Exercise 7.8 In a large class the students call out their birthdays until someone in the class finds that his or her birthday has been called. Technically, this is not a random variable since it is possible that no pair of students have the same birthday. However, if we assume that a match will eventually be found, then it is a random variable which is approximately exponentially distributed. Compute the mean of this exponential distribution. Compare with Exercise 3.25.

Exercise 7.9 A sociobiologist wishes to test whether or not birds of a certain species practice territorial spacing of their nest locations. Compute the distribution of the distance of a given nest from its nearest neighbor. Use this to formulate a statistical test.

Exercise 7.10 Let $Y_1, Y_2, \ldots, Y_n$ be n independent, exponentially distributed random variables, each of which has intensity α. Compute the order statistics $Y_{(1)} \leq Y_{(2)} \leq \ldots \leq Y_{(n)}$ of these random variables. One can do this in two ways. Either change variables and convert to a Uniform process (see Section 6.7) or use a modification of the reasoning used in Exercise 4.20 which was made rigorous in Section 6.6 ("needles on a stick problem").

Exercise 7.11 Compute the expectations $E(Y_{(i)})$ in Exercise 7.10 above. Compare with the expectations of the order statistics of the gaps in the Uniform process (Exercise 4.20).

Exercise 7.12 (Feller) Three persons, A, B and C, arrive at a post office simultaneously. There are two counters, and these are taken immediately by A and B. Assume the service time of a given individual is exponentially distributed with rate parameter α. Assume also that different individuals have independent service times. Now C will take whichever of the two counters becomes free first. Use the answers to Exercise 7.10 and Exercise 7.11 (if necessary) to answer the following questions:

1. What is the probability that C is the last to leave the post office?

2. What is the distribution of the total time (i.e., both waiting time and service time) spent by C in the post office? How long does C spend on average?

3. When does the last of the three persons leave the post office? When on average?

Exercise 7.13 (Bad luck) Two persons A and B arrive simultaneously at the post office in Exercise 7.12 above. They find that one person is being served at each of the two counters. Of course, A waits for one counter to become free and B waits for the other. Assume that line changing is forbidden. What is the distribution of the ratio between A's waiting time and B's waiting time? Show that the average value is infinite! Thus we have a situation where both A and B, on average, experience bad luck indeed. How can this be?

Exercise 7.14 A textbook has 250 pages and has many misprints. If you assume that the misprints are uniformly distributed, would you be surprised to find 4 misprints on one page? How about 8 misprints on a single page? Why and/or why not?

Exercise 7.15 Ladislaus Bortkiewicz in 1898 published the number of deaths from horse kicks among Prussian cavalry soldiers from 1875 to 1894. The observed frequencies were the following:

Number of deaths in one month	Frequency
0	109
1	65
2	22
3	3
4	1
5 or more	0

Fit these data with a Poisson distribution. The parameter for the Poisson distribution can be estimated by computing the sample mean of the number of deaths.

Exercise 7.16 On average, a secretary makes two errors per page. What is the probability that he makes five or more errors on the next page he types? A random page of this secretary's work is examined and found to have five errors. Would this be sufficient evidence for believing (at a 5% significance level) that he makes more than two errors per page on average?

Exercise 7.17 The IRS finds that numerical mistakes occur in 1% of all tax returns on average. A random sample of 1000 forms is chosen from the forms submitted by taxpayers in one particular tax bracket. Only three are found to have a numerical mistake. Is there a reason to believe (at a 5% significance level) that taxpayers in that bracket are more careful in preparing their tax returns than the average taxpayer?

Exercise 7.18 A small insurance company sells just one kind of insurance. Every year they charge each client \$100, and each client receives \$1,000,000 with probability 5×10^{-5}. The company has 20,000 clients and assets of only \$1,000,000. What is the probability that the company will go bankrupt in a given year?

Exercise 7.19 In a racially integrated school, 15% of the students are non-white, and the rest are white. What is the probability that a randomly chosen class of 50 students is all-white?

Exercise 7.20 The manager of a pharmacy orders aspirin once each week. Experience shows that 50 bottles of aspirin are sold each week on average. If she wants to minimize her stock of aspirin but also wants to be 80% confident that she will not run out in a given week, how much aspirin should she have on hand at the beginning of each week? Use a Poisson model.

Exercise 7.21 An exterminator company claims that its insecticide will kill a given flea with probability 99.9% and a flea egg with probability 99%. If there are about 100 fleas and 100 flea eggs in a house, what is the probability that the exterminator company will solve the flea problem for the house?

Exercise 7.22 In a city of 2,000,000, a man was brought to court on a change of larceny. Although the prosecution could not establish that the defendant was certainly the guilty party, it did establish that a number of traits were common to the defendant and to the thief. These included the make and model of his car, the color of his and his girlfriend's hair, etc. The prosecution estimated the probability of each of the traits in question. Although there is very little reason to believe that such traits will be independent of one another, the court allowed the prosecutor to assume this. The prosecutor then multiplied the various probabilities and concluded that the chance of anyone having these traits is 1 in 5,000,000. The jury was duly impressed and the defendant was convicted.

The case was appealed, and the defense brought in a statistician. The statistician pointed out that the prosecutor had asked the wrong question. One should ask what the probability is that this is the guilty party beyond a reasonable doubt. The statistician computed the answer to be about 5/6, which is not overwhelming. Calculate this probability exactly.

Exercise 7.23 (Luria and Delbruck) When a bacterial culture is attacked by a bacteriophage almost all the bacteria are killed, but a few resistant bacteria survive. This resistance is hereditary, but it might be due either to random, spontaneous mutations occurring before the attack by the bacteriophage, or to mutations directly induced by this attack. In the latter case, the number of resistant bacteria in different, but similar, cultures should follow the Poisson distribution, while in the former case, the variance should be larger than the mean since the number of resistant bacteria will depend markedly on whether a mutation occurs early or late in the life of the culture. The following numbers of resistant bacteria were found in ten samples from the same culture: 14, 15, 13, 21, 15, 14, 26, 16, 20, 13, while the following were found in ten samples from different cultures grown in similar conditions: 6, 5, 10, 8, 24, 13, 165, 15, 6, 10. Compare the mean with the variance in the two sets of observations. Which hypothesis does the experiment tend to support?

Exercise 7.24 An entomologist is studying the habits of a leaf-eating caterpillar. He wishes to test whether the insects choose the leaves on which they eat in a random (Poisson) manner. Accordingly, he chooses 200 leaves at random and counts the number of insects on each leaf. This is what he found:

Number of insects per leaf	0	1	2	3	4 or more
Number of leaves	140	48	10	2	0

Does this fit a Poisson distribution? Compare with the horse-kick problem of Exercise 7.15.

Exercise 7.25 A statistician looking at the data in Exercise 7.24 notices that it would fit a Poisson distribution if the number of leaves having no insects on them were slightly smaller. She suggests that one should use a modified Poisson model in which some leaves have no insects on them because they are not suitable for the insect to feed on and not just because of random effects. Fit such a model to the data.

Exercise 7.26 (Gosset) A sample of 400 unit volumes containing various numbers of yeast cells were examined and the number of yeast cells in each volume was counted. Here are the counts:

0	1	2	3	4	5	6	7	8	9	10	11	12	13+
0	20	43	53	86	70	54	37	18	10	5	2	2	0

The first line is the number of cells per unit volume. The second line is the number of unit volumes with the count on the first line. Compare these data with a suitable Poisson distribution.

Exercise 7.27 The fish population problem (Exercise 2.22 and Exercise 5.46) can also be modeled with the Poisson process. If we suppose that there are n fish in the lake and that we have tagged 50 of them, then we can model our subsequent catch of 50 fish with a Poisson process in which $\alpha = \frac{50}{n}$ is the density of tagged fish in the lake, $t = 50$, and $\lambda = \alpha t = \frac{2500}{n}$. Compare this model with the one in Exercise 5.46.

Exercise 7.28 (Mosteller and Rourke) In a series of essays called the *Federalist Papers*, Alexander Hamilton and James Madison discussed many of the most important political questions facing the United States at that time. Although scholars had been able to agree on the authorship of most of the *Federalist Papers*, there remained a small set of them which remained in dispute until recent times. The issue was finally settled by using statistical rather than traditional scholarly methods. We will give some of the flavor of these methods in this problem. In the actual analysis, the frequency counts of many different words were compared.

Suppose that the number of occurrences of the word "also" in 2000-word blocks in the writings of Hamilton and Madison are known to have approximately Poisson distributions with means 0.6 and 1.3, respectively. A 2000-word passage, known to have been written by one of them, is examined. Formulate a statistical test which will distinguish at the 10% level between whether Hamilton or Madison wrote the passage.

Exercise 7.29 (The waiting time paradox) A Geiger counter is set up to examine a small radioactive sample which produces 1 click every hour on average. In addition, the machine has a timer which shows how long it has been since the last click or since the timer has last been reset manually. The Geiger counter has been running unobserved for some days when you first observe it. At this time, you push the reset button but at the same time you write down what the timer indicated (call this T_0). You then wait until the next time the Geiger counter clicks, and you also write down what the timer indicates when this happens (call this T_1).

There are 2 ways to compute the expectations of T_0 and T_1. First, $T_0 + T_1$ is the total waiting time between 2 successive clicks of the Geiger counter, so that $E(T_0 + T_1)$ should be 1 hour. Since your arrival may be regarded as randomly dividing in 2 the total time between the 2 successive clicks, $E(T_0)$ should be the same as $E(T_1)$, and hence, $E(T_0)$ and $E(T_1)$ both should be 30 minutes. The second analysis begins by noting that because the exponential distribution is memoryless, the waiting time T_1 is itself exponentially distributed with mean 1 hour. Therefore, whatever the distribution of T_0 is, we have $E(T_1) = 1$ hour and $E(T_0 + T_1) > 1$ hour. Thus, the 2 analyses disagree. Compute the actual distributions of T_0 and T_1, and find their expectations. Which of the above analyses is correct? Refer to the "reference point paradox" in Section 4.5, and formulate an explanation of the waiting time paradox.

Exercise 7.30 Write a program in R to test the 2 explanations of the waiting time problem in Exercise 7.29. Assume that you arrive at the Geiger counter exactly 100 hours after it has been turned on. Run the same experiment many times and graph your results.

Exercise 7.31 Prove that the gaps of the Poisson process are independent exponentially distributed random variables with intensity α.

7.6 Answers to Selected Exercises

Answer to 7.1. Assume that the *Bicentennial Eagle* enters ordinary space at the "edge" of the debris, so that it must travel along a diameter of the region without colliding with a rock. Also ignore the motion of the debris so that it can be regarded as a uniform distribution of rocks in a spherical region. To get through the debris, no rocks may be in a cylindrical region 2×10^6 km long and 10 m in radius. Because the rocks have a 20 m radius, this means that no centers of rocks may be in a cylindrical region 2×10^6 km long and 30 m in radius. The volume of this region is 5.65×10^{12} m^3. To compute the density of the centers of rocks in the entire spherical region we first compute the number of rocks. This is easily computed by dividing the volume of the original planet by the volume of one rock:

$$\frac{\frac{4}{3}\pi \left(6.4 \times 10^6 \text{m}\right)^3}{\frac{4}{3}\pi (20\text{m})^3} = \frac{\left(6.4 \times 10^6\right)^3}{(20)^3}$$
$$= 3.28 \times 10^{16}$$

The density of the rocks in the region is then the ratio of this number to the volume of the region:

$$\frac{3.28 \times 10^{16}}{\frac{4}{3}\pi (1 \times 10^9 \text{m})^3} = 7.83 \times 10^{-12} \text{ rocks/m}^3$$

We use the Poisson process with intensity $\alpha = 7.83 \times 10^{-12}$ rocks/m^3 to model this situation. Thus, the probability of finding no rocks in a volume v is $e^{-\alpha v}$. In this case, we can compute the probability of avoiding a collision to be

$$\exp\left(-\left(7.83 \times 10^{-12} \text{ rocks/m}^3\right)\left(5.65 \times 10^{12}\text{m}^3\right)\right) \approx e^{-44} \approx 10^{-19}$$

Things look very bleak for the *Bicentennial Eagle*. The average number of collisions would be about 44, assuming that the ship could survive a collision at such a high speed.

Answer to 7.2. Following the reasoning in the answer to Exercise 7.1, the

chances of a collision reach 10% when $e^{-\alpha v} = 0.90$ or when $\alpha v \approx 0.1$. This
gives $v \approx 1 \times 10^{10} \mathrm{m}^3$, which seems like a lot of room. It represents a cylinder
of radius 30 m and length 3500 km. Unfortunately, the *Bicentennial Eagle*
is moving very rapidly: 5×10^4 km/hour. It will travel 3500 km in around 4
minutes or so.

Note that our conclusions were based on an unconditional probability. If 4
minutes go by without a collision, then the probability of traveling another
4 minutes without a collision is again 90%. In any case, these probabilities
"mount up," and, as we saw in our first computation above, there is essentially
no chance for the ship to travel all the way through the debris field without
a collision. We strongly advise the crew to repair the ship quickly: in about
half an hour the probability of collision reaches 50%.

Answer to 7.3. Let X be the distance a proton travels into the pool. As
in Exercise 7.1, the event $(X > x)$ means that a cylindrical region of volume
$\pi(2r)^2 x$ is devoid of proton centers, where r is the radius of a proton. Thus,

$$P(X > x) = \exp\left(-\alpha \pi (2r)^2 x\right)$$

where α is the density of protons in the pool.

Unfortunately, this reasoning assumes that protons are tiny hard spheres,
a description we know is wrong. On the other hand, we may simply define r
by this formula, and then measure r experimentally. In practice, one speaks
of the "cross-sectional area" of a proton with respect to such a collision.

A further complication is that the cross-sectional area of a proton depends
on its velocity. Fortunately, accelerators deliver protons in a very narrow energy range so this is not a problem. We do, however, have to consider that
the protons lose energy (i.e., slow down) in the pool. In any case, one begins
to see that while there are many technical details to dispose of, probability
allows one to compute a very subtle quantity: the size of a proton.

Answer to 7.4. If we assume that the nucleotide sequence occurs randomly,
then the Uniform process is a good model, but this model can be too difficult
for computational purposes. The Poisson process is an excellent approximation
when the number of such sequences in one chromosome is large (as it generally
would be). Thus, we could test randomness by allowing the enzyme to cleave
the chromosome at every appropriate sequence. Then we collect the fragments
and compare their distributions with an exponential distribution. Of course,
many other tests can be devised. This is just one possible test.

Answer to 7.10. Let $Y_1, Y_2, \ldots, Y_n$ be independent, exponentially distributed random variables with rate parameter α. Let $Y_{(1)} \leq Y_{(2)} \leq \cdots \leq Y_{(n)}$
be the order statistics of the Y_j's. Now it is easy to verify that if Y is exponentially distributed with rate parameter α, then $X = 1 - e^{-\alpha Y}$ is uniformly
distributed on the interval $[0, 1]$ (cf. Section 6.7). This transformation is order-
preserving: $Y_1 \leq Y_2$ if and only if $1 - \exp(-\alpha Y_1) \leq 1 - \exp(-\alpha Y_2)$. Thus, if we
set $X_i = 1 - \exp(-\alpha Y_i)$, then $X_{(i)} = 1 - \exp(-\alpha Y_{(i)})$. Let $F_{(i)}(x)$ be the

distribution of $X_{(i)}$. Then

$$P\big(Y_{(i)}{\leq}y\big) = P\big(1 - \exp\big(-\alpha Y_{(i)}\big){\leq}1 - e^{-\alpha y}\big)$$
$$= P\big(X_{(i)}{\leq}1 - e^{-\alpha y}\big) = F_{(i)}\big(1 - e^{-\alpha y}\big)$$

So the density of $Y_{(i)}$ can be computed by the chain rule:

$$
\begin{aligned}
\mathrm{dens}\big(Y_{(i)} = y\big) &= F'_{(i)}\big(1 - e^{-\alpha y}\big)\alpha e^{-\alpha y} \\
&= n\binom{n-1}{i-1}\big(1 - e^{-\alpha y}\big)^{i-1}\big(e^{-\alpha y}\big)^{n-i}\alpha e^{-\alpha y} \\
&= \alpha n\binom{n-1}{i-1}\big(1 - e^{-\alpha y}\big)^{i-1}\exp\big(-\alpha(n-i+1)y\big)
\end{aligned}
$$

Answer to 7.12. By the memorylessness of the exponential distribution, the probability that C is the last to leave is $1/2$. The expected waiting time for C to get service is easily seen to be $\dfrac{1}{2\alpha}$. The expected service time is $\dfrac{1}{\alpha}$, so the expected time in all is $\dfrac{3}{2\alpha}$. The expected time for the last person to leave is $\dfrac{2}{\alpha}$.

Answer to 7.14. This problem is rather vaguely worded, so there are several possible interpretations. If we assume that there are exactly 250 misprints on the 250 pages, then the sampling process of placing 250 objects in 250 containers is the appropriate model. But if the assumption is that the average density of misprints is 1 misprint per page (and not that there are exactly 250 misprints), then the appropriate model is the Poisson process. Another ambiguity is whether the question is asking for the probability of exactly 4 misprints on one page or the probability of 4 or more misprints. Yet another ambiguity is whether the question is about a particular page or about any page in the book. Both the sampling process and the Poisson process are available in R, so we can use either one for the computations. Let M_n be the number of misprints on page n. First, consider the question of the misprints on a specific page. In the sampling process, every M_n is the same as S_{250} in the Bernoulli process with bias $1/250$:

```
> dbinom(4, 250, 1/250)
[1] 0.01517449
```

To compute $P(M_n{\geq}4)$, we write it in terms of the probability distribution function of S_{250}:

$$P(M_n{\geq}4) = 1 - P(S_{250}{\leq}3)$$

This is computed in R with the `pbinom` function:

```
> 1 - pbinom(3, 250, 1/250)
[1] 0.01874274
```

The computations for the Poisson process interpretation are:

```
> dpois(4, 1)
[1] 0.01532831
> 1 - ppois(3, 1)
[1] 0.01898816
```

Notice that the results are nearly the same for either process, as we expected. Furthermore, it also does not matter too much whether we interpret the question as asking for exactly 4 misprints or for at least 4 misprints. In all cases, finding this many misprints is a significant result (i.e., an event with probability less than 5%) but not very significant (i.e., an event with probability less than 1%). For the case of 8 errors the probabilities are:

```
> dbinom(8, 250, 1/250)
[1] 8.396622e-06
> 1-pbinom(7, 250, 1/250)
[1] 9.399465e-06
> dpois(8,1)
[1] 9.123994e-06
> 1-ppois(7,1)
[1] 1.024920e-05
```

Again, all of the values are nearly the same. In all cases finding 8 errors on a page would be surprising in the sense of being a very significant result.

Now consider the alternative interpretation in which we are asking whether *at least* 1 page has 4 misprints. Let A be this event:

$$A = \bigcup_{n=1}^{250} (M_n = 4)$$

In the sampling process, the numbers of misprints on different pages are not independent so the computation must use the Inclusion-Exclusion Principle in Section 3.5:

$$
\begin{aligned}
P(A) &= \sum_n P(M_n = 4) \\
&\quad - \sum_{n_1 < n_2} P(M_{n_1} = 4, M_{n_2} = 4) + \cdots \\
&= \binom{250}{1} P(M_1 = 4) \\
&\quad - \binom{250}{2} P(M_1 = 4, M_2 = 4) + \cdots \quad \text{by exchangeability} \\
&= \binom{250}{1}\binom{250}{4}\left(\frac{1}{250}\right)^4 \left(\frac{249}{250}\right)^{246} \\
&\quad - \binom{250}{2}\binom{250}{4}\binom{246}{4}\left(\frac{1}{250}\right)^8 \left(\frac{249}{250}\right)^{242} + \cdots
\end{aligned}
$$

Even with the use of exchangeability in the computation above, it is clear that the formula is going to be very complicated. It gets much worse for the computation of the event

$$B = \bigcup_{n=1}^{250} (M_n \geq 4)$$

that at least 1 page has at least 4 misprints. Accordingly, we will only consider the Poisson approximation. The fact that the pages are now independent simplifies the computation enormously:

$$P\left(\bigcup_{n=1}^{250} (M_n = 4)\right) = 1 - P\left(\bigcap_{n=1}^{250} (M_n \neq 4)\right)$$

$$= 1 - P\left(\bigcap_{n=1}^{250} ((N(n) - N(n-1)) \neq 4)\right)$$

$$= 1 - \prod_{n=1}^{250} P((N(n) - N(n-1)) \neq 4)$$

$$= 1 - \prod_{n=1}^{250} P(N(1) \neq 4)$$

$$= 1 - P(N(1) \neq 4)^{250}$$

$$= 1 - (1 - P(N(1) = 4))^{250}$$

Now compute this in R:

```
> 1 - (1-dpois(4,1))^250
[1] 0.9789689
```

The computation for at least 4 misprints is similar so we just show the result:

$$P\left(\bigcup_{n=1}^{250} (M_n \geq 4)\right) = 1 - P(N(1) \leq 3)^{250}$$

Now compute this in R:

```
> 1 - ppois(3,1)^250
[1] 0.9917098
```

Therefore, it is nearly certain that some page will have at least 4 misprints. In the case of 8 misprints, the computations are:

```
> 1 - (1-dpois(8,1))^250
[1] 0.002278409
> 1 - ppois(7,1)^250
[1] 0.002559032
```

Unlike the case for 4 errors, we would be surprised to find 8 errors. It would be a reason for rejecting our model.

Answer to 7.18. The company receives $20,000 \times 100 = \$2,000,000$ in income annually, and has assets of $1,000,000, so it goes bankrupt (i.e., has no assets) if 3 or more clients receive the $1,000,000. The probability of this is

```
> 1-pbinom(3,20000,.00005)
[1] 0.01898509
```

Answer to 7.19. Using a Poisson process with $\alpha = .15$, we find that

$$P(N(20) = 0) = e^{-20(0.15)} = e^{-3} \approx 0.0498$$

Thus, there is a 1 in 20 chance for such a class to occur, which is not over-whelmingly likely. The Poisson process, of course, is only an approximation. Either the Bernoulli process or a sampling process would be more accurate. For example, using the Bernoulli process, the probability is $(0.85)^{20} \approx 0.0388$, which is roughly the same.

Answer to 7.20. She should stock

```
> qpois(0.8,50)
[1] 56
```

bottles.

Answer to 7.21. About 82%.

Answer to 7.22. The key question is the interpretation of "beyond a reasonable doubt." We will take this to mean that there is a high probability that the defendant *is* the guilty party, not that there is a high probability that the defendant *matches the description* of the guilty party. Thus, if there are 2 or more people who were in or near the city at the time of the crime and who match the description of the guilty party, then the defendant is *not* guilty beyond a reasonable doubt, because we do not know which of them committed the crime. In a city of 2,000,000, the density of people matching the description will be

$$\frac{2 \times 10^6}{5 \times 10^6} = 0.4$$

Using a Poisson process with $\alpha = 0.4$, the chance that there are two or more people matching the description, given that at least one exists, is

$$
\begin{aligned}
P(N(1) \geq 2 \mid N(1) \geq 1) &= \frac{P(N(1) \geq 2)}{P(N(1) \geq 1)} \\
&= \frac{1 - P(N(1) \leq 1)}{1 - P(N(1) = 0)}
\end{aligned}
$$

We now compute it:

```
> (1-ppois(1,0.4))/(1-ppois(0,0.4))
[1] 0.1867021
```

Accordingly, the defendant is guilty beyond a reasonable doubt with probability 0.8132979, which is not very convincing.

Answer to 7.28. One can never reject the hypothesis that Madison wrote the passage, but if there are three or more occurrences of "also" one can reject the hypothesis that Hamilton did.

CHAPTER 8

Randomization and Compound Processes

Pierre-Simon, Marquis de Laplace was a French mathematician who made outstanding contributions to pure mathematics, probability theory and astronomy. During his lifetime, France went through a series of upheavals starting with the French Revolution, and including a series of revolutionary governments, the republic, the empire and the Bourbon restoration. Laplace saw this as an opportunity and took advantage of the chaos by shifting his political allegiance whenever it was convenient. Each successive government rewarded his services with numerous honors. Napoleon made him a count, and Louis XVII made him a marquis. While he was politically inconsistent, Laplace's devotion to science was unwavering. His contribution to celestial mechanics was so substantial that Siméon Poisson described him as the Isaac Newton of France, for Laplace's results were a brilliant vindication of the principle of universal gravitation. Among some of his accomplishments, he proved our solar system is a stable dynamical system. When Laplace began working in probability theory, it was largely combinatorial and was only applied to games of chance. Laplace was the first to apply calculus to probability theory. The basic concept employed by Laplace in his analysis was the generating function. Generating functions were used by Laplace not only in the solution of probability problems but also in applications to problems in pure mathematics, such as interpolation, the solution of finite difference equations, and the expression of functions in terms of definite integrals, such as what we now call the Laplace transform. Another important contribution of Laplace was the extension of the work of the Reverend Thomas Bayes.

Laplace summarized probability theory as being, "basically nothing but common sense reduced to calculus; it enables us to appreciate with exactness that which accurate minds feel with a sort of instinct for which ofttimes they are unable to account." However, as we have seen, and will continue to see,

241

there are many paradoxes of probability that arise from having a false intuition. To reach a level in which probability is truly "common sense" requires that we "train" our instinct until it becomes intuition.

In this chapter we examine the close connection between stochastic processes and transforms such as the Laplace transform. The connection uses a method for creating new stochastic processes. Randomization is a general term for the method whereby new stochastic processes are created by allowing a parameter of a given stochastic process to be chosen randomly according to some distribution. In this section, we consider the randomized forms of the fundamental stochastic processes. However, the possibilities for randomizations are endless: any parameter of any process can be randomized, including the parameters of randomized processes.

8.1 Randomized Bernoulli Process

The simplest fundamental process is the Bernoulli process. To randomize this process, we allow the bias to be a random variable P instead of a constant p. This is the appropriate process to use when we do not know the bias of the coin. The bus stop problem in Exercise 2.15 is an example of this. Since we do not know anything about the bus schedule, we suppose that the probability of the man going uptown will be a random variable P, where P is chosen uniformly from $[0, 1]$.

Let X_i be the ith toss of a coin in the **randomized Bernoulli process**. By the Continuous Law of Alternatives,

$$
\begin{aligned}
P(X_i = 1) &= \int_0^1 P(X_i = 1 \mid P = p) \, \mathrm{dens}(P = p) \, dp \\
&= \int_0^1 p \, dp = \frac{1}{2}
\end{aligned}
$$

This comes as no surprise. If the bias is completely random, we would intuitively expect that the toss of a coin will be also. For once, our intuition is correct.

We now ask whether the individual tosses are independent. This is true for the unrandomized Bernoulli process, so our intuition suggests that it should remain true after randomization. Consider the first two tosses,

$$
\begin{aligned}
P(X_1 = 1, X_2 = 1) &= \int_0^1 P(X_1 = 1, X_2 = 1 \mid P = p) \, \mathrm{dens}(P = p) \, dp \\
&= \int_0^1 p^2 \, dp = \left[\frac{p^3}{3} \right]_0^1 = 1/3
\end{aligned}
$$

Since the individual tosses have probability $1/2$ of showing heads, the first two tosses are *not* independent! We will later see more precisely why this happens in Section 8.7. For now we just give an intuitive explanation. If the first toss is a head, then it gives us some information about the bias. In particular, observing a head on the first toss tells us that a very low bias is less likely than a high bias. In other words, when we condition on getting a head on the first toss, the bias random variable P is no longer uniformly distributed. Consequently, on the second toss, the conditional probability of seeing a head will be higher than $1/2$. From the computation above, the conditional probability is $2/3$.

While the tosses are dependent, when we condition on a specific bias, the resulting sample space becomes the ordinary, unrandomized Bernoulli process where the tosses are independent. This will be the case whenever we randomize a process constructed from independent random variables. The random variables in the randomized process will usually be dependent, but they will be conditionally independent.

Another important property of the tosses in the randomized Bernoulli process is that they are exchangeable. In general, if one randomizes a sequence of independent, equidistributed random variables, the result is a sequence of exchangeable random variables. Curiously, the converse is also true: essentially every sequence of exchangeable random variables arises from a sequence of independent, equidistributed random variables by randomization.

8.2 Randomized Uniform Process

Suppose that instead of sampling a fixed number n of points in the Uniform process, we sample a *random* number N of points. That is, we consider the two-step process:

1. Choose a number N of points to be sampled, according to some probability distribution: $P(N = n) = p_n$.

2. Once N is known, sample that many points from $[0, a]$ according to the Uniform process.

We have replaced the fixed number n of points by the integer random variable N, having the probability distribution $P(N = n) = p_n$. The result is a new stochastic process, the **randomized Uniform process**.

Every question we have asked about the ordinary Uniform process can now be asked for the randomized process. To compute the answers in this new process, we use the Law of Alternatives. For example, we might ask what the probability is for exactly k points to be in the interval $[0, t]$. Call this event $A_{k,t}$. We want to compute $P(A_{k,t})$. By the Law of Alternatives.

$$P(A_{k,t}) = \sum_{n=0}^{\infty} P(A_{k,t} \mid N = n) P(N = n)$$

Now $P(A_{k,t} \mid N = n)$ is the probability that exactly k points are in $[0, t]$ but in the *ordinary* Uniform process of sampling a fixed number n of points from $[0, a]$. We computed this probability back in Section 7.2, where we denoted it by $P(U_{n,a}(t) = k)$, and we computed it to be $\binom{n}{k} \left(\frac{t}{a}\right)^k \left(1 - \frac{t}{a}\right)^{n-k}$.

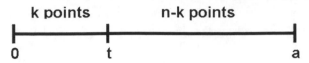

Therefore, the unconditional probability $P(A_{k,t})$ in the randomized uniform process is

$$P(A_{k,t}) = \sum_{n=0}^{\infty} \binom{n}{k} \left(\frac{t}{a}\right)^k \left(1 - \frac{t}{a}\right)^{n-k} p_n$$

If the probabilities p_n have a nice form, then it may be possible to simplify this expression, but normally the answer to a question about a randomized process will be in the form of an infinite series.

Randomizing stochastic processes allows one to define stochastic models which can be more realistic reflections of the phenomena being studied. This gives us a very powerful and effective tool for modeling phenomena. We will see many examples in the exercises and later sections.

Consider another example in the randomized Uniform process. Let B_t be the event "in the process of sampling N points uniformly from $[0, 1]$, all the points appear in $[0, t]$." Then

$$B_t = \sum_{n=0}^{\infty} P(B_t \mid N = n) \, P(N = n)$$

by the Law of Alternatives. Now $P(B_t \mid N = n)$ is the probability that all the points of the ordinary Uniform process of sampling n points from $[0, 1]$ occur in $[0, t]$. Therefore, $P(B_t \mid N = n) = t^n$, since we have chosen the length of the interval to be 1. Therefore, $P(B_t) = \sum_{n=0}^{\infty} t^n p_n$. This is a function of t that is usually called the **generating function** of the sequence $\{p_n\}$.

The technique of generating functions is important both in probability theory and in other branches of mathematics, science and engineering. For example, if we substitute $t = z^{-1}$ in the generating function, then the series is called the z-transform, an important technique in modern engineering. Unfortunately, generating functions and z-transforms are usually defined by fiat with little motivation behind saying that they are useful. Using probability theory we see more intuitively how it arises. Namely, given a sequence $\{p_n\}$ that defines a probability distribution, the generating function $f(t) = \sum_{n=0}^{\infty} t^n p_n$ of the sequence is the probability that a random number of points, the number chosen according to the distribution p_n when sampled from the unit interval, all occur in the subinterval $[0, t]$. In other words, the generating function is a way of studying a sequence $\{p_n\}$ by setting up a certain experiment and studying the properties of this experiment. This is the underlying reason why this technique turns out to be so useful.

8.3 Randomized Poisson Process

Consider the Poisson process. Since this process may be regarded as being the Uniform process as n and a approach infinity but with $\alpha = \dfrac{n}{a}$ fixed, we see that the intensity α is the analog in the Poisson process of the number of points sampled in the Uniform process. The **randomized Poisson process** is a Poisson process but with a random intensity A (capital alpha) instead of a fixed intensity. More precisely, this is a two-step process:

1. Choose an intensity α according to the density $g(\alpha) = \text{dens}(A = \alpha)$ of the positive continuous random variable A.

2. Observe a rare sequence of blips in the Poisson process having the chosen intensity.

Let T be the waiting time for the first blip in the randomized Poisson process. To compute the distribution of T, we use the Law of Alternatives, but this time, the continuous version.

$$P(T{>}t) = \int_0^\infty P(T{>}t \mid A = \alpha)\,\text{dens}(A = \alpha)\,d\alpha$$

The conditional probability $P(T{>}t \mid A = \alpha)$ is computed in the ordinary Poisson process with intensity α. Therefore:

$$P(T{>}t) = \int_0^\infty e^{-\alpha t} g(\alpha)\,d\alpha$$

The function

$$\hat{g}(t) = \int_0^\infty e^{-\alpha t} g(\alpha)\,d\alpha$$

is called the **Laplace transform** of the function $g(\alpha)$. The Laplace transform can be written more succinctly using expectations:

$$\hat{g}(t) = E\!\left(e^{tX}\right)$$

The Laplace transform is an important technique in engineering and in the sciences as well as in mathematics. We now see why. If we are given a function $g(\alpha)$ forming the probability density of a positive random variable, we can study $g(\alpha)$ by setting up an experiment and then studying the properties of the experiment. The experiment consists of waiting for the first blip of a Poisson process whose intensity is chosen according to the density $g(\alpha)$. The probability distribution of this experiment is $1 - \hat{g}(t)$ where $\hat{g}(t)$ is the Laplace transform of $g(\alpha)$. The Laplace transform can also be applied to random variables that are not positive random variables. Such a transform is sometimes called the *two-sided Laplace transform*. Amazingly, the Fourier transform, as well as two other commonly used transforms, can all be regarded as "perturbations" of the two-sided Laplace transform. Each one is in a different location in the complex plane:

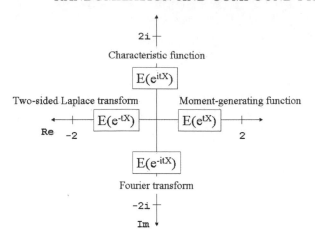

Note that there are other definitions of the Fourier transform. In one definition, the expectation is multiplied by $\frac{1}{2\pi}$, and in another it is multiplied by $\frac{1}{\sqrt{2\pi}}$. The characteristic functions of the most important distributions are in this table:

Distribution	$\varphi_X(t)$	Distribution	$\varphi_X(t)$
Bernoulli	$q + pe^{it}$	Uniform	$\dfrac{1 - e^{iat}}{iat}$
Binomial	$(q + pe^{it})^n$	Exponential	$\left(1 - \dfrac{it}{\alpha}\right)^{-1}$
Geometric	$\dfrac{pe^{it}}{1 - qe^{it}}$	Gamma	$\left(1 - \dfrac{it}{\alpha}\right)^{-k}$
Neg. Binomial	$\left(\dfrac{pe^{it}}{1 - qe^{it}}\right)^k$	Normal	$\exp\left(mit - \dfrac{\sigma^2 t^2}{2}\right)$
Poisson	$\exp(\alpha(e^{it} - 1))$	Beta	$_1F_1(k; k + l; it)$
		Cauchy	$\exp(mit - d|t|)$

Table of Characteristic Functions

As a simple example of this point of view, we can explain an important property of all of the transforms: the transform of the convolution of functions is the product of their transforms:

$$\widehat{f*g} = \hat{f}\hat{g}$$

Suppose that $f(\alpha)$ and $g(\alpha)$ are the densities of independent random variables A and B. Their convolution is the density of the sum $A + B$. Let T be the waiting time for the first gap in the Poisson process with random intensity $A + B$. We can compute $P(t>t)$ in two ways. Since the density of $A + B$ is $f*g$, we know that $P(T>t) = \widehat{f*g}$. On the other hand, we may view the event $(T>t)$ in another way. Sprinkle blips on $[0, \infty)$ with intensity A and then with intensity B. Then $(T>t) = (T_A>t) \cap (T_B>t)$, where T_A is the waiting time for the first A-blip and T_B is the waiting time for the first B-blip. Since these two kinds of blips were sprinkled independently,

$$P(T>t) = P_A(T>t)P(T_B>t)$$

Therefore $\widehat{f*g} = \hat{f}\hat{g}$.

Randomization Principle. Randomizing by an integer or a continuous random variable results in a "generating function" or a "transform" of the distribution or density, respectively.

Transforms can be regarded as a different way to look at probability distributions because they actually *characterize* distributions. More precisely, different probability distributions have different characteristic functions. This is a consequence of the Fourier inversion formula applied to probability densities:

$$\text{dens}(X = x) = \frac{1}{2\pi} \int_{-\infty}^{\infty} e^{-isx} \varphi(s) ds,$$

So one can go back and forth between the ordinary representation of probability distributions and the transformed representation, performing operations in whichever representation is most convenient. One can even perform limit operations in either representation because transforms are continuous operations, i.e., they preserve limits. The use of transforms is analogous to our use of logarithms to perform computations, switching from ordinary numbers to their logarithms and back to avoid overflows. We can have confidence in the accuracy of the result of such operations because the logarithm and exponential functions are continuous functions.

8.4 Laplace Transforms and Renewal Processes

Aside from the many applications in science and engineering for Laplace transforms, there are important applications for computing information about probability distributions. To introduce the use of Laplace transforms, we consider a specific problem: how long it takes to cross the street. We will model this problem with a Poisson process with intensity α. Now one can cross the street only if there is a pause or gap in traffic that is long enough for one to cross. Let b be the size of the gap necessary for crossing the street safely. This is a waiting time, but it is not continuous. There is a positive probability for the waiting time to be *exactly* b. This will happen if the first gap is already sufficiently long for one to cross. However, whenever a car does go by, one must "start again" to wait for a pause in traffic. For this reason, such a waiting time is called a **renewal time**, and the process is called a **renewal process**.

Let W be the waiting time for the beginning of a gap of size b (or more) in the Poisson process with intensity α. Write $F(t) = P(W \leq t)$ for the distribution of W. To compute $F(t)$, we condition on the time T_1 of the first blip, then apply the Continuous Law of Alternatives,

$$P(W \leq t) \quad = \quad \int_0^\infty P(W \leq t \mid T_1 = s) \operatorname{dens}(T_1 = s) \, ds$$

$$= \quad \int_0^\infty P(W \leq t \mid T_1 = s) \alpha e^{-\alpha s} ds$$

There are two cases for s. If $s \geq b$, then the first gap is long enough to satisfy the requirement, so $P(W \leq t \mid T_1 = s) = 1$, in this case. If, on the other hand, $s < b$, then the first gap is not long enough, and the conditional probability $P(W \leq t \mid T_1 = s)$ is the same as $P(W \leq t - s)$. In other words, it is the same as if one started at the time s of the first blip; one must "start over from scratch," the previous waiting time being irrelevant. Note that in this case, if $s > t$, then $P(W \leq t \mid T_1 = s) = P(W \leq t - s) = 0$, for now $t - s$ is negative. Put another way, if the first blip occurs at time s and the first gap is not long enough, then the waiting time for a long enough gap will necessarily be at least as long as s. We now apply these two cases to the previous formula:

$$F(t) = P(W \leq t) \quad = \quad \int_0^b P(W \leq t - s) \alpha e^{-\alpha s} ds + \int_b^\infty \alpha e^{-\alpha s} ds$$

$$= \quad \int_0^b P(W \leq t - s) \alpha e^{-\alpha s} ds + e^{-\alpha b}$$

One should immediately recognize this as being a convolution. Let G be the function:

$$G(s) = \begin{cases} \alpha e^{-\alpha s} & \text{if } 0 \leq s \leq b \\ 0 & \text{otherwise} \end{cases}$$

Then,

$$F(t) = (F * G)(t) + e^{-\alpha b}$$

This is a nice formula, but how can it help us compute $F(t)$? The trick, that engineers may recognize as one used for computing the transform of a feedback loop, is to take the Laplace transforms and solve for the transform of $F(t)$. One can then apply the reverse transform to obtain $F(t)$. Accordingly, write $\hat{F}(s)$ for the Laplace transform of $F(t)$. Then

$$\hat{F}(s) \quad = \quad \widehat{F * G}(s) + \frac{e^{-\alpha b}}{s}$$

$$= \quad \hat{F}(s)\hat{G}(s) + \frac{e^{-\alpha b}}{s}$$

Hence,

$$\hat{F}(s) = \frac{e^{-\alpha b}}{s\left(1 - \hat{G}(s)\right)}$$

Now,

$$\hat{G}(s) \quad = \quad \int_0^\infty e^{-st} G(t) dt$$

$$= \int_0^b e^{-st}\alpha e^{-\alpha t}dt$$

$$= \alpha\int_0^b \exp\left(-(\alpha+s)t\right)dt$$

$$= \alpha\left[\frac{\exp\left(-(\alpha+s)t\right)}{\alpha+s}\right]_0^b$$

$$= \frac{\alpha}{\alpha+s}(1-\exp\left(-(\alpha+s)b\right))$$

Therefore,

$$\hat{F}(s) = \frac{(\alpha+s)e^{-\alpha b}}{(\alpha+s)s - \alpha s(1-\exp\left(-(\alpha+s)b\right))}$$

$$= \frac{(\alpha+s)e^{sb}}{(\alpha+s)s\exp\left((\alpha+s)b\right) - \alpha s(\exp\left(-(\alpha+s)b\right)-1)}$$

$$= \frac{(\alpha+s)e^{sb}}{s^2\exp\left((\alpha+s)b\right) + \alpha s}$$

We now have the Laplace transform of the distribution for W. In principle, one could proceed to apply the reverse transform to obtain the distribution itself. However, we will see that, as in engineering, it can be more useful just to leave the answer in the form of a transform. In particular, it can be much easier to compute information such as the mean and variance by using the Laplace transform than to use the density. It is also much easier to perform convolution. We begin by computing the Laplace transform of the density from the Laplace transform of the distribution. To get this, one simply multiplies by s. To see why, recall that:

$$\hat{F}(s) = \int_0^\infty e^{-st}F(t)dt$$

Multiplying this by s and integrating by parts we get:

$$s\hat{F}(s) = \int_0^\infty se^{-st}F(t)dt$$

$$= \left[-e^{-st}F(t)\right]_0^\infty + \int_0^\infty e^{-st}F'(t)dt$$

$$= F(0) + \int_0^\infty e^{-st}F'(t)dt$$

Now if $F(t)$ is the distribution for a *positive* random variable, i.e., $F(0) = 0$, we get that $s\hat{F} = \hat{f}$, where $f(t)$ is the density function. However, if there is a positive probability that the random variable will be exactly equal to 0, as is the case for the renewal waiting time W, then the situation is a little more complicated. For one thing, we have not yet considered how to deal with random variables that are both continuous and discrete. The usual notation for this is to use the **Dirac delta function**. It indicates that there is a discontinuous "jump" in the distribution function at a value that has a positive probability. If the value is v and the probability that X will take this value is $P(X = v) = p_v$, one should add $p_v \delta(x - v)$ to the density function. However, we will avoid dealing with such functions by doing our computations in the "Laplace transform space." In this case, the Laplace transform of the density of the renewal time W is

$$\hat{f}(s) = \frac{(\alpha + s)e^{sb}}{s \exp((\alpha + s)b) + \alpha}$$

We now use another important fact about Laplace transforms:

Laplace Transform Expectation Formula. If $\hat{f}(s)$ is the Laplace transform of the density of a nonnegative random variable X, then for every power $n = 1, 2, \ldots$,

$$E(X^n) = (-1)^n \left[\frac{d^n}{ds^n} \hat{f}(s) \right]_{s=0}$$

In other words, the nth derivative of $\hat{f}(s)$, evaluated at zero, is the expectation of X^n, except that the sign alternates.

Proof. To see why this is true, we start with the definition of the Laplace transform:

$$\hat{f}(s) = \int_0^\infty e^{-st} f(t)dt$$

and differentiate under the integral sign to get:

$$\hat{f}'(s) = \int_0^\infty -te^{-st} f(t)dt$$

Now set s to 0:

$$\hat{f}'(0) = -\int_0^\infty tf(t)dt = -E(X)$$

For a general power n, differentiate n times, and we have proved the expectation formula.

We now apply the expectation formula to the renewal time W:

$$
\begin{aligned}
E(W) &= -\left[\frac{d}{ds}\left(\frac{(\alpha + s)e^{sb}}{s \exp((\alpha + s)b) + \alpha} \right) \right]_{s=0} \\
&= -\left[\frac{\exp(sb) + (\alpha + s)be^{sb}}{s \exp((\alpha + s)b) + \alpha} \right. \\
&\quad \left. - \frac{(\alpha + s)e^{sb}}{(s \exp((\alpha + s)b) + \alpha)^2}(\exp((\alpha + s)b) + sb \exp((\alpha + s)b)) \right]_{s=0} \\
&= -\left(\frac{1 + \alpha b}{\alpha} - \frac{\alpha e^{\alpha b}}{\alpha^2} \right) \\
&= \frac{e^{\alpha b} - 1 - \alpha b}{\alpha}
\end{aligned}
$$

For example, suppose that, on average, 10 vehicles travel down the street every minute and that one can cross the street only if there is a gap in traffic that is 15 seconds or more. Then the average time one waits for a gap in traffic is 52 seconds. By applying the Laplace Transform Expectation Formula for the case $n = 2$, one can compute the variance:

$$Var(W) = \frac{e^{2ab} - 1 - 2abe^{ab}}{a^2}$$

While the average time to cross the street in our example seems reasonable, the variance is quite high, with the standard deviation being about 56 seconds. So the wait will sometimes be quite long.

8.5 Proof of the Central Limit Theorem

We now show how one can use transforms to prove the Central Limit Theorem. Let $X_1, X_2, \ldots$ be a sequence of independent equidistributed random variables having finite variance. Without loss of generality, we may assume that they are standard. Set $S_n = X_l + X_2 + \ldots + X_n$. We wish to show that the distribution of $S_n/\sqrt{n}$ tends to the standard normal distribution.

Recall from Section 8.4 above that the expectations $E(X^n)$ can be computed from the Laplace transform by differentiation. The same is true, by a change of variables, for the characteristic function, except that the derivative must be multiplied by $(i)^n$ rather than $(-1)^n$. Consequently, if $\varphi(s)$ is the characteristic function of a standard random variable X, then:

1. $\varphi(0) = E(X^0) = 1$
2. $\varphi'(0) = iE(X^1) = 0$
3. $\varphi''(0) = -E(X^2) = -1$

We now apply the Taylor expansion to $\varphi(s)$:

$$
\begin{aligned}
\varphi(s) &= \varphi(0) + \varphi'(0)s + \frac{\varphi''(0)}{2}s^2 + o(s^2) \\
&= 1 - \frac{1}{2}s^2 + o(s^2)
\end{aligned}
$$

where $o(s^2)$ represents a term that approaches 0 as $s \to 0$. By the convolution property for transforms, the characteristic function of S_n is $\varphi(s)^n$. By a change of variables, the random variable $S_n/\sqrt{n}$ has characteristic function $\varphi(s/\sqrt{n})^n$. Utilizing the Taylor expansion computed above, we find that:

$$
\begin{aligned}
\varphi(s/\sqrt{n})^n &= \left(1 - \frac{1}{2}\left(\frac{s}{\sqrt{n}}\right)^2 + o\left(\frac{s^2}{n}\right)\right)^n \\
&= \left(1 - \frac{s^2/2}{n} + o\left(\frac{s^2}{n}\right)\right)^n
\end{aligned}
$$

Now we know from Calculus that:

$$\lim_{n \to \infty}\left(1 - \frac{s^2/2}{n}\right)^n = \exp\left(-s^2/2\right)$$

and we leave it as an exercise to show that this also works when we have the extra $o\left(\frac{s^2}{n}\right)$ term. Therefore, the characteristic function of the standardized sum $S_n/\sqrt{n}$ approaches $\exp\left(-s^2/2\right)$ as $n\to\infty$, which we know is the characteristic function of the standard normal distribution. As we saw in Section 8.3, different probability distributions have different characteristic functions, so the standard normal distribution is the only distribution with this characteristic function. Since transforms are continuous operations, the limit of the random variables $S_n/\sqrt{n}$ must be the standard normal distribution.

8.6 Randomized Sampling Processes

A common model in statistical mechanics consists of a collection of particles each of which is randomly in one of a set of states. There are usually many constraints on how many particles can be in each state, and these constraints can involve several states. Each state has an energy level, and the expectation of the total energy is an important physical quantity. The stochastic model for this system is the finite sampling process, where the states are the containers, and the constraints are on the occupancy numbers. Since the number of particles in such a system can be large and the constraints can be complicated, it can be difficult to compute the average energy. Erwin Schrödinger made the observation that the computation can be greatly simplified by randomizing the finite sampling process. Namely, choose a random number K of objects and then place them randomly into the n containers. Schrödinger's idea was that if K has the Poisson distribution, then the computations will be much easier.

The reason that the Poisson distribution works so well is the fact about the Poisson process noted earlier: if we combine two independent Poisson processes with intensities α and β, the result is a Poisson process with intensity $\alpha + \beta$. In terms of the Poisson distribution, this says that if X and Y are independent Poisson random variables of parameters λ and μ, then $X+Y$ has Poisson distribution with parameter $\lambda + \mu$. We simply reverse this. Suppose that K has Poisson distribution with parameter α. Then $K = K_1 + K_2 + \cdots + K_n$, where the K_i are independent Poisson random variables each with parameter α/n. The randomized finite sampling process then "splits up" into n independent randomized finite sampling processes. Each of these randomized finite processes consists of placing a random number K_i of objects into *one* container. For example, if K has the same distribution as $N(1)$, the number of points occurring in $[0,1)$ in the Poisson process of intensity α, then

$$N(1) = N_1\left(\frac{1}{n}\right) + N_2\left(\frac{1}{n}\right) + \cdots + N_n\left(\frac{1}{n}\right)$$

is a sum of n independent Poisson random variables each of intensity α/n, where $N_i\left(\frac{1}{n}\right)$ is the number of points occurring in $\left[\frac{i-1}{n}, \frac{i}{n}\right)$, i.e., $N_i\left(\frac{1}{n}\right) = N\left(\frac{i}{n}\right) - N\left(\frac{i-1}{n}\right)$.

Here is a pictorial representation of the sampling process randomized with Poisson random variables:

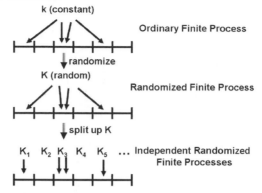

By randomizing, we make nonindependent random variables (the occupancy numbers) independent. We return to the nonrandomized process by conditioning the randomized one, using the Law of Alternatives.

A generalization of this process immediately comes to mind. We could just as easily place objects into containers of different sizes, i.e., the particles are not equally likely to fall into the various containers. The Schrödinger technique works just as well in this case; the only change required is that K be split into a sum $K_1 + \cdots + K_n$, where K_i is Poisson with parameter $p_i \alpha$. The parameter $p_i \alpha$ is the probability that any given object falls in container i.

8.7 Prior and Posterior Distributions

We began our study of randomized processes with the Bernoulli process in Section 8.1. We now address the question of why the tosses are not independent. As we already noted, if we observe that the first toss is heads, then we have some information about the bias. How can this information be expressed? Well, we ask the question "What is the bias given that the first toss was heads?" The answer is the conditional probability distribution $\mathrm{dens}(P = p \mid X_1 = 1)$, where P is the random bias. We know that P is uniformly distributed, but that is the unconditional distribution. After we have observed an event, we must use the conditional distribution, and this will be different unless the event is independent of P. To compute this new distribution we use the Continuous Bayes' Law,

$$\mathrm{dens}(P = p \mid X_1 = 1) = \frac{P(X_1 = 1 \mid P = p)\,\mathrm{dens}(P = p)}{\int_0^1 P(X_1 = 1 \mid P = p)\,\mathrm{dens}(P = p)\,dp}$$

$$= \frac{p}{\int_0^1 p\,dp} = \frac{p}{\left[\frac{p^2}{2}\right]_0^1} = \frac{p}{1/2} = 2p$$

So observing the first toss has changed the distribution of P from being uniformly distributed to having density $2p$. This new distribution is called the a **posteriori** or **posterior distribution** to distinguish it from the original uniform distribution of P, which is called the **a priori** or **prior distribution**. When one is computing a posterior distribution, the denominator in Bayes' Law serves the purpose of ensuring that the probabilities or densities add to 1. So if the numerator is a distribution that we recognize, then it is not necessary to evaluate the denominator.

We now see precisely how the tosses affect each other. After observing the first toss, the posterior distribution becomes the randomization distribution for the rest of the randomized Bernoulli process. If the first toss had been tails, it is easy to check that the posterior distribution of P would be $2q$. This is very different from the posterior distribution if we had observed heads, so the probability of getting heads on the second toss will be affected by the observation of the first toss.

Now suppose that we observe a sequence of tosses. What is the conditional distribution of P given all of this information? If the observed tosses are $i_1, \ldots, i_n$, and consist of h heads and t tails, then the numerator in Bayes' Law will be

$$P(X_1 = i_1, \ldots, X_n = i_n \mid P = p) \operatorname{dens}(P = p) = p^h (1 - p)^t$$

This depends only on the number of heads and tails, and we can recognize the density as the beta distribution whose standard model is the order statistic $X_{(h+1)}$ in the Uniform process of dropping $n + 1$ points on the interval $[0, 1]$. See Section 4.6.

Having determined the distribution of the bias P, we can compute the probability that the next toss will be heads by applying the Continuous Law of Alternatives. The result is:

$$\int_0^1 P(X_{n+1} = 1 \mid P = p) \operatorname{dens}(P = p) \, dp = \int_0^1 p \operatorname{dens}(P = p) \, dp$$

which we recognize as being the expectation of P. Since this random variable has the beta distribution with parameters $h + 1$ and $t + 1$, its expectation is $\frac{h+1}{h+t+2}$. So we conclude that if we initially have no information about the bias of a coin, and if we observe h heads and t tails, then the probability that the next toss will be heads is $\frac{h+1}{h+t+2}$. This is called **Laplace's rule of succession**. Laplace is said to have been willing to wager that if the sun has risen n times in a row then it will rise again with probability $\frac{n+1}{n+2}$.

We can graphically observe how our knowledge of the distribution of the bias P changes, as we observe tosses, by plotting the beta distribution. Here is a program that displays the beta distribution as a function of the number of heads and tails. The **dbeta** function computes the beta density for all values in a range from 0 to 1. After drawing the graph, we pause so that we can observe the graph more easily. The earlier graphs change much more dramatically, so we will spend more time watching them. We use the **ifelse** function to prevent division by zero. The **ifelse** function can have one of two values.

If the first parameter is TRUE, then it uses the second parameter. If the first parameter is FALSE, then it uses the the third parameter.

```
> plot.beta <- function(h, t) {
+    support <- (0:100)/100
+    density <- dbeta(support, h+1, t+1)
+    plot(support, density, ylim=c(0,8), type="l")
+    Sys.sleep(ifelse(h+t==0, 1, 1/(h+t)))
+ }
```

Next, we write a function that simulates the Bernoulli process with a specified number of trials. We need the number of heads for every n, so we apply the cumsum function to the Bernoulli sample point. We append 0 to the sequence of heads so that we will also see the initial uniform distribution. The number of tails is computed by subtracting the number of heads from every n in the range from 1 to the total number of trials. Since plot.beta has two parameters, we need to use the mapply function to apply the sequences of heads and tails.

```
> run.model <- function(trials, p) {
+    heads <- c(0,cumsum(rbinom(trials, 1, p)))
+    mapply(plot.beta, heads, (0:trials)-heads)
+ }
> ignore <- run.model(100, 0.6)
```

The animation will show how the posterior distribution starts from a uniform distribution and then gradually converges to the bias. It is interesting to compare the posterior distribution with the corresponding normal density function for the sample mean. The program is almost the same, except that we can no longer include the initial uniform distribution because there is no notion of a sample mean of a sample of size 0. The program looks like this:

```
> plot.beta <- function(h, t) {
+    support <- (0:100)/100
+    bdensity <- dbeta(support, h+1, t+1)
+    plot(support, bdensity, ylim=c(0,8), type="l")
+    ndensity <- dnorm(support, h/(h+t), sqrt(h*t)/(h+t)^(3/2))
+    points(support, ndensity, type="l")
+    Sys.sleep(1/(h+t))
+ }
> run.model <- function(trials, p) {
+    heads <- cumsum(rbinom(trials, 1, p))
+    mapply(plot.beta, heads, (1:trials)-heads)
+ }
> ignore <- run.model(100, 0.6)
```

Try running this animation with various values for the bias. As n increases the two graphs gradually coincide, but for small values of n, or for a bias near 0 or 1, the two distributions are substantially different. This animation gives an especially compelling case for using Bayesian methods rather than frequentist methods because one has much more precise information for smaller samples or at an extreme bias.

8.8 Reliability Theory

One of the prominent features of technological development is that increasingly complex machines are being built. Cars, computers and spacecraft are all getting more and more intricate. The problem of reliability theory is to measure the reliability of such large systems when the reliability of individual components is known. This is particularly relevant in spacecraft, for example, because they incorporate a high degree of redundancy among their subsystems. It is not often easy to determine to what extent reliability is enhanced by the presence of backup systems.

There are, of course, many ways to interpret the notion of reliability. We will describe one approach which provides a pattern for developing concepts of reliability more suited to a particular application. Our approach is concerned with the probability distribution of the time until failure of a system made up of components that fail, not because they "wear out," but rather because they are subject to a chance malfunction. The probability model that is most appropriate for such components is the exponential distribution. More precisely, we have a component C, for which the time T_C until C fails (or equivalently, the length of time that C functions properly) has a probability distribution given by $P(T_C \leq t) = 1 - e^{-\alpha t}$, where α is some constant, characteristic of the component. We would use this model for a component which is not subject to significant deterioration during the operating lifetime of the system and which fails as a result of chance causes, e.g., lightning, power surges, meteorites, etc.

Each component will be viewed as having inputs and outputs connected to other components; the whole system forming a network or circuit. Components in series must all function for the system to function, and if components are in parallel then the system will function if any one of these components operates. For example, suppose that our components are three power lines joining a power plant to the city it serves. The network may be represented like this:

Assume that the times until failure, T_1, T_2 and T_3, of the three lines are independent random variables, each exponentially distributed with parameter α. How reliably is the power plant joined to the city? Well, the city will have its power supply if and only if all of the lines function properly. Therefore the time T until failure of the whole network satisfies

$$
\begin{aligned}
P(T>t) &= P(T_1>t, T_2>t, T_3>t) \\
&= P(T_1>t)\, P(T_2>t)\, P(T_3>t) \\
&= \left(e^{-\alpha t}\right)^3 = e^{-3\alpha t}
\end{aligned}
$$

The distribution and density of T are given by:

$$P(T\leq t) = 1 - e^{-3\alpha t} \quad \text{and} \quad \text{dens}(T = t) = 3\alpha e^{-3\alpha t}$$

Thus, the network acts precisely as if it consisted of a single line whose average time until failure is one-third that of an individual line.

As another example, suppose that there are three cooling systems at the power plant and that only one is required to cool the plant. Thus, we may view the three cooling systems as being in parallel rather than in series. This system may be represented like this:

How reliably is the power plant being cooled? Well, for the power plant to lose its coolant, all three systems must have failed. Using the same assumptions about the systems as before, we find that the time T until failure of the network has distribution:

$$
\begin{aligned}
P(T\leq t) &= P(T_1\leq t, T_2\leq t, T_3\leq t) \\
&= P(T_1\leq t)\, P(T_2\leq t)\, P(T_3\leq t) \\
&= \left(1 - e^{-\alpha t}\right)^3 \\
&= 1 - 3e^{-\alpha t} + 3e^{-2\alpha t} + e^{-3\alpha t}
\end{aligned}
$$

Therefore, the density of T is given by:

$$\text{dens}(T = t) = 3\alpha e^{-\alpha t} - 6\alpha e^{-2\alpha t} + 3\alpha e^{-3\alpha t}$$

In this case, we find that T is *not* exponentially distributed. Its mean and variance are given by

$$
\begin{aligned}
E(T) &= \frac{3}{\alpha} - \frac{3}{2\alpha} + \frac{1}{3\alpha} = \frac{11}{6\alpha} \approx \frac{1.83}{\alpha} \\
Var(T) &= E(T^2) - E(T)^2 = \frac{31}{6\alpha^2} - \left(\frac{11}{6\alpha}\right)^2 = \frac{65}{36\alpha^2} \\
\sigma(T) &\approx \frac{1.34}{\alpha}
\end{aligned}
$$

In other words, the average time until failure is about 1.83 times as long as the average time until failure of a single line. However, the standard deviation is smaller than that of an exponential distribution whose mean is $1.83/\alpha$. Here are the graphs of the distribution of T:

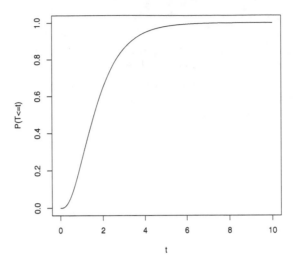

and the density of T:

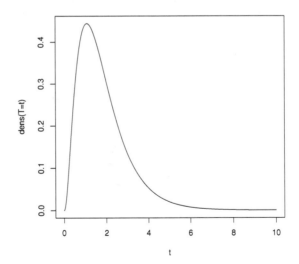

For more complicated networks, a similar kind of reasoning may be applied. We give some examples in the exercises. Moreover, one can use more complex distributions to model the behavior of the components of a system. By building larger networks from components which are, in turn, networks of subcomponents, one can give a precise model for the reliability of very complicated systems.

8.9 Bayesian Networks

By now you should be convinced that randomization is a powerful technique for constructing new stochastic processes. The basis for randomization is the notion of a random function which parametrizes a family of random variables. When the parameters are themselves random variables, the result is a randomized process. However, one might naturally wonder where the random parameters come from. The answer is simple: they come from other processes. For example, when we randomized the Poisson process, the intensity A was a random variable. This random variable could, for example, be chosen uniformly from the interval $[0, a]$. In other words, A is the result of the Uniform process. Using the picture of a random function we introduced in Section 3.1, we can depict a randomized process like this:

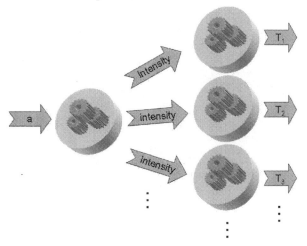

The parameter a of the process that produces the random variable A is a parameter at a "higher" level. Terms like **metaparameter** or **hyperparameter** are sometimes used for such a parameter.

Since the process that produces the random variable A is itself a random function, its parameter could also be randomized. This leads us to define a **stochastic program** to be a description of random variables that are specified by random functions whose parameters are constants or other random variables. For example, the randomized Poisson process would be a program something like this:

```
A <- unif(1,0,5)
X <- exp(*,A)
```

where **unif** is the random function for the Uniform process, and **exp** is the random function for the Poisson process as a sequence of independent exponential random variables. The asterisk means that the vector of exponentially distributed random variables is infinite.

If we combine several independent processes, the result is a special kind of stochastic program in which the random functions are not dependent on each other. Such a stochastic program is called a **compound process**. For example, here is a stochastic program in which we have combined a Poisson process with intensity 5 with an independent Bernoulli process with bias 1/4:

```
X <- exp(*,5)
Y <- bern(*,1/4)
```

We can interpret this compound process as a *multicolored Poisson process*. When each blip is observed, we flip a coin with 1/4. If the toss is heads, we paint the blip with one color, and if the toss is tails we paint the blip with a different color. A multidimensional random walk is another example of a compound process. See Exercise 6.56.

Of course, for a stochastic program to be meaningful, the parameters must be *type compatible*. In other words, the support of a random variable must be a subset of the allowed values for any input parameter where the random variable is used. For example, the following stochastic program is not type consistent:

```
A <- norm(1,0,1)
X <- exp(*,A)
```

because the normal distribution has negative values in its support, and the intensity of an exponentially distributed random variable cannot be negative. Another issue is whether a random variable could, directly or indirectly, be defined in terms of itself. In computer programs, this is called *recursion*. We will see an example of a recursive stochastic program in Chapter 10.

A **Bayesian network** (or **belief network**) is a stochastic program which has a finite number of random variables and no recursion. Other than these restrictions, the main difference between stochastic programs and Bayesian networks is the terminology used in each case. Bayesian networks use graph terminology instead of function terminology. For example, a random variable is called a **node**, and when a random variable is used to randomize another random variable, one says that the two nodes are **connected** by an **edge**. The requirement that the stochastic program be nonrecursive means that the graph of the Bayesian network has no cycles. The node producing the output is called the *parent node*, and the node receiving the input is called the *child node*. The parent/child terms were borrowed from the terminology for hierarchies and taxonomies.

Bayesian networks have emerged as a popular technique for constructing stochastic models of phenomena. By requiring Bayesian networks to be finite and to have no cycles, one can give a general definition of its probability measure, which is not easily done for a general stochastic program, as we will see in Chapter 10. To show how to compute probability measures, we give an example of a Bayesian network for medical diagnosis. Although this network is very simple, it illustrates the kind of networks that are used in actual medical diagnosis. The purpose of the network is to help determine whether a patient has a skin disease which we will call simply D. We presume that we are only including patients that have been screened for the potential to have the disease, and not the general population. The disease is known to cause some symptoms, such as a particular rash R, and a high fever F. However, D does not always cause one or the other symptom. For simplicity, we will assume that the disease D and the symptoms R and F either occur or do not occur. Each one will be regarded as a random variable that can only take values 0 or 1. In other words, they are Bernoulli random variables. The Bayesian network looks like this:

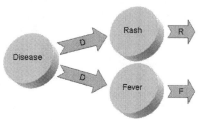

In this network, node D has no inputs so it is an ordinary Bernoulli random variable. Since a Bayesian network is finite and has no cycles, there must always be at least one node that has no inputs. The nodes R and F each have one input, so they are randomized Bernoulli trials as in Section 8.1, except that there are only two possibilities for the bias of each node, one for each of the outputs of node D. Here are the probability distributions that define the random functions of this Bayesian network:

Node	Input	p_0	p_1	Reason for the Distribution
D	–	0.99	0.01	A priori probability of the disease
R	0	0.90	0.10	Probability of rash without the disease
R	1	0.20	0.80	Probability of rash with the disease
F	0	0.98	0.02	Probability of fever without the disease
F	1	0.40	0.60	Probability of fever with the disease

Diagnostic Bayesian Network Distributions

The distribution for node D is the probability distribution of the disease, prior to observing whether a person has either of the symptoms. It is the only unconditional probability distribution in the model. The other probability distributions are conditional in the sense that one must specify an input parameter to determine whether the distribution applies.

To use this Bayesian network, we observe the patient to determine which symptoms are exhibited. Suppose that we observe that a patient has a fever but does not have a rash. Does the patient have the disease? The answer is the conditional probability $P(D = 1 \mid R = 0, F = 1)$. By Bayes' Law, we can write this as

$$P(D = 1 \mid R = 0, F = 1) \, \frac{P(R = 0, F = 1 \mid D = 1)\, P(D = 1)}{\sum_{i=0}^{1} P(R = 0, F = 1 \mid D = i)\, P(D = i)}$$

From the table that defines the Bayesian network, we know that

$$P(R = 0 \mid D = 1) = 0.20 \text{ and } P(F = 1 \mid D = 1) = 0.60$$

However, we need the probability

$$P(R = 0, F = 1 \mid D = 1)$$

To find this probability, we use the fact that the Bayesian network is a randomization, so that, as we discussed in Section 8.1, the random variables R and F are conditionally independent. Therefore,

$$
\begin{aligned}
P(R = 0, F = 1 \mid D = 1) &= P(R = 0 \mid D = 1)\, P(F = 1 \mid D = 1) \\
&= (0.20)(0.60) = 0.12
\end{aligned}
$$

We similarly compute:

$$
\begin{aligned}
P(R = 0, F = 1 \mid D = 0) &= P(R = 0 \mid D = 0)\, P(F = 1 \mid D = 0) \\
&= (0.90)(0.02) = 0.018
\end{aligned}
$$

Putting these together, we find the probability that the patient has the disease:

$$P(D = 1 \mid R = 0, F = 1) = \frac{(0.12)(0.01)}{(0.12)(0.01) + (0.018)(0.99)} \approx 0.02$$

So observing symptom F has approximately doubled the likelihood of the disease.

By the definition of conditional probability, the numerator in our application of Bayes' Law above is a joint probability:

$$
\begin{aligned}
P(R = 0, F = 1, D = 1) &= P(R = 0, F = 1 \mid D = 1)\, P(D = 1) \\
&= P(R = 0 \mid D = 1)\, P(F = 1 \mid D = 1)\, P(D = 1) \\
&= (0.20)(0.60)(0.01) = 0.0012
\end{aligned}
$$

By varying the values of the random variables, one can compute the entire joint probability distribution in the same manner. More generally, one can compute the joint probability distribution of any Bayesian network by multiplying together all of the conditional probabilities of all of the nodes, where all of the random variables are equated to one set of values. We now give the general definition of the sample space and probability measure of a Bayesian network. For simplicity, we will only consider the case of a network of discrete random variables. The definition for general random variables requires one to deal with densities, so it is more complicated, but the idea is the same.

Definition. Let $X_1, X_2, \ldots, X_n$ be the random variables of a Bayesian network. For each X_i, let S_i be its set of values (i.e., the support of X_i), and let parents(X_i) be the sequence of subscripts of the inputs to the node of X_i. The sample space Ω for the Bayesian network is the Cartesian product $\prod_{i=1}^{n} S_i$.

The probability measure is defined for any element $x = (x_1, x_2, \ldots, x_n) \in \Omega$ by the product:

$$P(\{x\}) \quad = \quad \prod_{i=1}^{n} P\left(X_i = x_i \mid \bigcap_{j \in \text{parents}(X_i)} (X_j = x_j)\right) \tag{8.1}$$

This definition can be regarded as a variation on the Law of Successive Conditioning, which has been simplified by using conditional independence so that the conditions only involve the parent random variables rather than all earlier random variables.

While we have given reasons for believing that the definition of the Bayesian network probability space is plausible, it is not obvious that it is well defined. For Equation 8.1 to define a probability measure, it is necessary to show that the probabilities add up to 1. We will do this by induction. It is trivial to show it when $n = 1$, for in that case there is just one node and it is unconditional, so its probabilities add to 1. We assume, by induction, that the result holds for a Bayesian network with $n - 1$ nodes. We must then show that it holds for a Bayesian network with n nodes. To do this, we rearrange the order of the nodes so that the first node is unconditional. This can always be done because a Bayesian network is finite and has no cycles. Our objective is to prove that:

$$\sum_{x_1, \ldots, x_n} P(X_1 = x_1) \cdots \quad = \quad 1 \tag{8.2}$$

The first probability is unconditional, so it can be factored out, and the sum becomes:

$$\sum_{x_1} P(X_1 = x_1) \sum_{x_2, \ldots, x_n} P(X_2 = x_2 \mid \cdots) \tag{8.3}$$

Now fix one of the values $x_1 \in S_1$. Because x_1 is now a constant, we can regard the random variables $X_2, \ldots, X_n$ as defining a Bayesian network with only $n - 1$ nodes. Of course, this will be a different Bayesian network for each choice of x_1. Nevertheless, for each one, we may use induction to prove that the products sum to 1, i.e., Equation (8.3) is equal to:

$$\sum_{x_1} P(X_1 = x_1)$$

Since X_1 is an unconditional random variable, this sum is 1. Therefore, by induction, Equation 8.2 holds, and the probability measure for a Bayesian network is well defined.

8.10 Exercises

Exercise 8.1 A man sitting on his front porch is watching the traffic going by. Some of the vehicles are cars and some are trucks. Assume that the cars and trucks pass by with independent Poisson distributions of intensities α and

β for cars and trucks, respectively. What is the probability that a run of 10 cars pass by before a run of 10 trucks do so?

Exercise 8.2 In the same situation as Exercise 8.1, suppose that the man first sits on his porch at time t_1 and later leaves at time $t_2 > t_1$. What is the probability that during this interval of time no run of two cars (or trucks) passes by (i.e., what vehicles he sees are alternating cars and trucks)?

Exercise 8.3 The cabs carrying passengers to an airport terminal arrive with a Poisson distribution having intensity α. Assume for simplicity that each cab brings exactly one passenger, who is either a man or a woman with probabilities p and $q = 1 - p$, respectively. What is the distribution of $N_w(t)$, the number of women arriving during the time interval $[0, t)$? What assumption must one make to answer this question?

Exercise 8.4 In a telephone network, it is known that phone calls are made with a Poisson distribution having intensity α. It is also found that the length of time of any given phone conversation is exponentially distributed with parameter β. The phone company charges customers $c + kt$ dollars for a phone call lasting t units of time, where c and k are constants. How much money does the company receive during a billing period $[t_1, t_2)$? The billing period of a conversation is determined by when it begins.

Exercise 8.5 Suppose that a suspension of cells contains, on average, λ cells per cm^3 so that the number of cells in 1 cm^3 follows a Poisson distribution with mean λ; and suppose that the probability that a cell will survive and form a visible colony is p, so that the number of colonies formed from a fixed number of cells follows a binomial distribution. Show that the number of colonies formed from 1 cm^3 of the suspension follow a Poisson distribution with mean λp.

Exercise 8.6 Suppose we have a molecular beam firing molecules one at a time at a small target crystal such that 10^{14} molecules/sec hit the crystal face. The target is a square crystal lattice having 10^{16} crystal sites. Each molecule strikes one of the crystal sites at random. If the site is empty, the molecule adheres to the crystal at that site. If there is already a molecule at that site, the new molecule rebounds and is lost. How long will it take for the beam to cover the surface of the target crystal? Use a Poisson model for this problem. See Exercise 3.22. Explicitly compute how long it takes before there is a 50% chance that the crystal is totally covered. Do the same for a 99.9% chance. Finally, compute the expected length of time for the crystal to be totally covered.

Exercise 8.7 Return to the jailer paradox of Exercise 6.5. Model this problem with a randomized Bernoulli process in which the bias p is replaced by a random variable P whose density is $f(p)$. Using the Continuous Law of Alternatives, give an expression for the conditional probability that A is to be set free given that the jailer says that B will go free. Compute this exactly for the special case for which the bias P is uniformly distributed on $[0, 1]$.

Exercise 8.8 In the jailer paradox of Exercise 6.5, one can give a very quick argument that what the jailer says does convey information to the prisoner. Suppose that the prisoner makes his plea to a sequence of jailers (say one for each shift). If each jailer gave the same answer, the prisoner would have reason to feel confident. If two jailers gave different answers, ... Model this process using a randomized Bernoulli process as in Exercise 8.7. Assume all the jailers have the same bias. Compute A's chances of being set free if n jailers say that B will go free.

Exercise 8.9 Show that the Cauchy distribution has the property that the sample mean of a sequence of independent, equidistributed Cauchy random variables has the same distribution as a single random variable in the sequence. Hint: Use characteristic functions.

Exercise 8.10 You wish to set up an electrical link between a city A and a power station B. You must use power lines that have a fixed length equal to one-third the distance between A and B. The length of time until a given line fails is exponentially distributed, i.e., failure of a line is due to chance causes (e.g., lightning) and not due to the wearing out of the line. All lines to be used are identical. The link is used until it can no longer send power from B to A, at which time the whole apparatus is overhauled. Discuss the relative merits of each of the following three designs:

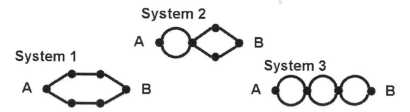

In each case, graph the distribution of the time until failure of the given design. Compare the mean time until failure and the standard deviation of the time until failure for each design. Discuss which design is better given that the first costs C_1, the second C_2, and the third C_3, where $C_1 > C_2 > C_3$.

Exercise 8.11 In a manned spacecraft, there are numerous built-in redundancies because it is usually not possible to service the craft during the mission and the lives of the astronauts depend on the proper functioning of the various subsystems. Consider the following simplified example. Suppose that two separate rocket thrusters are provided. Each has its own fuel supply. In addition, a bypass is provided for moving fuel from one fuel supply to the other rocket thruster. Diagrammatically, the system looks like this:

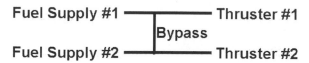

Assume that the time until failure of each of the five subsystems in the diagram above is exponentially distributed with mean α for each of the fuel supplies, β for each of the rockets and γ for the bypass. The overall rocket thruster system is functioning if one of the thrusters is functioning, one of the fuel supplies is functioning, and, if necessary, the bypass is functioning. What is the distribution of the time until failure of the rocket thruster system? Compare it with a system in which there is no bypass. Compute the mean time until failure of the system. If α, β and γ are all twice as long as the mission itself, what is the probability that the rocket thruster system works throughout the mission?

8.11 Answers to Selected Exercises

Answer to 8.1.   The fact that the sum of independent Poisson random variables is again Poisson tells us that we may view this experiment in one of two equivalent ways:

1. Two independent Poisson processes of cars and trucks that produce a Poisson process of vehicles having intensity $\alpha + \beta$, and

2. a Poisson process of vehicles, having intensity $\alpha + \beta$ which is made into a compound Poisson process by tossing a biased coin with bias $p = \frac{\alpha}{\alpha+\beta}$ and changing each vehicle into a car if heads occurs and a truck if tails occurs.

Thus, for this problem one should use the Bernoulli process: coin tossing model, with bias $p = \frac{\alpha}{\alpha+\beta}$, rather than the Poisson process. From Section 6.2, we find that the probability of a run of 10 cars before a run of 10 trucks is:

$$\frac{p^9\left(1 - q^{10}\right)}{p^9 + q^9 - p^9 q^9} = \frac{\alpha^9(\alpha + \beta)^{10} - \alpha^9 \beta^{10}}{(\alpha^9 + \beta^9)(\alpha + \beta)^{10} - \alpha^9 \beta^9(\alpha + \beta)}$$

Answer to 8.2.   If we use a Poisson process for this problem, then $N(t_2) - N(t_1)$ represents the number of cars that pass the man between times t_1 and t_2. This has the same distribution as $N(t_2 - t_1)$. Let $t = t_2 - t_1$. We use the Law of Alternatives. Let A be the event that the vehicles the man sees are alternating cars and trucks. Then,

$$P(A) = \sum_{n=0}^{\infty} P(A \mid N(t) = n)\, P(N(t) = n)$$

Now $P(A \mid N(t) = n)$ may be viewed in the Bernoulli process having bias $p = \frac{\alpha}{\alpha+\beta}$ as the event that the first n coin tosses are alternately heads and tails. This probability is easy to compute. If $n = 0$, then $P(A \mid N(t) = n)$ is trivially 1. If n is even, and $n > 0$, then,

$$P(A \mid N(t) = n) = 2p^{\frac{n}{2}} q^{\frac{n}{2}}$$

because there are just two ways for this event to happen. If n is odd, we also have just two ways for the event to happen. These two ways have different formulas, but their sum can be simplified:

$$P(A \mid N(t) = n) = p^{\frac{n-1}{2}} q^{\frac{n+1}{2}} + p^{\frac{n+1}{2}} q^{\frac{n-1}{2}} = p^{\frac{n-1}{2}} q^{\frac{n-1}{2}}$$

Let $\lambda = (\alpha + \lambda)t$. Then,

$$
\begin{aligned}
P(A) &= \sum_{n=0}^{\infty} P(A \mid N(t) = n)\, P(N(t) = n) \\
&= P(A \mid N(t) = 0)\, e^{-\lambda} \\
&\quad + \sum_{m=1}^{\infty} P(A \mid N(t) = 2m)\, \lambda^{2m} \frac{e^{-\lambda}}{(2m)!} \\
&\quad + \sum_{m=1}^{\infty} P(A \mid N(t) = 2m - 1)\, \lambda^{2m-1} \frac{e^{-\lambda}}{(2m-1)!} \\
&= e^{-\lambda} + 2 \sum_{m=1}^{\infty} p^m q^m \lambda^{2m} \frac{e^{-\lambda}}{(2m)!} \\
&\quad + \sum_{m=1}^{\infty} p^{m-1} q^{m-1} \lambda^{2m-1} \frac{e^{-\lambda}}{(2m-1)!} \\
&= e^{-\lambda} + 2 e^{-\lambda} \sum_{m=1}^{\infty} \frac{\left(\lambda\sqrt{pq}\right)^{2m}}{(2m)!} + \frac{e^{-\lambda}}{\sqrt{pq}} \sum_{m=1}^{\infty} \frac{\left(\lambda\sqrt{pq}\right)^{2m-1}}{(2m-1)!} \\
&= e^{-\lambda} + 2 e^{-\lambda} \left(\cosh\left(\lambda\sqrt{pq}\right) - 1\right) + \frac{e^{-\lambda} \sinh\left(\lambda\sqrt{pq}\right)}{\sqrt{pq}}
\end{aligned}
$$

Replacing λ by $(\alpha + \beta)$, p by $\frac{\alpha}{\alpha+\beta}$ and q by $\frac{\beta}{\alpha+\beta}$, we find that

$$P(A) = \left(2 \cosh\left(\sqrt{\alpha\beta} t\right) + \frac{(\alpha + \beta) \sinh\left(\sqrt{\alpha\beta} t\right)}{\sqrt{\alpha\beta}} - 1\right) e^{-(\alpha+\beta)t}$$

Answer to 8.3. To answer the question, one must assume that the men and women are independent from one another. If this is so, then $N_w(t)$ has a Poisson distribution with parameter $q\alpha t$. Here is how that is derived. By the Law of Alternatives,

$$P(N_w(t) = k) = \sum_{n=0}^{\infty} P(N_w(t) = k \mid N(t) = n)\, P(N(t) = n)$$

Now $P(N_w(t) = k \mid N(t) = n)$ is the same as the probability that the number of tails in n tosses of a coin is k, where the coin has bias p. Thus,

$$P(N_w(t) = k \mid N(t) = n) = \binom{n}{k} p^{n-k} q^k$$

and hence,

$$
\begin{aligned}
P(N_w(t) = k) &= \sum_{n=k}^{\infty} \binom{n}{k} p^{n-k} q^k \frac{(\alpha t)^n}{n!} e^{-\alpha t} \\
&= \frac{q^k e^{-\alpha t}}{k! p^k} \sum_{n=k}^{\infty} \frac{n! p^n}{(n-k)!} \frac{(\alpha t)^n}{n!} \\
&= \frac{q^k e^{-\alpha t}}{k! p^k} \sum_{n=k}^{\infty} \frac{(p\alpha t)^n}{(n-k)!} \\
&= \frac{q^k e^{-\alpha t} (p\alpha t)^k}{k! p^k} \sum_{n=k}^{\infty} \frac{(p\alpha t)^{n-k}}{(n-k)!} \\
&= \frac{q^k e^{-\alpha t} (p\alpha t)^k}{k! p^k} \sum_{m=0}^{\infty} \frac{(p\alpha t)^m}{(m)!} \\
&= \frac{q^k e^{-\alpha t} (p\alpha t)^k}{k! p^k} e^{p\alpha t} \\
&= \frac{q^k e^{p\alpha t - \alpha t} (\alpha t)^k}{k!} \\
&= \frac{\left(q\alpha t^k e^{-q\alpha t}\right)}{k!}
\end{aligned}
$$

Answer to 8.4.   This is a compound Poisson process. As usual, the relevant distribution is $N(t)$ where $t = t_2 - t_1$. Write D for the amount received by the phone company in this billing period. The number of phone calls made in this period is, by assumption, $N(t)$, where the Poisson distribution has parameter αt. Then D may be written as

$$
\begin{aligned}
D &= (c + kT_1) + (c + kT_2) + \cdots + \left(c + kT_{N(t)}\right) \\
&= cN(t) + k\left(T_1 + T_2 + \cdots + T_{N(t)}\right)
\end{aligned}
$$

where $T_1, T_2, \ldots$ are independent, exponentially distributed random variables with parameter β. Recall that the sum $T_1 + \cdots + T_n$ of a fixed number of such T_i's has the gamma distribution. So this problem is essentially the computation of a randomized gamma distribution. Write $F_n(t)$ for the gamma probability distribution with parameters n and β.

To compute anything about D one uses the Law of Alternatives:

$$
\begin{aligned}
P(D \le d) &= \sum_{n=0}^{\infty} P(D \le d \mid N(t) = n) \, P(N(t) = n) \\
&= \sum_{n=0}^{\infty} P(cn + k(T_1 + \cdots + T_n) \le d) \, P(N(t) = n)
\end{aligned}
$$

$$
= \sum_{n=0}^{\infty} P\left(T_1 + \cdots + T_n \leq \frac{d - cn}{k}\right) P(N(t) = n)
$$

$$
= \sum_{n=0}^{\infty} F_n\left(\frac{d - cn}{k}\right) \frac{(\alpha t)^n}{n!} e^{-\alpha t}
$$

$$
= \sum_{n \leq d/c} F_n\left(\frac{d - cn}{k}\right) \frac{(\alpha t)^n}{n!} e^{-\alpha t}
$$

The last step above follows from the fact that $F_n(x) = 0$ when $x < 0$. The density of D is

$$
\mathrm{dens}(D = d) = \sum_{n \leq d/c} F_n'\left(\frac{d - cn}{k}\right) \frac{1}{k} \frac{(\alpha t)^n}{n!} e^{-\alpha t}
$$

$$
= \sum_{n \leq d/c} \frac{\beta^n (d - cn)^{n-1}}{k^n (n-1)!} \exp\left(-\beta\left(\frac{x - cn}{k}\right)\right) \frac{(\alpha t)^n}{n!} e^{-\alpha t}
$$

$$
= \sum_{n \leq d/c} \frac{(\alpha \beta t)^n (d - cn)^{n-1}}{k^n n! (n-1)!} \exp\left(-\alpha t - \beta\left(\frac{x - cn}{k}\right)\right)
$$

This answers the problem as it was stated. However, the formula above is very complicated. How would one compute the expectation, for example? Again we use the Law of Alternatives, but now one uses conditional *expectations*,

$$
E(D) = \sum_{n=0}^{\infty} E(D \mid N(t) = n) P(N(t) = n)
$$

$$
= \sum_{n=0}^{\infty} E(cn + k(T_1 + \cdots + T_n)) P(N(t) = n)
$$

$$
= \sum_{n=0}^{\infty} \left(cn + \frac{kn}{\beta}\right) \frac{(\alpha t)^n}{n!} e^{-\alpha t}
$$

$$
= \left(c + \frac{k}{\beta}\right) e^{-\alpha t} \sum_{n=0}^{\infty} \frac{(\alpha t)^n}{(n-1)!}
$$

$$
= \left(c + \frac{k}{\beta}\right) e^{-\alpha t} (\alpha t) e^{\alpha t}
$$

$$
= \left(c + \frac{k}{\beta}\right) (\alpha t)
$$

Answer to 8.6. Let T be the waiting time for the crystal to be *completely* covered. If there were just one crystal site, then the distribution of T would be given by $1 - e^{-\alpha t}$, where α is the intensity of the molecular beam in molecules per unit time. In this case, there are 10^{16} crystal sites and the beam is firing 10^{14} molecules per second at the crystal. Thus

$$
\alpha = \frac{10^{14}}{10^{16}} = 10^{-2}
$$

The crucial property of the Poisson process, not shared by the finite sampling process, is that what happens at each crystal site is independent of what happens elsewhere. Therefore,

$$P(T{\le}t) = \left(1 - \exp\left(-10^{-2}t\right)\right)^{10^{16}}$$

We wish to find t such that the probability is 50% and such that it is 99.9%. If p is the desired probability, then

$$\left(1 - \exp\left(-10^{-2}t\right)\right)^{10^{16}} = p$$

In theory, one could just solve for t in this equation and compute the answer with R. However, this is likely to cause overflows during a computation. One should not trust any answer the computer might give in this case. The usual way to deal with this situation is to take logarithms. This gives us the equation

$$10^{16}\log\left(1 - \exp\left(-10^{-2}t\right)\right) = \log(p)$$

which is equivalent to

$$\log\left(1 - \exp\left(-10^{-2}t\right)\right) = \log(p)10^{-16}$$

Unfortunately, this does not seem to help us solve for t. However, there is a trick that is frequently useful for probabilistic computations. The trick is to use the the Taylor series for $\log(1-x)$:

$$\log(1-x) \approx -x - \frac{x^2}{2} - \frac{x^3}{3}$$

If x is very close to zero, then the terms in this series will go to 0 so quickly that one can often use just the first term alone. In this case, $\exp\left(-10^{-2}t\right)$ will be around 10^{-16}, so using the first term of the Taylor series will give an answer that is correct to 16 places. Thus

$$\exp\left(-10^{-2}t\right) \approx -\log(p)10^{-16}$$

and hence

$$t \approx -100(\log(-\log(p)) - 16\log(10))$$

So we compute

```
> -100*(log(-log(c(0.5,0.999)))-16*log(10))
[1] 3720.787 4374.862
```

or about 62 minutes for 50% confidence and 73 minutes for 99.9%. Note that it only takes 11 minutes longer to go from 50% chance of full coverage to 99.9%.

Be careful not to confuse the probability of full coverage with the percentage coverage, which is a very different quantity. We are not considering the latter quantity.

We now compute the expected time until full coverage. Let T_n be the waiting time until full coverage when there are n crystal sites and molecular beam intensity α. The particular case we wish to solve in this problem is $n = 10^{16}$ and $\alpha = 0.01$. Let

$$F_n(t) = P(T_n{\le}t) = \left(1 - e^{-\alpha t}\right)^n$$

Then, using integration by parts,

$$
\begin{aligned}
E(T_n) &= \int_0^\infty t F_n'(t)\,dt \\
&= \int_0^\infty (1 - F_n(t))\,dt \\
&= \int_0^\infty \left(1 - \left(1 - e^{-\alpha t}\right)^n\right)dt
\end{aligned}
$$

Using the substitution $u = 1 - e^{-\alpha t}$, we compute that $du = e^{-\alpha t}\alpha\,dt$, and $dt = \dfrac{du}{\alpha(1 - u)}$, so that

$$
E(T_n) = \frac{1}{\alpha}\int_0^1 (1 - u^n)\frac{du}{1 - u}
$$

Now the integrand above is well known as the sum of the powers of u:

$$
\frac{1 - u^n}{1 - u} = 1 + u + u^2 + \cdots + u^{n-1}
$$

Therefore,

$$
\begin{aligned}
E(T_n) &= \frac{1}{\alpha}\int_0^1 \left(1 + u + u^2 + \cdots + u^{n-1}\right)du \\
&= \frac{1}{\alpha}\left[u + \frac{u^2}{2} + \frac{u^3}{3} + \cdots + \frac{u^n}{n}\right]_0^1 \\
&= \frac{1}{\alpha}\sum_{k=1}^n \frac{1}{k}
\end{aligned}
$$

The series $\displaystyle\sum_{k=1}^n \frac{1}{k}$ is called the **harmonic series**, and by Euler's approximation,

$$
\sum_{k=1}^n \frac{1}{k} \approx \log(n) + \gamma \approx \log(n) + 0.57721
$$

Substituting $n = 10^{16}$ and $\alpha = 0.01$, we get

$$
E(T) = E(T_{10^{16}}) = 100\sum_{k=1}^{10^{16}} \frac{1}{k} \approx 100\left(\log\left(10^{16}\right) + 0.57721\right) \approx 3742 \text{ seconds}
$$

which is almost the same as the time until there is a 50% chance of coverage.

Answer to 8.10. There are many ways to interpret the notion of being a "better" system. Here is one interpretation. It is not the only one nor even the best one, but it has the advantage of being relatively simple. The memoryless-ness of the lines means that the time until failure of one line is exponentially distributed. All the lines are the same, so the parameter of this distribution is the same one for all of them. Call it α. Let T_i be the time until failure of system i. From reliability theory, we know how to compute the distributions of the T_i's:

Single line $P(T \leq t) = 1 - e^{-\alpha t}$ and $P(T > t) = e^{-\alpha t}$

Series of n lines $P(T > t) = (e^{-\alpha t})^n = e^{-n\alpha t}$

Parallel set of n lines $P(T \leq t) = (1 - e^{-\alpha t})^n$

System 1   This system has two 3-line series in parallel:

$$P(T_1 \leq t) = \left(1 - e^{-3\alpha t}\right)^2 = 1 - 2e^{-3\alpha t} + e^{-6\alpha t}$$

System 2   This has a parallel pair in series with two 2-line series in parallel:

$$
\begin{aligned}
P(T_2 \leq t) &= 1 - \left(1 - (1 - e^{-\alpha t})^2\right)\left(1 - (1 - e^{-2\alpha t})^2\right) \\
&= 1 - 4e^{-3\alpha t} + 2e^{-4\alpha t} + 2e^{-5\alpha t} - e^{-6\alpha t}
\end{aligned}
$$

System 3   This is a series of 3 parallel pairs:

$$
\begin{aligned}
P(T_3 \leq t) &= 1 - \left(1 - (1 - e^{-\alpha t})^2\right)^3 \\
&= 1 - 8e^{-3\alpha t} + 12e^{-4\alpha t} - 6e^{-5\alpha t} + e^{-6\alpha t}
\end{aligned}
$$

Graphs of the distributions are:

```
> a <- 1
> F1 <- function(t) 1 - 2*exp(-3*a*t) + exp(-6*a*t)
> F2 <- function(t) 1 - 4*exp(-3*a*t) + 2*exp(-4*a*t) +
+ 2*exp(-5*a*t) - exp(-6*a*t)
> F3 <- function(t) 1 - 8*exp(-3*a*t) + 12*exp(-4*a*t) -
+ 6*exp(-5*a*t) + exp(-6*a*t)
> t <- (0:2000)/1000
> plot(t, sapply(t, F1), type="l", xlab="t", ylab="P(T<=t)")
> text(0.3, 0.6, labels="P(T1<=t)")
> points(t, sapply(t, F2), type="l")
> text(0.6, 0.55, labels="P(T2<=t)")
> points(t, sapply(t, F3), type="l")
> text(0.8, 0.5, labels="P(T3<=t)")
```

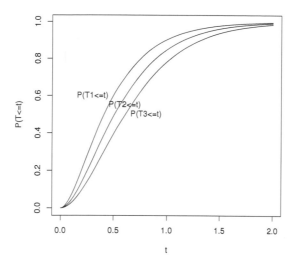

To distinguish the three distribution functions from each other, we labeled them using the `text` function. This function places text on a graph. The first two parameters are the x and y coordinates where the text should begin. The third parameter is the text string. The corresponding densities are:

```
> a <- 1
> f1 <- function(t) 6*a*exp(-3*a*t) - 6*a*exp(-6*a*t)
> f2 <- function(t) 12*a*exp(-3*a*t) - 8*a*exp(-4*a*t) -
+ 10*a*exp(-5*a*t) + 6*a*exp(-6*a*t)
> f3 <- function(t) 24*a*exp(-3*a*t) - 48*a*exp(-4*a*t) +
+ 30*a*exp(-5*a*t) - 6*a*exp(-6*a*t)
> t <- (0:2000)/1000
> plot(t, sapply(t, f1), type="l", xlab="t", ylab="dens(T=t)")
> text(0.6, 1.5, labels="dens(T1=t)")
> points(t, sapply(t, f2), type="l")
> text(0.8, 1.1, labels="dens(T2=t)")
> points(t, sapply(t, f3), type="l")
> text(1.4, 0.4, labels="dens(T3=t)")
```

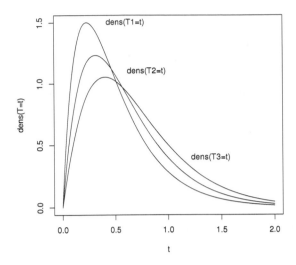

The expectations are:

$$E(T_1) = \frac{2}{3\alpha} - \frac{1}{6\alpha} = \frac{0.5}{\alpha}$$

$$E(T_2) = \frac{4}{3\alpha} - \frac{2}{4\alpha} - \frac{2}{5\alpha} + \frac{1}{6\alpha} = \frac{0.6}{\alpha}$$

$$E(T_3) = \frac{8}{3\alpha} - \frac{12}{4\alpha} + \frac{6}{5\alpha} - \frac{1}{6\alpha} = \frac{0.7}{\alpha}$$

We see, therefore, that each system is better than the next in terms of the mean time until failure. The variances are also increasing. The standard deviations are:

$$\sigma(T_1) \approx \frac{0.37}{\alpha} \quad \sigma(T_2) \approx \frac{0.50}{\alpha} \quad \sigma(T_3) \approx \frac{0.55}{\alpha}$$

In spite of the larger variance, it seems clear that the third system is the best one. This is reinforced by the fact that it is also the least expensive one.

CHAPTER 9

Entropy and Information

Ludwig Eduard Boltzmann was an Austrian physicist who founded the branch of physics known as statistical mechanics and who established the relationship between entropy and probability. He coined the word "ergodic" now used for a certain kind of Markov chain, which we will introduce in Chapter 10. He was one of the most important advocates for atomic theory when that scientific model was still highly controversial. His work established the basis for quantum mechanics, which is inherently a stochastic theory. Boltzmann was subject to rapid alternation between depression and elevated moods. It seems likely that he suffered from what we now call bipolar disorder. During one of his extreme periods of depression, he committed suicide in 1906.

That probability is closely connected with information should come as no surprise after problems such as Exercise 6.5 (the jailer paradox). What entropy does is to make this connection precise. We begin with finite-valued random variables. The notion of entropy is quite clear in this case, and it forms the basis for one of the most dramatic applications of the law of large numbers to information theory: The Shannon Coding Theorem. We then consider continuous random variables. Entropy is much harder to define in this case, but the reward is that we can then prove that essentially all the interesting distributions we have seen in probability theory may be defined by entropy considerations.

9.1 Discrete Entropy

A random variable is said to be a **finite-valued random variable** if it takes finitely many values. For example, S_n in the Bernoulli process is finite-valued since it can only take on values from 0 to n. If X is a finite-valued random variable whose values are $1, 2, \ldots, n$, then X determines the events $(X = 1), (X = 2), \ldots, (X = n)$. Moreover, every outcome of Ω is in exactly one of these events. We call this situation a **partition** of Ω as shown in this drawing:

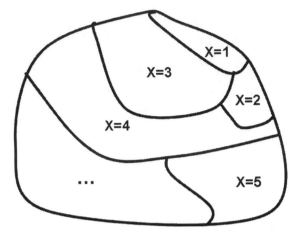

In general, a *partition* π of Ω is a collection of nonempty events $B_1, B_2, \ldots, B_n$, called the *blocks* of π such that:

Disjointness   No two blocks intersect.

Completeness   Every sample point is in some block, i.e., $\bigcup_i B_i = \Omega$.

The only difference between a random variable X and a partition π is that a random variable consists not only of a partition but also of a *label* (the value X takes on that block) for each block. The partition $\pi(X)$ *defined by* X is the partition whose blocks are $(X = 1), (X = 2), \ldots, (X = n)$, i.e., $\pi(X)$ is obtained by ignoring the particular labels that X attaches to the events it defines.

More generally, suppose that we have several finite-valued random variables $X_1, X_2, \ldots, X_r$. The smallest events that one can define by these random variables are:

$$((X_1 = i_1) \cap (X_2 = i_2) \cap \ldots \cap (X_r = i_r))$$

and any event definable by the random variables $X_1, X_2, \ldots, X_r$ is necessarily a union of some of the above events. The partition whose blocks are the above events is called the **joint partition** $\pi(X_1, X_2, \ldots, X_r)$ *defined by* $X_1, X_2, \ldots, X_r$. The partition $\pi(X_1, X_2, \ldots, X_r)$ is related to the partitions $\pi(X_1), \pi(X_2), \ldots, \pi(X_r)$ by means of the operation on partitions called the *meet*. In general, if σ and τ are two partitions, whose blocks are $C_1, C_2, \ldots, C_m$ and $D_1, D_2, \ldots, D_n$, respectively, then the meet of σ and τ, written $\sigma \wedge \tau$ is the partition whose blocks are $C_i \cap D_j$, whenever they are nonempty. In terms of the meet,

$$\pi(X_1, X_2, \ldots, X_r) = \pi(X_1) \wedge \pi(X_2) \wedge \ldots \wedge \pi(X_r)$$

As the joint distribution of random variables determines everything about the set of random variables, so the joint partition of a set of partitions determines the set of partitions. In particular, it is easy to see that independence of random variables is really a property of the partitions defined by them. Let σ and τ be two partitions. We say σ and τ are *independent* if and only if

$$P(C \cap D) = P(C)\, P(D)$$

for all blocks C of σ and D of τ. When σ and τ are independent, we can display the sample space Ω as a "checkerboard" like this

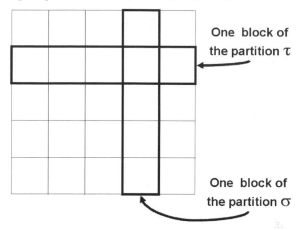

whose rows are blocks of τ, whose columns blocks of σ, and such that the "area" is proportional to the probability.

The meet is the analog for partitions of the intersection of sets. There is a whole algebra of partitions analogous to that for sets. For example, there is an analog of set union called the *join* of partitions and written $\sigma \vee \tau$. We leave it as an exercise to decide how this ought to be defined. We will not have need of this particular operation.

Another notion from sets is that of *subset*, and its analog for partitions will be very important for us. We say that a partition σ, with blocks $C_1, C_2, \ldots, C_m$, is *finer than* a partition τ, with blocks $D_1, D_2, \ldots, D_n$, if every block C_i of σ is contained in some block D_j of τ. We write $\sigma \leq \tau$ for this relation. If X and Y are finite-valued random variables, then $\pi(X) \leq \pi(Y)$ means that an observation of X is *sufficient* to determine anything one might ask about Y. The technical term for this relation is that X is a **sufficient statistic** for Y. More generally, if $X_1, X_2, \ldots, X_r$ are a collection of finite-valued random variables such that $\pi(X_1, X_2, \ldots, X_r)$ is finer than $\pi(Y)$, we say that $X_1, X_2, \ldots, X_r$ are *sufficient* for Y. In practice, one often finds that in a particular experiment one wants the value of Y but that the random variables one actually measures form a sequence $X_1, X_2, \ldots$ If for some r, $X_1, X_2, \ldots, X_r$ are sufficient for Y, then one can, in principle, compute Y from the measurements of the X_i's. One also says that $X_1, X_2, \ldots, X_r$ *code for* Y.

The reason for introducing partitions is that the "information content" of a finite-valued random variable X is a property of the collection of events defined by X and not by the particular labels X happens to assign to these events. We now make this precise. The **entropy of a partition** π whose blocks are the events $B_1, B_2, \ldots, B_n$ is defined by:

$$H_2(\pi) = \sum_i P(B_i) \log_2 \left(\frac{1}{P(B_i)} \right) = -\sum_i P(B_i) \log_2(P(B_i))$$

where by convention $0 \cdot \log_2 \frac{1}{0}$ is defined to be 0. The **entropy of a finite-valued random variable** X is the entropy of its partition:

$$H_2(X) = H_2(\pi(X))$$

We remark that $\log_2$ could be replaced by $\log_b$ for any base $b > 0$. The only effect on $H_2(\pi)$ is to multiply by the scale factor $\log_b(2)$, i.e., we merely alter the units in which the entropy is measured. The use of $\log_2$ is traditional. In this case, we say that $H_2(\pi)$ is measured in **bits**. More generally, we will write $H_b(\pi)$ for $-\sum_i P(B_i) \log_b(P(B_i))$. If we use $H(\pi)$ without a subscript, we mean that the base b should be taken to be e, the base of the natural logarithms. We say that $H(\pi)$ is measured in **nats** (natural digits).

Consider the example of tossing a biased coin with bias p, i.e., consider a partition consisting of one or two blocks. The function $H_2(X)$, as a function of the bias p, is:

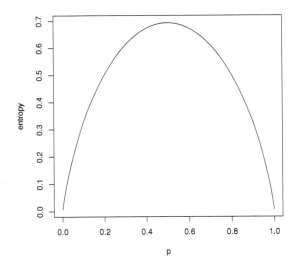

If p is 1, then we know for certain that the coin will always show heads. In this case,

$$H_2(X) = -0 \cdot \log_2 0 - 1 \cdot \log_2 1 = 0$$

Entropy zero corresponds to *total certainty*. Now suppose that p is somewhat less than 1. The toss is now somewhat less predictable, and we find that the entropy is a small positive number. As p decreases, the entropy gradually increases, reaching a maximum at $p = 1/2$. For a fair coin,

$$H_2(X) = -\frac{1}{2}\cdot\log_2\left(\frac{1}{2}\right) - \frac{1}{2}\cdot\log_2\left(\frac{1}{2}\right) = \frac{1}{2} + \frac{1}{2} = 1 \text{ bit}$$

Finally, as p decreases from $1/2$ to 0, the entropy again decreases to zero; for now the toss is becoming increasingly predictable.

More generally, suppose that π has n blocks. Shannon proved that $H_2(\pi)$ takes its maximum value precisely when all n outcomes are equally likely. In this case, the entropy is $H_2(\pi) = \log_2(n)$ bits or $H(\pi) = \log(n)$ nats. We will now prove this. All of our later characterizations of distributions having maximum entropy rely on the same basic technique we will use in this case. The key fact is this inequality:

$$\log(u) \quad \leq \quad u - 1 \qquad \text{for all } u > 0$$

$$\log(u) \quad = \quad u - 1 \qquad \text{if and only if } u = 1$$

This fact is intuitively obvious when one looks at the graphs of the functions $x - 1$ and $\log(x)$:

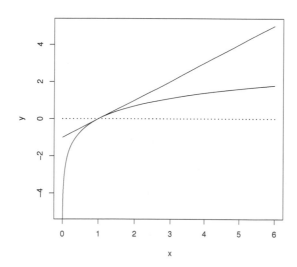

and it is easy to prove using Calculus. The function $f(u) = \log(u) - u + 1$ has the derivative $f'(u) = \dfrac{1}{u} - 1$ so $f'(u) > 0$ for $u < 1$ and $f'(u) < 0$ for $u > 1$. Therefore, $f(u)$ takes its maximum value at $u = 1$.

Now compare $H(\pi)$ to $\log(n)$ using the above inequality:

$$H(\pi) - \log(n) \quad = \quad -\sum_{i=1}^{n} P(B_i)\left(\log(P(B_i)) + \log(n)\right)$$

$$= \sum_{i=1}^{n} P(B_i) \log\left(\frac{1}{P(B_i)\,n}\right)$$

$$\leq \sum_{i=1}^{n} P(B_i)\left(\frac{1}{P(B_i)\,n} - 1\right)$$

$$= \sum_{i=1}^{n}\left(\frac{1}{n}\right) - \sum_{i=1}^{n} P(B_i)$$

$$= 1 - 1 = 0$$

The fact that the probabilities of the n blocks add up to 1, $\sum_{i=1}^{n} P(B_i) = 1$, is used twice above: in the first equality and in the second last one. In any case, we find that if π has n blocks, then $H(\pi) \leq \log(n)$.

When is equality possible? In our derivation of $H(\pi) \leq \log(n)$, equality can fail in only one of the steps:

$$\log\left(\frac{1}{P(B_i)\,n}\right) \leq \frac{1}{P(B_i)\,n} - 1$$

for all i. Now the basic logarithmic inequality tells us that this will be an equality if and only if $\frac{1}{P(B_i)\,n} = 1$ for all blocks, i.e., $P(B_i) = \frac{1}{n}$ for all i. This completes our proof.

When X has a partition π, all of whose blocks have the same probability, we say that X is **completely random** or **totally random**, although this is not quite the best terminology. One should really say that X has **maximum uncertainty**. Equivalently, the measurement of X gives one the **maximum information** about the outcome of an experiment, of any random variable having the same number of outcomes. It is unfortunate that it has become standard terminology to describe such a random variable as being simply "random." For example, one often says "choose a card at random" rather than "choose a card completely at random," as if there were no other way to choose a card from a deck. In fact, most "random" shuffles of a deck are far from being completely random (see Exercise 9.4); as a result, choosing a card or dealing a hand is not totally random, and the probabilities computed in Exercise 3.1 would seldom be achieved in a game played with physical cards (although one can achieve complete randomness by using computers). On the other hand, the terminology suggests that we always do have complete randomness. This is the price one pays for using a vague, imprecise language to describe probabilistic concepts.

Properties of Entropy

So far we have discussed examples of the entropy of some random variables. Although these examples provide some motivation for our definition of entropy, they leave unanswered the basic question of why this formula and not some other is the one we use to define entropy. We will now consider why our formula is the only possible one. We will do this by finding three self-evident properties that ought to hold for any reasonable measure of information (or entropy). It then turns out that our definition of entropy is the only one that satisfies all these properties.

We begin with the most obvious of properties. As we have defined it, H is a function of partitions of the sample space. However, it should be clear that we want H to depend only on the set of probabilities of the blocks of the partition. In fact, we want H to depend only on the *positive* probabilities which occur. Moreover, we want H to be a continuous function of these probabilities. This is a convenience only. We could, with a great deal of effort, derive continuity from other more complex conditions; but we would rather concentrate on the important issues. We summarize the conditions on H we have just described before going on to the difficult question of conditional entropy.

Entropy Property 1. An entropy is a function H defined on sets $\{p_1, p_2, \ldots, p_n\}$ of nonnegative real numbers which satisfy $p_1 + p_2 + \cdots + p_n = 1$.

Entropy Property 2. If H is an entropy function, then for any set $\{p_1, p_2, \ldots, p_n\}$ on which H is defined, H satisfies:

$$H(p_1, p_2, \ldots, p_n, 0) = H(p_1, p_2, \ldots, p_n)$$

In other words, H depends only on the nonzero p_i's in a given set.

Entropy Property 3. An entropy function is continuous.

Entropy Property 4. If H is an entropy function, then for any set $\{p_1, p_2, \ldots, p_n\}$ on which H is defined, H satisfies:

$$H(p_1, p_2, \ldots, p_n) \leq H\left(\frac{1}{n}, \frac{1}{n}, \ldots, \frac{1}{n}\right)$$

The last property of entropy we consider requires the concept of conditional entropy. There are two ways to think of the concept of conditional entropy, and the fact that they are equivalent is our next property of entropy. To illustrate the ideas involved, we consider the following simple weighing problem. We have three coins, some of which may be counterfeit (but not all). Counterfeit coins are distinguishable from normal coins by the fact that they are lighter. We are given a balance scale, and we wish to find out which, if any, of the coins are counterfeit. The sample space for this problem consists of seven sample points, one for each possible set of good coins. We denote them as follows:

$$\Omega = \{1, 2, 3, 12, 13, 23, 123\}$$

Now what happens when we put the first two coins on each side of the scale? The sample space is partitioned into three blocks corresponding to the three possible outcomes of the weighing: $\sigma = \{12, 123, 3\}, \{2, 23\}, \{1, 13\}$. After recording the result of this weighing, we then place the second and third coins on the two sides of the scale. The result of this second weighing is to partition each of the blocks of the first weighing:

$$\{12, 123, 3\} \quad \text{becomes} \quad \{12\}, \{123\}, \{3\}$$

$$\{2, 23\} \quad \text{becomes} \quad \{2\}, \{23\}$$

$$\{1, 13\} \quad \text{becomes} \quad \{1\}, \{13\}$$

The combined information of the two weighings is represented by the partition into seven blocks, each with one sample point. Call this partition π. Conditional entropy is concerned with the effect of the second weighing, *given* that the first has occurred. One way to analyze this is to look at each block σ_i of the partition of the first weighing and to analyze the situation as if σ_i were the whole sample space. In general, for a event A and a partition τ, we define the **conditional entropy** of τ given A, written $H(\tau \mid A)$, to be the entropy of the partition $\tau_1 \cap A, \tau_2 \cap A, \ldots$ that τ induces on A. Thus in the above weighing problem, we have three conditional entropies, one for each possible outcome to the first weighing: $H(\pi \mid \sigma_1)$, $H(\pi \mid \sigma_2)$ and $H(\pi \mid \sigma_3)$. The conditional entropy of π given σ is then defined to be the average of these. More precisely, if π and σ are any two partitions of a sample space Ω such that π is finer than σ, we define the *conditional entropy* of π given σ to be the average value of $H(\pi \mid \sigma_i)$ over all blocks σ_i of σ; namely,

$$H(\pi \mid \sigma) = \sum_i P(\sigma_i) H(\pi \mid \sigma_i)$$

On the other hand, we would like to think of information as a "quantity" that increases as we ask more and more questions about our experiment. Therefore, the conditional entropy of π given σ ought to be the net increase in entropy from σ to π. In other words, we require our entropy function to satisfy this property:

Entropy Property 5. If π is a finer partition than σ, then

$$H(\pi \mid \sigma) = H(\pi) - H(\sigma)$$

Remarkably, these five properties of the entropy uniquely determine the function H, except for the units used to measure entropy. This was shown by Shannon, and we give the proof in Section 9.4

We conclude by giving an interpretation of independence of partitions in terms of conditional entropy. Intuitively, if π and σ are independent, then their joint entropy $H(\pi \wedge \sigma)$ is the sum of the individual entropies: $H(\pi) + H(\sigma)$. In terms of conditional entropy, this says that $H(\pi \wedge \sigma \mid \sigma) = H(\pi)$.

9.2 The Shannon Coding Theorem

Twenty questions is a popular parlor game in which one person thinks of a subject and the other persons ask yes-no questions in order to guess the subject. Sometimes the questions also allow "maybe" as an answer. Let us write X for the subject to be guessed and $S_1, S_2, \ldots, S_n$ for the questions that the players can ask. We view all of these as random variables on a probability space of potential subjects. One consequence of Entropy Property 4 of Section 9.1 is that the joint entropy of the S_i's must be at least as large as the entropy of X. In particular, if the S_i's are yes-no questions, then $H_2(S_i) \leq 1$, and we must have that $H_2(S_1 \wedge S_2 \wedge \ldots \wedge S_n) \leq n$. The problem of finding a set of sufficient statistics for a random variable X is called the **coding problem** for X, and the sequence $S_1, S_2, \ldots, S_n$ is said to *code* X. In practice, the kinds of questions one may ask are usually restricted to some class of questions. As a result, devising particular codes is a highly nontrivial task.

Here is an example. Suppose that X takes value 0 with probability 0.85 and takes values 1 through 200 each with probability 7.5×10^{-4}. Then $H_2(X)$ is less than 1. However, because $201 > 2^7$, one can see that at least eight yes-no questions will be needed to achieve a sufficient statistic for X, even though the entropy suggests that one should be able to determine X with a single yes-no question.

A further complication of the coding problem is that one is usually required to answer a whole sequence of questions $X_1, X_2, \ldots$ produced by some process, rather than just a single question. Shannon's Theorem states that for any finite-valued random variable X, it is possible to efficiently encode a sequence of independent copies of X provided that:

- A block $X_1, X_2, \ldots, X_n$ is encoded all at one time.

- One is willing to accept a small probability of error, $\epsilon > 0$, that a block is incorrectly coded, such that ϵ can be made arbitrarily small.

Since one frequently encounters sequences of random variables in actual practice, it is not unreasonable to encode them in blocks. The small probability of error is also acceptable since it can be made arbitrarily small. Consider for example, the random variable X mentioned in the preceding paragraph. Since $H_2(X) < 1$, Shannon's Theorem says that there is a block of size n such that a sequence of n independent copies of X can be encoded with a sequence of n yes-no questions $S_1, S_2, \ldots, S_n$. Consider that the sequence of X_i's can take one of 201^n values, while the sequence of S_i's can take on at most 2^n possible values, and you will begin to appreciate Shannon's Theorem.

To give a statement of Shannon's Theorem we need to make precise the idea that a sequence of random variables "almost" codes for another sequence. Let $X_1, X_2, \ldots, X_n$ and $S_1, S_2, \ldots, S_r$ be two sequences of random variables. We say that $S_1, S_2, \ldots, S_r$ is **almost sufficient** for $X_1, X_2, \ldots, X_n$ with *confidence* $1 - \epsilon$ if there is an event A such that

1. $P(A) = 1 - \epsilon$, and

2. the conditional random variables $S_1 \mid A, \ldots, S_r \mid A$ are sufficient for $X_1 \mid A, \ldots, X_n \mid A$.

Put another way, condition (2) says that the joint partition $\pi(S_1) \wedge \ldots \wedge \pi(S_r)$, when restricted to A, is finer than $\pi(X_1) \wedge \ldots \wedge \pi(X_n)$, when also restricted to A.

The only other technical detail we need for the Shannon Coding Theorem is the **ceiling function**. For a real number x, the ceiling of x, written $\lceil x \rceil$, is the smallest integer larger than x. In R this function is called `ceiling`. We can now state the Shannon Coding Theorem. The proof is rather technical, so it is given later on in Section 9.4.

Shannon Coding Theorem. Let $X_1, X_2, \ldots$ be a sequence of independent, equidistributed, finite-valued random variables such that $H_2(X_i) = h$. For any $\epsilon > 0$, no matter how small, and any $\delta > 0$, no matter how small, there is an integer N such that for any block size $n \geq N$, one can find a sequence $S_1, S_2, \ldots, S_r$ of $r = \lceil hn + \delta n \rceil$ random variables each taking two values, which is almost sufficient for $X_1, X_2, \ldots, X_n$ with confidence $1 - \epsilon$.

The confidence $1 - \epsilon$ represents the probability that the S_i's are able to code for a particular sequence of values of $X_1, X_2, \ldots, X_n$. By entropy considerations, we know that at least $\lceil hn \rceil$ S_j's will be needed to code for the X_i's. The additional δn S_j's represent an extra set of S_j's beyond those required by entropy, but they can be chosen to be as small a fraction of the total set of S_j's as we please.

The usual form in which one sees this theorem is called the Shannon Channel Coding Theorem. The problem here is to transmit information through a noisy channel. The channel we consider is called the **binary symmetric channel**. Each bit of information one transmits through the BSC is either left alone or changed. The probability that it is changed is p, the same for all bits, and each bit is altered or not independently of the others.

Transmission through the BSC proceeds as follows. A message k bits long is first sent through an *encoder* where it is changed into a string of n bits. This string of bits is then transmitted through the BSC to a *decoder* that converts the received n bits into a string of k bits, which we hope is the same as the original message. The channel looks like this:

The problem is to design the encoder and decoder so that the probability of error per transmitted bit is smaller than some preassigned value and so that the *redundancy* n/k is as small as possible. Equivalently, we want the **rate of transmission** k/n to be as high as possible.

We may think of the noise as a sequence of Bernoulli random variables $X_1, X_2, \ldots$ that are added to the signal. Let $h = H_2(X_i) = -p \log_2(p) - q \log_2(q)$. Then the input signal plus the noise constitute a total entropy of $k + nh$ bits. The decoder can ask at most n questions about the data it receives, since it receives just n bits of data. From these n questions it must determine both the noise and the original signal, hence $k + nh \leq n$. Put another way, the decoder may ask just $n - k$ questions in order to determine the noise and eliminate it. Thus, $k + nh \leq n$ or $1 - h \geq k/n$. This says that the rate of communication through the BSC can never be greater than $1 - h$. One calls $1 - h$ the **capacity** of the channel.

The Shannon Channel Coding Theorem says that for any rate r less than the channel capacity $1 - h$, it is possible to choose k and n so that $k/n > r$, and to design an encoder and decoder so that the (average) probability of error per message bit is as small as we please. The proof is very similar to the proof of the Shannon Coding Theorem.

9.3 Continuous Entropy

We now consider what entropy means for continuous random variables. The concepts in this case are by no means as self-evident as in the case of a finite-valued random variable. We will need two new ways to measure entropy. The first is called **relative entropy**, and it is appropriate for random variables with bounded support. When a random variable has unbounded support, we will need yet another way to measure entropy called **Boltzmann entropy**.

The most obvious way to begin is to try to "finitize." Let X be a continuous random variable taking values in some finite interval. For simplicity, take this interval to be $[0, a]$. Now exactly as in Calculus, we *partition* (or subdivide) this interval into n *blocks* $B_1, B_2, \ldots, B_n$. The ith block is the subinterval $[(i - 1)a/n, ia/n)$. Define a new random variable Y_n that takes value $(i-1)a/n$ whenever X takes a value in the block B_i. We call Y_n the nth **truncation** of X. We show why we use this name by means of an example. Suppose that $a = 1$ and that $n = 1000$. Now imagine that we perform our experiment and that the outcome X is

$$0.1415926\ldots$$

The value of Y_{1000} in this case would be

$$0.1410000\ldots$$

i.e., we *truncate* the value of X to 3 decimal places. Clearly, the truncations of X will be better and better approximations as $n \to \infty$. Moreover, the truncations are finite-valued random variables. One might add that, in practice, one always uses a truncation in an actual experiment. It is only in our idealized mathematical models that one can speak of an arbitrary real number.

Now compute the entropy of Y_n. By definition of Y_n,

$$P\left(Y_n = \frac{(i - 1)a}{n}\right) = P\left(\frac{(i - 1)a}{n} \leq X < \frac{ia}{n}\right)$$

$$= F\left(\frac{ia}{n}\right) - F\left(\frac{(i - 1)a}{n}\right)$$

where $F(t)$ is the probability distribution of X. Thus:

$$H(Y_n) = -\sum_{i=1}^{n} P\left(Y_n = \frac{(i - 1)a}{n}\right) \log\left(P\left(Y_n = \frac{(i - 1)a}{n}\right)\right)$$

$$= -\sum_{i=1}^{n}\left(F\left(\frac{ia}{n}\right) - F\left(\frac{(i-1)a}{n}\right)\right)\log\left(F\left(\frac{ia}{n}\right) - F\left(\frac{(i-1)a}{n}\right)\right)$$

The crucial step in the computation is the mean value theorem of Calculus: if F is differentiable on the interval $[s,t]$, then for some x between s and t:

$$F'(x)(t-s) = F(t) - F(s)$$

We apply this to each block B_i. Each block has length a/n, so

$$
\begin{aligned}
H(Y_n) &= -\sum_{i=1}^{n}F'(x_i)\frac{a}{n}\log\left(F'(x_i)\frac{a}{n}\right) \\
&= -\sum_{i=1}^{n}F'(x_i)\frac{a}{n}\log(F'(x_i)) + \sum_{i=1}^{n}F'(x_i)\frac{a}{n}\log\left(\frac{n}{a}\right)
\end{aligned}
$$

where each x_i is some point in the block B_i. If we write $f(t) = F'(t)$ for the density of X, then the first term above is:

$$-\sum_{i=1}^{n}f(x_i)\log(f(x_i))\frac{a}{n}$$

This is just the Riemann sum for our partition of $[0,a]$. so as $n\to\infty$ this approaches:

$$-\int_{0}^{a}f(x)\log(f(x))dx$$

Next, consider the second term above. We may write this as:

$$\log\left(\frac{n}{a}\right)\sum_{i=1}^{n}f(x_i)\frac{a}{n}$$

Except for the factor $\log(n/a)$, we would have a Riemann sum for $\int_{0}^{a}f(x)dx$. Consequently, as $n\to\infty$,

$$H(Y_n) \approx \log(n) - \log(a) - \int_{0}^{a}f(x)\log(f(x))dx$$

So $H(Y_n)$ diverges to ∞.

The difficulty is easily seen. As we partition $[0,a]$ into finer and finer blocks, the random variable Y_n is taking an enormous number of values, some fractions of which are roughly equally likely. This is an artifact of our subdivision process and ought to be eliminated. We do this by measuring not the absolute entropy of Y_n but rather the difference between the entropy of Y_n and the maximum possible entropy of a random variable taking n values. We call this the **relative entropy** of Y_n:

$$RH_n(Y_n) = \text{ Relative entropy of } Y_n = H(Y_n) - \log(n)$$

In other words, instead of measuring how far Y_n is from being completely certain, we measure how close Y_n is to being completely random. For finite-valued random variables these two ways of measuring entropy are equivalent, but when we take the limit as $n \to \infty$, only the relative entropy converges. We therefore define:

$$RH(X) = \lim_{n \to \infty} RH_n(Y_n)$$

$$= -\int_0^a f(x)\log(f(x))dx - \log(a)$$

For this notion of entropy, the case of total randomness will be represented by a relative entropy of zero. Less uncertain random variables will have a negative relative entropy. Continuous random variables can have arbitrarily large negative entropy: complete certainty is impossible for continuous random variables.

Which continuous random variables will have entropy zero? In other words, what is the continuous analog of the equally likely probability distribution? To answer this we proceed as we did for finite-valued random variables. Let X be any continuous random variable taking values in $[0, a]$. Then

$$RH(X) = -\int_0^a f(x)\log(f(x))dx - \log(a)$$

$$= -\int_0^a f(x)\log(f(x))dx - \log(a)\int_0^a f(x)dx$$

$$= -\int_0^a f(x)(\log(f(x)) + \log(a))dx$$

$$= \int_0^a f(x)\log\left(\frac{1}{f(x)a}\right)dx$$

$$\leq \int_0^a \left(\frac{1}{f(x)a} - 1\right)dx$$

$$= \int_0^a \frac{dx}{a} - \int_0^a f(x)dx = 1 - 1 = 0$$

Now the logarithmic inequality tells us that the above inequality is an equality if and only if $\frac{1}{f(x)a} = 1$ or $f(x) = \frac{1}{a}$. In other words, the maximum entropy occurs precisely when X has the uniform distribution on $[0, a]$.

The notion of relative entropy is fine for random variables taking values in a finite interval, but most continuous random variables we have seen do not have this property. The most natural way to try to extend entropy to arbitrary continuous random variables is to use a limiting process similar to what we used for extending entropy from finite-valued random variables to finite-interval random variables. We will do this first for positive random variables before going on to the general case.

Let T be a positive continuous random variable, i.e., $P(T \leq 0) = 0$. For $a > 0$, we define the **restriction** of T to $[0, a]$ to be the random variable:

$$T_a = T \mid (T \leq a)$$

By this we mean that T_a takes the value of T *conditioned* on the occurrence of $(T \leq a)$. The probability distribution of T_a is given by:

$$P(T_a \leq t) = P(T \leq t \mid T \leq a)$$

The density of T_a is the same as that of T within the interval $[0, a]$, except that it must be "normalized" so that the total probability is 1. The normalization constant is:

$$C_a \;\; = \;\; \frac{1}{P(0 \leq X \leq a)}$$

$$= \;\; \frac{1}{\int_0^a f(x)dx}$$

The density of T_a is then given by:

$$\mathrm{dens}(T_a = t) = \begin{cases} C_a f(t) & \text{if } 0 \leq t \leq a \\ 0 & \text{if } t < 0 \text{ or } t > a \end{cases}$$

where $f(t) = \mathrm{dens}(T = t)$. Clearly, T_a will be a better and better approximation to T as $n \to \infty$. As with truncations, restrictions are always used in an actual experiment.

It would be nice if we could define the relative entropy of T to be the limit of the relative entropy of T_a as $n \to \infty$, but unfortunately this diverges. As before, the difficulty is that we are not measuring entropy properly. The case of total randomness, entropy zero, is the uniform distribution on $[0, a]$; but as $a \to \infty$, this distribution ceases to make sense. So we are attempting to measure the entropy of T relative to that of a nonexistent distribution! What should we do? We no longer have either total certainty or total uncertainty from which to measure entropy.

What we do is to "renormalize" our measurement of entropy so that the entropy of the uniform distribution on $[0, a]$ is $\log(a)$ rather than 0. We do this by analogy with the "equally likely" distribution on n points, whose entropy is $\log(n)$. There is no really convincing justification for this choice of normalization. The entropy defined in this way is called the **Boltzmann entropy** or **differential entropy**:

$$H(T) \;\; = \;\; \lim_{a \to \infty} \left(RH(T_a) + \log(a) \right)$$

$$= \;\; \lim_{a \to \infty} -\int_0^a C_a f(t) \log(C_a f(t)) dt$$

$$= \;\; -\int_0^\infty f(t) \log(f(t)) dt, \qquad \text{because } C_a \to 1 \text{ as } a \to \infty$$

if this improper integral exists. The Boltzmann entropy can also be defined for continuous random variables that are not positive. To do this, we must allow both the upper and lower bound of the interval to increase to infinity. The bounds of the improper integral above are then both infinite. Note that we use the same letter "H" to denote both discrete and continuous entropy. The formulas are analogous, and they apply to different kinds of random variable, so there should not be any confusion about this ambiguity.

We now ask which positive continuous random variables take maximum Boltzmann entropy. Let T be such a random variable, and let $\mu = E(T)$ be its expectation. To maximize the entropy of T we use a method known as the **Lagrange multiplier method**. This method is appropriate wherever we wish to maximize some quality subject to constraints. In this case, the constraints that the density function $f(t)$ of T must satisfy are:

$$\int_0^\infty f(t)dt = 1 \quad \text{and} \quad \int_0^\infty tf(t)dt = \mu$$

Multiply the constraints by constants α and β to be determined later, and subtract both from the entropy of T. Then proceed as in all our previous maximum entropy calculations:

$$
\begin{aligned}
H(T) - \alpha - \beta\mu &= -\int_0^\infty f(t)\log(f(t))dt - \alpha\int_0^\infty f(t)dt - \beta\int_0^\infty tf(t)dt \\
&= \int_0^\infty f(t)(-\log(f(t)) - \alpha - \beta t)dt \\
&= \int_0^\infty f(t)\log\left(\frac{1}{f(t)e^{\alpha+\beta t}}\right)dt \\
&\leq \int_0^\infty f(t)\left(\frac{1}{f(t)e^{\alpha+\beta t}} - 1\right)dt \\
&= \int_0^\infty e^{-\alpha-\beta t}dt - \int_0^\infty f(t)dt \\
&= \left[\frac{e^{-\alpha-\beta t}}{\beta}\right]_0^\infty - 1 \\
&= \frac{e^{-\alpha}}{\beta} - 1, \quad \text{if } \beta > 0
\end{aligned}
$$

By the basic logarithmic inequality, the above inequality is an equality if and only if $f(t) = e^{-\alpha-\beta t}$. We now use the constraints to solve for α and β:

$$1 = \int_0^\infty f(t)dt = \int_0^\infty e^{-\alpha-\beta t}dt = \frac{e^{-\alpha}}{\beta}$$

$$\mu = \int_0^\infty tf(t)dt = \int_0^\infty te^{-\alpha-\beta t}dt = \frac{e^{-\alpha}}{\beta^2}$$

Now the first equation implies that $\beta = e^{-\alpha}$, and the second equation gives us that $\beta^2\mu = e^{-\alpha}$. Combining these, we obtain that $\beta^2\mu = \beta$, and hence that $\beta = e^{-\alpha} = 1/\mu$. Substituting these into our formula for $f(t)$ from above, we have:

$$f(t) = e^{-\alpha - \beta t} = e^{-\alpha} e^{-\beta t} = \frac{1}{\mu} e^{-t/\mu},$$

i.e., T is exponentially distributed with parameter $1/\mu$. Moreover, the entropy of T is $H(T) = \alpha + \beta\mu = 1 + \log(\mu)$. Therefore, we see that as μ gets large, T can have arbitrarily high entropy. So there is no positive random variable having maximum entropy among all such random variables.

The reason we had to specify the expectation of a positive random variable in order to find the one having maximum entropy arises from an important distinction between finite entropy and Boltzmann entropy: the choice of units in which we measure our random variable alters the Boltzmann entropy but has no effect on the finite entropy. For example, if X is uniformly distributed on $[0, 1]$, then $Y = 2X$ is uniformly distributed on $[0, 2]$. Although Y represents the same phenomenon as X, the difference being the units with which we measure distance, obviously an observation of X is more certain than an observation of Y (one bit more certain to be precise). More generally, for any continuous random variable X, we have that $H(cX) = H(X) + \log(c)$, when $c > 0$.

In order to speak of the entropy of the phenomenon represented by a random variable, independent of scale changes, we introduce yet one more notion of entropy. The **standard entropy** of a random variable is the Boltzmann entropy of its standardization. Using the notion of standard entropy we can ask an important question. Which continuous random variables have the maximum standard entropy? The answer is the normally distributed random variables.

We now show how this is done. Since we want the random variable to have maximum standard entropy, we may assume it is standard. Let X be such a random variable, and let $f(x)$ be its density. We maximize $H(x)$ by the method of Lagrange multipliers we used above, but now there are three constraints:

$$\int_0^\infty f(x)dx = 1, \quad \int_0^\infty xf(x)dx = 0 \quad \text{and} \quad \int_0^\infty x^2 f(x)dx = 1$$

One of the constraints being zero, we need just two parameters α and β. Using the same technique as before, it is easy to see that the maximum entropy will occur for a distribution having the form

$$\text{dens}(X = x) = \exp\left(-\alpha - \beta x^2\right)$$

We now use the constraints to solve for α and β.

$$1 = \int_{-\infty}^\infty f(x)dx \;=\; \int_{-\infty}^\infty \exp\left(-\alpha - \beta x^2\right)dx = e^{-\alpha}\sqrt{\pi/\beta}$$

$$1 = \int_{-\infty}^\infty x^2 f(x)dx \;=\; \int_{-\infty}^\infty x^2 \exp\left(-\alpha - \beta x^2\right)dx$$

$$= -\left[\frac{x\exp\left(-\alpha - \beta x^2\right)}{2\beta}\right]_{-\infty}^\infty$$

$$+ \frac{1}{2\beta} \int_{-\infty}^{\infty} \exp\left(-\alpha - \beta x^2\right) dx$$

$$= 0 + \frac{1}{2\beta} e^{-\alpha} \frac{\pi}{\beta}$$

Now the first equation implies that $e^{-\alpha} = \sqrt{\beta/\pi}$, while the second equation gives us that $e^{-\alpha} = 2\beta\sqrt{\beta/\pi}$ from which we conclude that $\beta = 1/2$ and $e^{-\alpha} = \frac{1}{2\pi}$. Hence, the maximum entropy among all standard continuous random variables is achieved precisely when

$$\text{dens}(X = x) = \frac{1}{\sqrt{2\pi}} \exp\left(-x^2/2\right)$$

which we recognize as the standard normal distribution.

It should be apparent that there is a general procedure for characterizing the distributions that have maximum entropy. This was first shown by Boltzmann. The idea is to define a collection of distributions by their support and by one or more expectation constraints. For example, the uniform distribution is supported on $[0, a]$, and the exponential distribution is supported on $[0, \infty)$. An expectation constraint usually specifies the expectation of a power or logarithm of the random variable. We always have the constraint that the density is a probability distribution, which is the constraint $\int f(x)dx = 1$, where the integral is over the support of the random variable. If we impose the constraint that $E(X) = m$, then all of the distributions in the collection have mean m. A constraint on both $E(X)$ and $E(X^2)$ will constrain all of the distributions in the collection to have the same mean and variance. Having defined the collection by its constraints, one can use Lagrange multipliers to determine the distribution that has the maximum entropy in the collection, if there is one. For example, among all distributions on $[0, 1]$ for which $E(\log(X)) = -n$, the one with the maximum entropy is the last order statistic (i.e., the maximum) of n points dropped on $[0, 1]$. We leave the details as an exercise.

The principal processes of probability theory are all determined by maximum entropy properties. Here is a summary of the most important ones.

Type	Support	Definition
Discrete	Finite set	$H(\pi) = -\sum_{i=1}^{n} P(B_i) \log(P(B_i))$
		$H(X) = H(\pi(X))$
Relative	$[0, a]$	$RH(X) = -\int_0^a f(x) \log(f(x)) \, dx - \log(a)$
Boltzmann	$[0, \infty)$	$H(X) = -\int_0^\infty f(x) \log(f(x)) \, dx$
Standard	$(-\infty, \infty)$	$SH(X) = H(\frac{X-m}{\sigma}) = H(X) - \log(\sigma)$

Types of Entropy

Process	Distribution/Model	Random Variables
Sampling	Finite uniform on n points	Discrete
	Sample of size 1 from a population of size n	Supported on n values
Uniform	Uniform on $[0, a]$	Continuous
	Drop one point completely at random in $[0, a]$	Supported on $[0, a]$ (Relative entropy or Boltzmann entropy)
Poisson	Exponential, intensity α	Positive continuous
	Continuous memoryless waiting time, intensity α	The mean is $1/\alpha$ (Boltzmann entropy)
Normal	Normal, $N(m, \sigma^2)$	Continuous
Sampling	Sample a member of a normal population with mean m and variance σ^2	The mean is m The variance is σ^2 (Boltzmann entropy)

Maximum Entropy Distributions

Distribution	H(X)	SH(X)
Finite uniform on n points	$\log_2(n)$ bits	–
Uniform on $[0, a]$	$\log(a)$	$\log(12)/2$ ≈ 1.2425 nats
Exponential intensity α	$1 - \log(\alpha)$	1
Normal, $N(m, \sigma^2)$	$\log(\sigma\sqrt{2\pi e})$	$\log(\sqrt{2\pi e})$ ≈ 1.4189 nats

Values of Entropy

9.4 Proofs of Shannon's Theorems

In this section, we give the proofs of two theorems due to Shannon. The first is the uniqueness of the entropy function, and the second is his Coding Theorem. The proofs are very technical, so this section is optional.

Uniqueness of Entropy. If H is a function satisfying the five properties of an entropy function in Section 9.1, then there is a constant C such that H is given by:

$$H(p_1, p_2, \ldots, p_n) = C\sum_{i=1}^{n} \log_2(p_i)$$

Proof. We first apply Entropy Property 4 to the partition consisting of just one block: Ω itself. By definition, $H(\Omega \mid \Omega)$ is the same as $H(\Omega)$. Therefore, $H(\Omega) = H(\Omega) - H(\Omega) = 0$.

We now define a function $f(n)$ by $H\left(\frac{1}{n}, \frac{1}{n}, \ldots, \frac{1}{n}\right)$. We have just shown that $f(1) = 0$, and we want to calculate $f(n)$ in general. By Entropy Properties 2 and 5, we show that $f(n)$ is an increasing function as follows:

$$\begin{aligned}
f(n) &= H\left(\frac{1}{n}, \frac{1}{n}, \ldots, \frac{1}{n}\right) \\
&= H\left(\frac{1}{n}, \frac{1}{n}, \ldots, \frac{1}{n}, 0\right) \\
&\leq H\left(\frac{1}{n+1}, \frac{1}{n+1}, \ldots, \frac{1}{n+1}\right) \\
&= f(n+1)
\end{aligned}$$

Next, we consider a partition σ consisting of n^{k+1} blocks, each of which has probability $\frac{1}{n^{k+1}}$. Then subdivide each of these into n parts, each of which has the same probability. Call the resulting partition π. The conditional entropy $H(\pi \mid \sigma_i)$ for each block σ_i is clearly given by $f(n)$. Thus the conditional entropy $H(\pi \mid \sigma)$ is $f(n)$. By Entropy Property 4, we know that:

$$f(n) = H(\pi \mid \sigma) = H(\pi) - H(\sigma) = f(n^k) - f(n^{k-1})$$

Now apply this fact k times to obtain: $f(n^k) = kf(n)$.

Let n and k be two positive integers. Since the exponential function is an increasing function, there is an integer b such that: $2^b \leq n^k \leq w^{b+1}$. By applying the two facts about $f(n)$ that we obtained above to this relation, we get

$$f(2^b) \leq f(n^k) \leq f(2^{b+1})$$

because f is increasing, and

$$bf(2) \leq kf(n) \leq (b+1)f(2)$$

because $f(n^k) = kf(n)$. Dividing these inequalities by $kf(2)$ gives us:

$$\frac{b}{k} \leq \frac{f(n)}{f(2)} \leq \frac{b+1}{k}$$

Now apply the increasing function $\log_2$ to the inequalities $2^b \leq n^k \leq w^{b+1}$ to get that $b \leq k\log_2(n) \leq b + 1$. If we divide these by k we obtain:

$$\frac{b}{k} \leq \log_2(n) \leq \frac{b+1}{k}$$

It therefore follows that both $f(n)/f(2)$ and $\log_2(n)$ are in the interval $\left[\frac{b}{k}, \frac{b}{k} + \frac{1}{k}\right]$. This implies that $f(n)/f(2)$ and $\log_2(n)$ can be no farther apart than $\frac{1}{k}$, the length of this interval. But n and k were arbitrary positive integers. So if we let k get very large, we are forced to conclude that $f(n)/f(2)$ coincides with $\log_2(n)$. Thus for positive integers n, we have:

$$f(n) = f(2)\log_2(n)$$

Define C to be the constant $-f(2)$. Since $f(2) \geq f(1) = 0$, we know that C is negative.

We next consider a set $\{p_1, p_2, \ldots, p_n\}$ of positive rational numbers such that $p_1 + p_2 + \cdots + p_n = 1$. Let N be their common denominator, i.e., $p_i = \frac{a_i}{N}$, for all i, where each a_i is an integer and $a_1 + a_2 + \cdots + a_n = N$. Let σ be a partition corresponding to the set of probabilities $\{p_1, p_2, \ldots, p_n\}$. Let π be a partition obtained by breaking up the ith block of σ into a_i parts. Then every block of π has probability $1/N$. By definition of conditional entropy, $H(\pi \mid \sigma_i) = f(a_i)$ and

$$H(\pi \mid \sigma) = \sum_i p_i H(\pi \mid \sigma_i) = \sum_i p_i f(a_i) = -C \sum_i p_i \log_2(a_i)$$

By Entropy Property 4, on the other hand, we have:

$$H(\pi \mid \sigma) = H(\pi) - H(\sigma) = f(N) - H(\sigma) = -C\log_2(N) - H(\sigma)$$

Combining the two expressions for $H(\pi \mid \sigma)$ gives us:

$$\begin{aligned}
H(\sigma) &= -C\log_2(N) + C\sum_i p_i \log_2(a_i) \\
&= C\left(\sum_i p_i \log_2(N) + \sum_i p_i \log_2(a_i)\right) \\
&= C\left(\sum_i p_i (\log_2(a_i) - \log_2(N))\right) \\
&= C\left(\sum_i p_i \log_2\left(\frac{a_i}{N}\right)\right) \\
&= C\sum_i p_i \log_2(p_i)
\end{aligned}$$

By Entropy Property 3, H must have this same formula for all sets $\{p_1, p_2, \ldots, p_n\}$ on which it is defined. This completes the proof.

We leave it as an exercise to show that the above formula for entropy actually satisfies the five postulated properties. We now show the Shannon Coding Theorem.

Shannon Coding Theorem. Let $X_1, X_2, \ldots$ be a sequence of independent, equidistributed, finite-valued random variables such that $H_2(X_i) = h$. For any $\epsilon > 0$, and any $\delta > 0$, there is an integer N such that for any block size $n \geq N$, one can find a sequence $S_1, S_2, \ldots, S_r$ of $r = \lceil hn + \delta n \rceil$ two-valued random variables, which are almost sufficient for $X_1, X_2, \ldots, X_n$ with confidence $1 - \epsilon$.

Proof. We begin by defining a sequence of random variables $Y_1, Y_2, \ldots$ by decreeing that if X_i takes value k then Y_i takes value $P(X_i = k)$. For example, if X_i takes values $1, \ldots, m$ each with probability $\frac{1}{m}$ then Y_i takes value $\frac{1}{m}$ with probability 1. These random variables have two properties we need. The first is that the Y_i's are independent. This is an immediate consequence of the fact that the X_i's are independent. The second fact is that the expected value of $\log_2(1/Y_i)$ is h, the entropy of X_i. To see this we compute:

$$E\left(\log_2\left(\frac{1}{Y_i}\right)\right) = \sum_{k=1}^{m} \log_2\left(\frac{1}{P(X_i = k)}\right) P(X_i = k)$$

$$= H_2(X_i) = h$$

because $\log_2\left(\frac{1}{Y_i}\right)$ takes value $\log_2\left(\frac{1}{P(X_i) = k}\right)$ when $(X_i = n)$.

The sequence $\log_2\left(\frac{1}{Y_1}\right), \log_2\left(\frac{1}{Y_2}\right), \ldots$ is a sequence of independent, equidistributed random variables each with mean h. By the Law of Large Numbers,

$$P\left(\lim_{n \to \infty} \left(\frac{\log_2\left(\frac{1}{Y_1}\right) + \log_2\left(\frac{1}{Y_2}\right) + \ldots + \log_2\left(\frac{1}{Y_n}\right)}{n}\right) = h\right) = 1$$

Now

$$\log_2\left(\frac{1}{Y_1}\right) + \log_2\left(\frac{1}{Y_2}\right) + \ldots + \log_2\left(\frac{1}{Y_n}\right) = \log_2\left(\frac{1}{Y_1 Y_2 \cdots Y_n}\right)$$

Therefore:

$$P\left(\lim_{n \to \infty} \left(\frac{1}{n}\log_2\left(\frac{1}{Y_1 Y_2 \cdots Y_n}\right)\right) = h\right) = 1$$

This says that for n large enough the expression

$$\frac{1}{n}\log_2\left(\frac{1}{Y_1 Y_2 \cdots Y_n}\right)$$

will be as close to h as we please with as high a probability as we please. The probability we want is $1 - \epsilon$, and we want the expression above to be within δ of h with this probability. In other words,

$$P\left(\left|\frac{1}{n}\log_2\left(\frac{1}{Y_1 Y_2 \cdots Y_n}\right) - h\right| < \delta\right) = 1 - \epsilon$$

As one might expect, the event A in the definition of a set of almost sufficient statistics will be the event:

$$A = \left(\left|\frac{1}{n}\log_2\left(\frac{1}{Y_1 Y_2 \cdots Y_n}\right) - h\right| < \delta\right)$$

$$= \left(\left|\log_2\left(\frac{1}{Y_1 Y_2 \cdots Y_n}\right) - nh\right| < n\delta\right)$$

$$= \left(-n\delta < \log_2\left(\frac{1}{Y_1 Y_2 \cdots Y_n}\right) - nh < n\delta\right)$$

Exponentiating every term in the above pair of inequalities preserves the inequalities so:

$$A = \left(s^{-n\delta} < \frac{2^{-nh}}{Y_1 Y_2 \cdots Y_n} < 2^{n\delta}\right)$$

$$= \left(2^{-n\delta+nh} < \frac{1}{Y_1 Y_2 \cdots Y_n} < 2^{n\delta+nh}\right)$$

$$= \left(2^{n\delta-nh} < Y_1 Y_2 \cdots Y_n < 2^{-n\delta-nh}\right)$$

We are now ready for the crucial step in the proof. We count how many blocks of the joint partition $\pi(X_1) \wedge \pi(X_2) \wedge \ldots \wedge \pi(X_n)$ are contained in the event A. Suppose that there are r such blocks; call them $B_1, B_2, \ldots, B_r$. Each of these blocks is of the form $(X_1 = i_1) \cap (X_2 = i_2) \cap \cdots \cap (X_n = i_n)$. If we sum the probabilities of *all* such events, we get 1. Since the X_i's were assumed to be independent, this means that:

$$\sum_{i_1, \ldots, i_n} P(X_1 = i_1) P(X_2 = i_2) \cdots P(X_n = i_n) = 1$$

Now each of the above factors is, by definition, the value that the corresponding Y_i takes. Therefore,

$$\sum_{i_1 : X_1 = i_1} \sum_{i_2 : X_2 = i_2} \cdots \sum_{i_n : X_1 = i_n} Y_1 Y_2 \cdots Y_n = 1$$

If we sum just over those blocks contained in A, we find that

$$\sum_{j=1}^{r} Y_1 Y_2 \cdots Y_n \leq 1$$

each Y_i taking the value appropriate to the blocks B_j. But for these blocks we know that $Y_1 Y_2 \cdots Y_n > 2^{-nh-n\delta}$. Hence,

$$\sum_{j=1}^{r} 2^{-nh-n\delta} < \sum_{j=1}^{r} Y_1 Y_2 \cdots Y_n \leq 1$$

Now the terms of the sum $\sum_{j=1}^{r} 2^{-nh-n\delta}$ do not depend on the block B_j. So we find that $r 2^{-nh-n\delta} < 1$ or that $r < 2^{nh+n\delta}$, i.e., there are fewer than $2^{nh+n\delta}$ blocks in A.

We are now ready to code the random variables $X_1, X_2, \ldots, X_n$. Number the blocks in A in binary using $\lceil nh + n\delta \rceil$ binary digits starting with $00\ldots01$ and ending with $11\ldots11$. Because $r < 2^{nh+n\delta}$, we will have at most $2^{\lceil nh+n\delta \rceil}$ blocks in A. We assign the binary number $00\ldots00$ to all blocks outside A.

The random variable S_i is defined to be the ith digit of the block in which the outcome occurs. By definition, the random variables $S_1, S_2, \ldots, S_{\lceil nh+n\delta \rceil}$, when restricted to A, are sufficient for $X_1, X_2, \ldots, X_n$, when restricted to A. When all the S_i's take the value 0, we are unable to determine the values of the X_i's, but when the S_i's take any other set of values, we can compute all the values of the X_i's. This completes the proof of the Shannon Coding Theorem.

9.5 Exercises

Exercise 9.1 A visitor to an imaginary country finds that the inhabitants of city A always tell the truth, while the inhabitants of city B always lie. The visitor wants to know which city he is in. He may ask only yes-no *direct* questions. (A question such as "If I were to ask you what city I am in, what would you say?" are indirect and would only confuse an inhabitant.) How many questions must the visitor ask? Note that an inhabitant of city A could be temporarily residing in city B and vice versa.

Exercise 9.2 You are given 12 coins, 1 of which is counterfeit, and a balance. The counterfeit coin is either light or heavy, you do not know which. How many weighings are necessary to determine which coin is counterfeit?

Exercise 9.3 You are given 5 coins, some of which may be counterfeit. A counterfeit coin is lighter than a good coin, and all counterfeit coins weigh the same. Again you are given a balance. How many weighings are necessary to find all of the counterfeit coins?

Exercise 9.4 A deck of 52 cards is said to have been *randomly shuffled* if all 52! permutations are equally likely. What we normally regard as being a random shuffle is in fact very far from random. For example, the cut-and-interlace shuffle (also called the *perfect shuffle*) has the following property. If a new deck (in the standard bridge order: 2 of clubs, 3 of clubs,..., ace of spades) is perfectly shuffled, "cut" at a point $4m$ cards from the top, and dealt as in a bridge game, then every player will receive all the cards of 1 suit. It is known that frequent bridge players are capable of consistently achieving a perfect shuffle. How much information is contained in a random shuffle? In a perfect shuffle? In a random cut? How many independent random cuts are needed to achieve a completely random shuffle?

Exercise 9.5 Let N be an integer between 1 and 2000. Divide N by 6, 10, 22 and 35, and find the remainders. How much information about N do these 4 remainders tell you?

Exercise 9.6 Show that the entropy of the normal distribution $N(m, \sigma^2)$ is $\log_2(\sigma\sqrt{2\pi e})$ bits. Use this to answer the following question. A coin is tossed 1000 times, getting 368 heads and 632 tails. How much additional information will one more toss of the coin give?

Exercise 9.7 Suppose that the imaginary country of Exercise 9.1 has another city C where the inhabitants alternately tell the truth and lie. What is the smallest number of questions the visitor must ask to find out which city he is in?

Exercise 9.8 Generalize problems 2 and 3 above to an arbitrary number of coins.

Exercise 9.9 Let $f(x)$ be a differentiable function defined on $[0, a]$. Assume that $f(0) = 0$ and that $|f'(x)| \leq b$ for all x in $[0, a]$. Find an upper bound on the amount of information necessary to determine the value of $f(x)$ at *every* $x \in [0, a]$ with an error not exceeding $\epsilon > 0$.

Exercise 9.10 You are playing a variation of "20 questions." A chooses a number between 1 and 1,000,000, and B must find this number by asking yes-no questions about it, except that B asks *random* questions. How long does it take for B to find the number?

Exercise 9.11 Let $k_1, k_2, \ldots$ be a sequence of numbers such that

$$\lim_{n \to \infty} (k_n - \log_2(n)) = c$$

Let T_n be the random variable in Exercise 9.10 above but for the problem of guessing a number between 1 and 2^n. Prove that

$$\lim_{n \to \infty} P(T_n = k_n) = \exp\left(\frac{-1}{2c}\right)$$

9.6 Answers to Selected Exercises

Answer to 9.1. The visitor wishes to know only which city he is in: not whether the inhabitant being questioned is a truth-teller or a liar. Therefore, he needs only *one bit* of information. Entropy suggests that there ought to be a single yes-no question that will determine the city. It is possible, however, that the restrictions placed on the kind of question that may be asked preclude the existence of such a question, so this is not such a trivial problem. Here is a question that will work: "Are you an inhabitant of this city?" If the answer is yes, then you are in city A, otherwise you are in city B. This is obvious for the truth tellers. The liars will lie about which city they inhabit, and so will give the same answer as the truth teller.

Answer to 9.2. There are really two questions here. One might just find the counterfeit coin or one might both find the bad coin and also determine whether it is heavy or light relative to a good coin. The first question requires $\log_3(12) \approx 2.26$ trits of information, while the second requires $\log_3(24) \approx 2.89$ trits. In either case, at least three weighings will be required to find the bad coin. Actually devising a procedure for finding the bad coin is far from easy, however, because there are great restrictions on the kinds of questions that may be asked (i.e., the partitions into which the sample space can be subdivided).

The sample space Ω in this problem has 24 sample points which we label Ah, Al, Bh, Bl, ..., Lh, Ll, where, for example, "Ah" means the first coin is too heavy. Let W_1, W_2, W_3 denote the three weighings, whose outcomes may be +1, -1 or 0, depending on whether the left side is heavier, lighter or the same, respectively, as the right side. The two questions we might ask correspond to two partitions of Ω. If we just want to know which coin is bad, the partition is

$$\{Ah, Al\}, \{Bh, Bl\}, \ldots, \{Lh, Ll\}$$

If we also want to know whether the bad coin is heavy or light, the partition is into all the sample points, i.e.,

$$\{Ah\}, \{Al\}, \{Bh\}, \ldots, \{Ll\}$$

which is finer than the first partition. We will find a weighing procedure that answers the second question, and hence also the first, in three or fewer weighings. We do this by always choosing the next weighing so as to maximize the information that it produces, For this purpose we assume that the probability on Ω is the equally likely probability measure. Other probability measures will have different results.

For example, suppose that one puts the first two coins on the scale. We write this A/B. If the scale tips left, then either the first coin is heavy or the second coin is light: $\{Ah, Bl\}$. If the scale tips right, then either the first coin is light or the second coin is heavy: $\{Al, Bh\}$. If the scale does not tip, then both coins are good coins which represent all of the other sample points in Ω. The entropy of this partition is

$$-\frac{2}{24}\log_3\left(\frac{2}{24}\right) - \frac{2}{24}\log_3\left(\frac{2}{24}\right) - \frac{20}{24}\log_3\left(\frac{20}{24}\right) \approx 0.52 \text{ trits}$$

We need to reach 2.89 trits to solve the problem so this is much too low. So we need to use some other weighing. The number of possibilities seem to be very large, but it is easy to reduce the possibilities by using symmetry. For example, any pair of coins will produce the same entropy as A/B. Clearly, for the first weighing we just need to consider how many coins to compare. If one compares k coins with k coins, then the partition is easily seen to consist of one block with the k coins on the left as heavy and the k coins on the right as light ($2k$ sample points in all), one block with the reverse (also $2k$ sample points), and one block with the rest of the sample points. If $k = 4$, then the three blocks will have the same size, so the information produced will be exactly 1 trit, which is the maximum entropy for a weighing. Any 8 coins can be used in the first weighing, so we use $ABCD/EFGH$. Based on what happens on this weighing, here is the rest of the procedure:

$(W_1 = +1)$	$(W_1 = -1)$	$(W_1 = 0)$
$\Rightarrow$ ABEF/CGIJ	$\Rightarrow$ ABEF/CDIJ	$\Rightarrow$ IJ/BK
$(W_2 = +1) \Rightarrow$ A/B	$(W_2 = +1) \Rightarrow$ E/F	$(W_2 = +1) \Rightarrow$ I/J
$(W_3 = +1) \Rightarrow$ Ah	$(W_3 = +1) \Rightarrow$ Eh	$(W_3 = +1) \Rightarrow$ Ih
$(W_3 = -1) \Rightarrow$ Bh	$(W_3 = -1) \Rightarrow$ Fh	$(W_3 = -1) \Rightarrow$ Jh
$(W_3 = 0) \Rightarrow$ Gl	$(W_3 = 0) \Rightarrow$ Cl	$(W_3 = 0) \Rightarrow$ Kl
$(W_2 = -1) \Rightarrow$ E/F	$(W_2 = -1) \Rightarrow$ A/B	$(W_2 = -1) \Rightarrow$ I/J
$(W_3 = +1) \Rightarrow$ Fl	$(W_3 = +1) \Rightarrow$ Bl	$(W_3 = +1) \Rightarrow$ Jl
$(W_3 = -1) \Rightarrow$ El	$(W_3 = -1) \Rightarrow$ Al	$(W_3 = -1) \Rightarrow$ Il
$(W_3 = 0) \Rightarrow$ Ch	$(W_3 = 0) \Rightarrow$ Gh	$(W_3 = 0) \Rightarrow$ Kh
$(W_2 = 0) \Rightarrow$ AB/DH	$(W_2 = 0) \Rightarrow$ AB/DH	$(W_2 = 0) \Rightarrow$ A/L
$(W_3 = +1) \Rightarrow$ Hl	$(W_3 = +1) \Rightarrow$ Dl	$(W_3 = +1) \Rightarrow$ Ll
$(W_3 = -1) \Rightarrow$ Dh	$(W_3 = -1) \Rightarrow$ Hh	$(W_3 = -1) \Rightarrow$ Lh

Note that the coins we use for each weighing, after the first one, depend on the outcomes of the previous weighings, since each weighing changes the sample space to the corresponding block of the partition. For the second weighing, the sample space will have 8 elements, so it cannot be partitioned into blocks having equal size. The best one can do is partition it into blocks of sizes 3, 3 and 2, which has entropy equal to 0.985 trits. One can find which weighings to use by trial and error, keeping in mind the objective of achieving the desired partition. Alternatively, one could write a program to examine all possible choices for weighings and select the first one that achieves the desired partition. For the third weighing, the sample space will be reduced to either 2 or 3 sample points, so it is quite easy to find an appropriate weighing to use in each case.

If we only wish to find the bad coin, not whether it is light or heavy, then the event $(W_1 = 0) \cap (W_2 = 0)$ is sufficient for determining that L is the bad coin (when using the procedure above). Thus, 3 weighings are not needed in this case, and only about 2.9 weighings are needed, on the average, to find the bad coin. This is much more than the 2.26 suggested by entropy considerations. Can you do better than 2.9?

An important point that is easily missed is that there are 24, not 12, sample points in Ω, regardless of which question is asked. This confusion is what leads one to think that the outcomes $(W_1 = +1)$ and $(W_1 = -1)$ only tell one that I, J, K and L are good coins. In fact, these outcomes tell one much more, as we saw above. This is crucial in subsequent weighings. For example, if $(W_1 = +1)$ and $(W_2 = 0)$, then $(W_3 = +1)$ does not seem to distinguish between the outcomes Dh and Hh. But if we look back at the first weighing, we see that D and H were on different sides of the scale, and the side of the scale with the D coin was heavier.

What about the case of 13 coins, one of which is bad? We will certainly need $\log_3(26) \approx 2.966$ trits of information, so that at least 3 weighings will be necessary. Unfortunately, one cannot find both the bad coin and whether it is light or heavy in only 3 weighings. The difficulty is that the same number of coins must be both sides of the balance. As a result, the partition of the first weighing, $\pi(W_1)$, can only have blocks with sizes that are even numbers, and hence 1 block must have 10 or more sample points. But for such a block, the entropy is $\log_3(10) \approx 2.096$ trits, so one cannot distinguish all sample points in the block with only 2 more weighings. It is possible to "break" this symmetry if one is also given 1 more coin that is known to be a good coin. This allows us to weigh the coins in such a way as to partition Ω into blocks of sizes 9, 9 and 8. We omit the details. The important point is that the restrictions on the kind of questions one may ask heavily influence whether a given question may be answered in the number of trials suggested by the entropy.

Answer to 9.4.

$$H_2(\text{random shuffle}) = \log_2(52!) \approx 226 \text{ bits}$$
$$H_2(\text{perfect shuffle}) = \log_2(1) = 0 \text{ bits}$$
$$H_2(\text{random cut}) = \log_2(52) \approx 5.7 \text{ bits}$$

The entropy for the random cut assumes that one can "cut" the deck by doing nothing to it as well as by dividing the deck into two parts and switching them. It also assumes that all 52 possibilities are equally likely.

No matter how many perfect shuffles one makes, one can never achieve a random shuffle, because there is no randomness at all in such a shuffle. As long as one performs a fixed number of perfect shuffles, there is no entropy.

At first it looks like $226/5.7 \approx 40$ random cuts would achieve a random shuffle, but this is not true. The reason is that random cuts are not independent. For example, a cut of 5 cards followed by one of 7 cards is the same as a single cut of 12 cards: a cut is a "circular shift" of the deck.

Answer to 9.5. It looks as though the remainders give one

$$\log_2(6) + \log_2(10) + \log_2(22) + \log_2(35) \approx 15.5 \text{ bits of information}$$

because there are 6 possibilities for the first remainder, 10 for the second, etc. However, the remainders are not independent. For example, if the division of N by 6 has remainder 3, then the division of N by 10 cannot have an even number as a remainder. The actual number of bits of information is $\log_2(2000) \approx 11$ bits, because N is determined by the remainders.

Answer to 9.6. Let X be $N(m, \sigma^2)$, and let $f(x)$ be the density of X. Then

$$H(X) = -\int_{-\infty}^{\infty} f(x)\log(f(x))dx$$

$$= \int_{-\infty}^{\infty} f(x)\log\left(\sigma\sqrt{2\pi}\exp\left(\frac{(x-m)^2}{2\sigma^2}\right)\right)dx$$

$$= \int_{-\infty}^{\infty} f(x)\log\left(\sigma\sqrt{2\pi}\right)dx + \int_{-\infty}^{\infty} \frac{(x-m)^2}{2\sigma^2}f(x)dx$$

$$= \log\left(\sigma\sqrt{2\pi}\right) + \frac{E\left((X-m)^2\right)}{2\sigma^2}$$

$$= \log\left(\sigma\sqrt{2\pi}\right) + \frac{\sigma^2}{2\sigma^2}$$

$$= \log\left(\sigma\sqrt{2\pi}\right) + 1/2$$

$$= \log\left(\sigma\sqrt{2\pi}\right) + \log(\sqrt{e})$$

$$= \log\left(\sigma\sqrt{2\pi e}\right)$$

For $\sigma = 1$, this is about 1.42 nats or about 2 bits.

CHAPTER 10

Markov Chains

Andrey Andreyevich Markov was a Russian mathematician. He is best known for his work on the theory of stochastic processes in which random variables are mutually dependent. His work founded a new branch of probability theory and launched the theory of stochastic processes. Markov lived through a period of great political unrest in Russia, and he became heavily involved, protesting actions taken by the Tsar that affected the Russian Academy of Sciences. Markov only escaped severe consequences because the authorities did not wish to make an example of a distinguished academician. When the Romanov dynasty celebrated their 300 years of power in 1913, Markov showed his disapproval by celebrating 200 years of the Law of Large Numbers! By coincidence, we are now nearing the 300th Anniversary of the Law of Large Numbers. In any case, when the Russian Revolution began early in 1917, Markov went to a small country town where he taught mathematics in the secondary school without any compensation. When he returned to St. Petersburg, his health was deteriorating, yet he continued to lecture on probability at the university until his death after months of severe suffering.

All the processes we have considered so far have been based on sequences of independent equidistributed random variables. We now consider processes which are based on sequences of dependent random variables but for which the dependence is of the simplest possible kind: the future depends on the present but not on the past.

10.1 The Markov Property

Let $X_0, X_1, X_2, \ldots$ be a sequence of integer random variables. We think of the values of the X_n's as being the *states* of the Markov chain. Thus, if $(X_n = i)$ we say the process is in *state i* at *time n*. Moreover, if $(X_n = i)$ and $(X_{n+1} = j)$, then we say there was a *transition* from *state i* to *state j* at *time n*.

Definition. A sequence $X_0, X_1, X_2, \ldots$ of integer random variables forms a **Markov chain** if for any integers $i_0, i_1, i_2, \ldots, i_n$,

$$P(X_n = i_n \mid X_{n-1} = i_{n-1}, X_{n-2} = i_{n-2}, \ldots, X_1 = i_1, X_0 = i_0)$$

$$= \quad P(X_n = i_n \mid X_{n-1} = i_{n-1})$$

In other words, the future states of the Markov chain are dependent only on the present state and not on how the Markov chain reached the present state. We call this condition the **Markov property**.

The conditional probability

$$p_{ijn} = P(X_{n+1} = j \mid X_n = i)$$

is called the **transition probability** from *state i* to *state j* at *time n*. By the Law of Alternatives, the probability distribution of X_{n+1} is determined by the transition probabilities and the probability distribution of X_n:

$$
\begin{aligned}
P(X_{n+1} = j) &= \sum_i P(X_{n+1} = j \mid X_n = i) \, P(X_n = i) \\
&= \sum_i p_{ijn} P(X_n = i)
\end{aligned}
$$

As a result, we see that all the probability distributions of the X_n's, as well as all their joint distributions, are determined by the distribution of X_0 and the transition probabilities.

We have seen several examples of Markov chains already. The Bernoulli process is a Markov chain having two states: heads and tails, or 1 and 0. In this case, the transition probabilities are given by

$$p_{00n} = q \quad p_{01n} = p$$
$$p_{10n} = q \quad p_{11n} = p$$

and the probability distribution of X_0 can be anything.

Another example is the sequential sampling process. Here the state is the number of red balls in the urn. For if we know the number of red balls in the urn, as well as the number of balls chosen so far (i.e., the time), we can compute how many black balls are in the urn. The sequential sampling process has the property that the transition probabilities depend not only on the states i and j but also on n, the number of balls chosen so far. In such a case, our process is continually changing or **inhomogeneous**. In this chapter, we will only study Markov chains such that the transition probabilities are independent of time.

Definition. A Markov chain $X_0, X_1, X_2, \ldots$ is said to be **homogeneous** if the transition probabilities

$$p_{ij} = P(X_{n+1} = j \mid X_n = i)$$

do not depend on n.

Many inhomogeneous Markov chains can be reinterpreted as homogeneous

Markov chains, so that this concept is not as special as it may at first appear. For example, if we define the "state" of the sequential sampling process to be the pair of numbers (number of red balls, number of black balls), then the sequential sampling process is a homogeneous Markov chain.

When we write the transition probabilities p_{ij} as a matrix we get a matrix M called the **transition probability matrix** of the Markov chain.

$$\begin{pmatrix} p_{11} & p_{12} & \cdot & \cdot & \cdot \\ p_{21} & p_{22} & \cdot & \cdot & \cdot \\ \cdot & & \cdot & \cdot & \cdot \\ \cdot & & \cdot & \cdot & \cdot \end{pmatrix}$$

The rows represent the starting states and the columns represent the ending states, during each unit of time. The transition probability matrix determines the Markov chain except for the probability distribution of X_0. The entries of the matrix must be between 0 and 1, and the sum of the entries of each row is 1. On the other hand, we can say nothing about the columns.

Definition. A row vector (with possibly infinitely many coefficients) is said to be a **stochastic vector**, if all entries are between 0 and 1 and the sum of all coefficients is 1. A matrix (with possibly infinitely many rows) is called a **stochastic matrix**, if its rows are all stochastic vectors.

The term "stochastic vector" is simply another way of looking at the probability distribution of an integer random variable, and a "stochastic matrix" is another way of looking at a random function whose input parameters are discrete. We see that distributions and Markov chains give rise to a new way of looking at vectors and matrices. A pair, consisting of a square stochastic matrix M and a stochastic vector $\vec{u_0}$, compatible with M, determines a unique Markov chain such that $\vec{u_0}$ is the row vector corresponding to the distribution of X_0 and M is the transition probability matrix.

The distribution of X_0 is called the **initial distribution** of the Markov chain. As we have already remarked, the distributions of $X_0, X_1, X_2, \ldots$ are determined successively by the formula

$$P(X_{n+1} = j) = \sum_i p_{ij} P(X_n = i)$$

In terms of matrices, this says that if $\vec{u_n}$ is the stochastic vector corresponding to the distribution of X_n, then

$$\vec{u_{n+1}} = \vec{u_n} \cdot M$$

is the stochastic vector corresponding to X_{n+1}, where $\vec{u_n} \cdot M$ is the product of the matrices $\vec{u_n}$ and M. In other words, the transition from time n to time $n+1$ in a Markov chain corresponds to matrix multiplication. More generally, we can iterate the above formula to get

$$\vec{u_n} = \vec{u_0} \cdot M^n$$

showing explicitly how the distributions of the X_n's depend on $\vec{u_0}$ and M.

One of the most intriguing ways to view a Markov chain is as a stochastic program. As we have already noted, a stochastic matrix is a random function. The stochastic matrix for a Markov chain is a square matrix, which means that its input parameters and output distribution have exactly the same support. Thus, the output of a square stochastic matrix can be used as the parameters for itself. The stochastic matrix can be viewed as a stochastic program like this:

```
X <- M(X)
```

In other words, a Markov chain is the simplest possible *recursive stochastic program*. Using Bayesian network notation, the stochastic matrix is a single node that is linked to itself by an edge that "loops back" like this:

If we now add the initial distribution, we obtain the stochastic program for the Markov chain:

```
X <- X0
X <- M(X)
```

It is easy to see how to interpret this. Set X to the initial value X0, and then iterate to obtain a sequence of distributions. We will give a few examples to show how this is done using R in the rest of this chapter.

The Bernoulli Process

The Bernoulli process, as the process of tossing a coin, is a Markov chain whose transition matrix is

$$M = \begin{pmatrix} q & p \\ q & p \end{pmatrix}$$

Notice that $M^n = M$ for all n and that $\vec{u_0} \cdot M_n = (q, p)$ no matter what the initial distribution is.

On the other hand, if we use the random walk interpretation of the Bernoulli process, we get a very different Markov chain. In this case, the states are the integers, both positive and negative. The state represents the position of the random walk at the given time. The transition probabilities are:

$$p_{ij} = \begin{cases} p & \text{if } j = i + 1 \text{ (move right)} \\ q & \text{if } j = i - 1 \text{ (move left)} \\ 0 & \text{in all other cases} \end{cases}$$

The matrix of this Markov chain is an infinite matrix,

$$
M = \begin{pmatrix}
\cdot & \cdot & \cdot & \cdot & \cdot & \cdot & \cdot & \cdot & \cdot & \cdot & \cdot \\
\cdot & \cdot & \cdot & 0 & p & 0 & 0 & 0 & \cdot & \cdot & \cdot \\
\cdot & \cdot & \cdot & q & 0 & p & 0 & 0 & \cdot & \cdot & \cdot \\
\cdot & \cdot & \cdot & 0 & q & 0 & p & 0 & \cdot & \cdot & \cdot \\
\cdot & \cdot & \cdot & 0 & 0 & q & 0 & p & \cdot & \cdot & \cdot \\
\cdot & \cdot & \cdot & \cdot & \cdot & \cdot & \cdot & \cdot & \cdot & \cdot & \cdot
\end{pmatrix}
$$

Unlike the coin-tossing manifestation of the Bernoulli process, the powers M^n of this transition matrix are progressively more complicated. Moreover, the long-term behavior of this Markov chain does depend on the initial distribution X_0. Typically, X_0 will be in state i with probability 1, in which case we say that i is the *starting point* or *starting state* of the random walk. If we start at $i = 0$, the successive distributions are:

$$
\begin{aligned}
\vec{u_0} &= (\ldots \quad 0 \quad 0 \quad 1 \quad 0 \quad 0 \quad \ldots) \\
\vec{u_1} &= (\ldots \quad 0 \quad q \quad 0 \quad p \quad 0 \quad \ldots) \\
\vec{u_2} &= (\ldots \quad q^2 \quad 0 \quad 2pq \quad 0 \quad p^2 \quad \ldots)
\end{aligned}
$$

When $p = q = 1/2$, we say the random walk is *symmetric*. In this case, the transition matrix M is symmetric. More generally, we say that a Markov chain is **symmetric** if its transition matrix is symmetric.

As we have already remarked, the random walk model and the coin-tossing model are both interpretations of a single process: the Bernoulli process. However, the two models correspond to very different Markov chains, and hence one asks completely different questions about the two models. For example, we will consider the question of how long it takes for the random walk to return to its starting point. One might also consider how many times the random walk crosses the origin. These questions will be considered not only for random walks but also for more general Markov chains. A great number of physical and chemical phenomena can be modeled using Markov chains and random walks in particular. For example, polymer growth can be modeled using two-and three-dimensional random walks. A two-dimensional random walk is just a pair of independent one-dimensional random walks proceeding simultaneously.

10.2 The Ruin Problem

Suppose that we are gambling in a casino. Suppose that we bet one dollar on each play and that we win another dollar with probability p and lose the dollar with probability q. This situation is modeled by a random walk. The starting point X_0 is our *initial fortune,* and the state at the time n is our fortune at that time. Unfortunately, the random walk model we have just considered does not take into consideration the fact that we cannot continue playing if we run out of money. Furthermore, there is a number c (possibly very large), such that if we ever succeed in reaching this state, the casino must stop allowing us to play because it has run out of money. Alternatively, we may simply choose to stop playing if our fortune ever reaches c.

The Markov chain corresponding to this situation is called a **random walk with absorbing barriers**. The *barriers* are the states 0 and c, and these have the property that once one of them occurs, the subsequent states of the Markov chain are all this same state. This Markov chain has only finitely many states, so the transition probability matrix M is an ordinary square matrix. All rows of M except the first and last have the same form as the rows of the barrierless random walk. The top and bottom rows have a 1 as the first and last entries, respectively, indicating that if either of these states is a starting state, then the ending state is the same state.

$$M = \begin{pmatrix} 1 & 0 & 0 & 0 & 0 & & & \\ q & 0 & p & 0 & 0 & & & \\ 0 & q & 0 & p & 0 & & & \\ 0 & 0 & q & 0 & p & & & \\ & & & & & \ddots & & \\ & & & & & q & 0 & p \\ & & & & & 0 & 0 & 1 \end{pmatrix}$$

Other kinds of barrier are possible. Suppose that if our fortune decreases to zero at any time, we are given a \$1 advance (or loan) from an outside source ("Daddy") so that we can continue to play. We call this a *reflecting barrier*. Still another possibility is the *elastic barrier* for which we are either reflected or remain in the same state depending on some probability. In other words, "Daddy" will give us a loan, but we may have to wait for it. The transition matrix for a random walk having a reflecting barrier at c, and an elastic barrier at 0, has this form:

$$M = \begin{pmatrix} s & r & 0 & 0 & & & \\ q & 0 & p & 0 & & & \\ 0 & q & 0 & p & & & \\ & & & & \ddots & & \\ & & & & q & 0 & p \\ & & & & 0 & 1 & 0 \end{pmatrix}$$

A problem of obvious relevance to any gambler is the probability, for a given initial fortune, that the random walk will reach state 0 before reaching state c. If the gambler's fortune ever reaches zero, we say the gambler is "ruined." For this reason, this problem has come to be called the **ruin problem**. This is only the beginning of the general question of how Markov chains behave in the long run, which we will consider later in this chapter.

Let A be the event "In the random walk with absorbing barriers, the walk reaches 0 before reaching c." Then the ruin problem is to compute $u_j = P(A \mid X_0 = j)$ for every initial fortune j. Now u_0 is 1 because $X_0 = 0$ means we are ruined from the start; u_c is 0 for the opposite reason. For the other cases, we use the Conditional Law of Alternatives (see Section 6.1), conditioning on the possible values of X_1. When $(X_0 = j)$, there are only two alternatives, $(X_1 = j - 1)$ and $(X_1 = j + 1)$, Therefore,

$$
\begin{aligned}
P(A \mid X_0 = j) &= P(A \mid X_0 = j, X_1 = j - 1)\, P(X_1 \mid X_0 = j) \\
&\quad + P(A \mid X_0 = j, X_1 = j - 1)\, P(X_1 \mid X_0 = j) \\
&= P(A \mid X_1 = j - 1)\, p + P(A \mid X_1 = j + 1)\, q
\end{aligned}
$$

by the Markov property and the definition of the transition probabilities. We now make the important observation that in a homogeneous Markov chain we may view any of the random variables X_n as the initial distribution of the sequence $X_n, X_{n+1}, X_{n+2}, \ldots$, which is itself a Markov chain having the same transition matrix as the original Markov chain $X_0, X_1, X_2, \ldots$ In other words, except for the numbering of the random variables and the initial distribution, this new Markov chain is the same as the old Markov chain. Since A is the event that the gambler is eventually ruined, it does not depend on the numbering of the random variables. That is, we do not care *when* the gambler is ruined. Hence,

$$
\begin{aligned}
P(A \mid X_1 = j - 1) &= u_{j-1} \\
P(A \mid X_1 = j + 1) &= u_{j+1}
\end{aligned}
$$

Therefore,

$$
u_j = P(A \mid X_0 = j) = u_{j-1} q + u_{j+1} p
$$

for $0 < j < c$. An equation of the above form is called a *difference equation*, while the conditions $u_0 = 1$ and $u_c = 0$ are its *boundary conditions*. A difference equation can be solved in a manner exactly analogous to a differential equation, except that instead of exponential functions, $u(x) = e^{\alpha x}$, we use the functions $u_j = a^j$, where α is a constant. We will proceed by steps to emphasize the similarity with differential equation techniques.

Step 1. Determine the possible values for α.

If we substitute α^j in the equation $u_j = u_{j-1} q + u_{j+1} p$, we get $\alpha^j = \alpha^{j-1} q + \alpha^{j+1} p$. Dividing by α^{j-1}, we find that $\alpha = q + \alpha^2 p$, a quadratic equation in α. Solving for α, we find that

$$
\alpha = \frac{1 \pm \sqrt{1 - 4pq}}{2p} = \frac{1 \pm \sqrt{1 - 4p + 4p^2}}{2p} = \frac{1 \pm (1 - 2p)}{2p} = \left\{ \frac{p}{q}, 1 \right\}
$$

Notice that there are two cases. When $p = q = 1/2$, there is a double root $\alpha = 1$; and when $p \neq q$ there are two distinct roots.

Step 2. Find the general solution to the difference equation.

When there are distinct roots, the general solution is just an arbitrary linear combination of the functions α^j, as α ranges over all roots. Thus when $p \neq q$, the general solution is:

$$
\begin{aligned}
u_j &= C_1\left(\frac{q}{p}\right)^j + C_2(1)^j \\
&= C_1\left(\frac{q}{p}\right)^j + C_2
\end{aligned}
$$

On the other hand, if there are multiple roots, we must use functions of the form α^j, $j\alpha^j$, $j^2\alpha^j,\ldots$, using as many as the multiplicity of α as a root. Therefore, the general solution when $p = q = 1/2$ is:

$$
\begin{aligned}
u_j &= C_1(1)^j + C_2 j(1)^j \\
&= C_1 + C_2 j
\end{aligned}
$$

Step 3. Use the boundary conditions to find the particular solution.

The boundary conditions are $u_0 = 1$ and $u_c = 0$. So when $p \neq q$ we have:

$$
\begin{aligned}
u_0 &= 1 = C_1\left(\frac{q}{p}\right)^0 + C_2 = C_1 + C_2 \\
u_c &= 0 = C_1\left(\frac{q}{p}\right)^c + C_2
\end{aligned}
$$

Solving for C_1 and C_2, we find that:

$$
C_1 = \frac{1}{1 - \left(\frac{q}{p}\right)^c}
$$

$$
C_2 = -\frac{\left(\frac{q}{p}\right)^c}{1 - \left(\frac{q}{p}\right)^c}
$$

Hence, the particular solution we seek is:

$$
u_j = \frac{\left(\frac{q}{p}\right)^j - \left(\frac{q}{p}\right)^c}{1 - \left(\frac{q}{p}\right)^c}
$$

On the other hand, when $p = q = 1/2$, we have:

$$
\begin{aligned}
u_0 = 1 &= C_1 + C_2 \cdot 0 = C_1 \\
u_c = 0 &= C_1 + C_2 \cdot c
\end{aligned}
$$

Solving for C_1 and C_2, we find that:

$$
\begin{aligned}
C_1 &= 1 \\
C_2 &= -\frac{1}{c}
\end{aligned}
$$

Hence, the solution in this case is:

$$u_j = 1 - \frac{j}{c}$$

Summarizing, we find that the probability of ruin starting from the initial fortune j is:

Solution to the Ruin Problem.

$$P(A \mid X_0 = j) = \begin{cases} 1 - j/c & \text{if } p = q = 1/2 \text{ (the game is fair)} \\ \frac{(q/p)^j - (q/p)^c}{1 - (q/p)^c} & \text{if } p \neq q \text{ (the game is unfair)} \end{cases}$$

The so-called **gambler's ruin paradox** refers to the fact that these probabilities are very close to 1 when perfectly reasonable values of p, q, j and c are used. For example, suppose that a gambler has an initial fortune of \$500, and the gambler decides to be smart and quit the moment his fortune reaches \$1000. He is playing \$1 bets on black or red in the game of roulette. In this game, $p = 18/38$ and $q = 20/38$. He reasons that although the game is unfair, the odds against his eventual win are only 10:9. This would be true if he would bet his entire \$500 on one turn of the wheel. However, by betting only \$1 at a time, his probability of ruin is, by the solution to the ruin problem

$$\begin{aligned} P(A \mid X_0 = 500) &= \frac{\left(\frac{20}{18}\right)^{500} - \left(\frac{20}{18}\right)^{1000}}{1 - \left(\frac{20}{18}\right)^{1000}} \\[2mm] &= 1 - \frac{\left(\frac{20}{18}\right)^{500} - 1}{\left(\frac{20}{18}\right)^{1000} - 1} \\[2mm] &\approx 1 - \left(\frac{10}{9}\right)^{-500} \\[2mm] &> 1 - 10^{-22} \end{aligned}$$

Therefore, the gambler has less than one chance in 10^{22} of eventually winning!

On the other hand, this says nothing about how long it will take for the gambler to be ruined nor whether the gambler will enjoy occasional "winning streaks." One can clearly see that it will take many more than 500 turns of the wheel, on the average, before the gambler is ruined. Moreover, one can show that "winning streaks" and "losing streaks" (when suitably defined) are actually probable events during long betting sessions. So the "structure" of the gambler's ruin is much more complicated than the solution to the ruin problem suggests. It is this complexity that the gambler is presumably paying for when he bets smaller bets instead of the one grand \$500 bet on a single turn of the wheel.

We end by considering what happens when $c \to \infty$. One can think of this as a random walk with just one absorbing barrier. It corresponds to the gambling situation in which the house has infinite resources, and the gambler sets no limit on how much he is willing to win. There are three cases.

$p<q$ *The game is unfair to the gambler.* In this case,

$$u_j = \frac{(q/p)^j - (q/p)^c}{1 - (q/p)^c} = \frac{(p/q)^{c-j} - 1}{(p/q)^c - 1} \rightarrow \frac{0 - 1}{0 - 1} = 1$$

as $c \rightarrow \infty$, because $p/q < 1$. So the gambler certainly loses in this case. This is no surprise.

$p = q = 1/2$ *The game is fair.* In this case,

$$u_j = 1 - \frac{j}{c} \rightarrow 1 \text{ as } c \rightarrow \infty$$

Therefore, the gambler eventually loses even in a fair game.

$p>q$ *The game is unfair to the house.* In this case, $q/p < 1$ so

$$u_j = \frac{(q/p)^j - (q/p)^c}{1 - (q/p)^c} \rightarrow (q/p)^j$$

Hence, there is a positive probability that the gambler continues winning forever. This follows from the fact that $p>q$ produces a "drift" of the random walk to the right as if there were a force acting in the positive direction.

10.3 The Network of a Markov Chain

The **network** or **graph** of a homogeneous Markov chain is an effective method of describing and picturing a Markov chain. Moreover, by using these networks we can view all homogeneous Markov chains as being "random walks" but on the network rather than on a straight line. The term "graph" is more commonly used for this concept, but to avoid a clash of terminology with graphs of functions, we will use the term "network" instead. One should also be careful not to confuse Markov chain networks with Bayesian networks.

Let us begin with a simple example. This is a simple model of machine operation. We suppose that there are two states $1 = $ "the machine runs" and $2 = $ "the machine is broken down." During each unit of time (say every hour), the machine either works or does not work. There is a certain probability p_{11} that a working machine will stay working and a probability p_{22} that a broken machine will still be broken. If we assume that these apply to the machine during each unit of time independently of previous states, then this model is a homogeneous Markov chain. Its transition matrix is

$$\begin{pmatrix} p_{11} & p_{12} = 1 - p_{11} \\ p_{21} = 1 - p_{22} & p_{22} \end{pmatrix}$$

To picture this Markov chain, we draw two points (vertices) to represent the states. We then draw lines (edges) between these vertices, with arrowheads to denote direction, indicating the possibility of passage from one state to another (or the same) state like this:

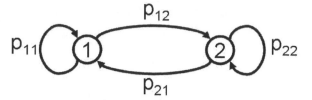

We think of the model as describing the motion of a point along the edges of this network. During each unit of time, the point follows exactly one of the edges in the indicated direction to its other end. The label on an edge denotes the probability that the edge will be chosen. We may think of a point as representing the position of someone "walking" on the network, in which case our model represents a "random walk on the network."

The random walk with no barriers has this network:

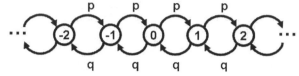

A random walk with absorbing barriers as in the ruin problem has a network like this:

A random walk with reflecting barriers has this network:

The only features that change in the various random walk models are the boundaries. The three kinds of boundary are:

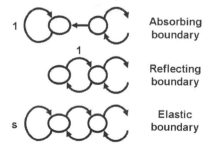

Absorbing boundary

Reflecting boundary

Elastic boundary

Definition. The **network of a homogeneous Markov chain** consists of:

1. One vertex for every state, and

2. A directed edge from state i to state j, for every pair of states i and j such that $p_{ij} > 0$.

Notice that we do *not* have an edge from state i to state j if $p_{ij} = 0$.

10.4 The Evolution of a Markov Chain

To illustrate how Markov chains evolve dynamically, we need to use some features of R that we have not used yet. Specifically, we need to construct matrices, and we need to multiply matrices. The `matrix` function can be used to build a matrix from a vector. The first parameter is a vector with the values of the matrix in "raster" fashion, i.e., as if one were reading the values one row at a time from top to bottom. The second and third parameters define the dimensions of the matrix. In our case, the stochastic matrix m is very large, so we start with a square matrix with all entries equal to 0, and then alter the values to define the stochastic matrix. The vector x is a row matrix that is also constructed as a 0 matrix, which is then altered to define the initial distribution of the Markov chain. Since vectors and matrices start their indexes at 1, we view index 1 as state 0, index 2 as state 1, etc. Matrix multiplication is done using the `%*%` operator. Note that one should not use the ordinary multiplication operator because this will just multiply element by element, which is very different from matrix multiplication.

The particular program we give here is for the gambler's ruin problem, with p chosen to be 18/38 as in betting on red or black in roulette. The Markov chain has absorbing barriers at both boundaries, and the initial fortune of the gambler is exactly halfway between the boundaries.

```
> p <- 18/38
> q <- 1-p
> n <- 100
> c <- 50
> m <- matrix(0, n+1, n+1)
> for (i in 2:n) {
+   m[i,i-1] <- q
+   m[i,i+1] <- p
+ }
> m[1,1] <- 1
> m[n+1,n+1] <- 1
> x <- matrix(0, 1, n+1)
> x[1,c+1] <- 1
> for (i in 1:2000) {
+   x <- x %*% m
+   plot((1:(n+1))[x>0], x[x>0], xlim=c(1,n+1),
+     ylim=c(0,1), xlab="n", ylab="P(X=n)")
+ }
```

Note that we did not include a `Sys.sleep` statement. The convergence is slow enough that it is not necessary. However, if your computer is very fast, or you have a lot of patience, you should add a pause as in the other animations. It is easy to experiment with this program. For example, a reflecting barrier at the left boundary can be introduced by changing the assignment of `m[1,1]` to `m[1,2] <- 1`, and similarly for the other boundary. The other examples of Markov chains in this chapter can also be simulated using a program like this, in principle. However, some of them (like the Ehrenfest diffusion model in the next section and the chemical models in the exercises) are much too large to be simulated using this technique.

The Ehrenfest Diffusion Model

The Ehrenfest model attempts to explain the following physical experiment. A container is divided into two equal parts by a removable wall. We place k gas molecules in one part and $r - k$ in the other. Then we remove the wall and wait for a time. If we now reinsert the wall, we will find almost the same number of particles in each part, no matter how many particles were initially placed in the two parts. Explaining this phenomenon is called the **diffusion problem**.

removable barrier

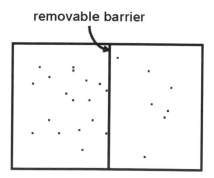

This model was one of the earliest successful attempts to explain the phenomenon of diffusion using probability. Much more sophisticated models now exist, but it is best to start with the simplest model. For this model, we imagine that we have two urns or containers filled with r balls or particles, k in urn 1 and $r - k$ in urn 2. The *state* of the model is the number of particles in urn 1. A *transition* of the model consists of transferring one particle from one urn to the other urn. Therefore, there are two possible transitions from state k: to state $k - 1$ or to state $k + 1$. Here is a typical transition:

The transition probabilities are assigned in such a way that every particle has the same probability of being transferred to the other urn as any other particle. Therefore, the transition probabilities are:

$$p_{k,k-1} = \frac{k}{r}$$

$$p_{k,k+1} = \frac{r-k}{r}$$

Using networks, the transitions from state k are:

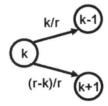

The network of this Markov chain is:

In other words, this Markov chain is a random walk with reflecting barriers but with a "central force" tending to keep the state near $r/2$. We will consider in Section 10.6 what it means to say that the state of this model "tends" to be near $r/2$.

Finite Sampling

Problems of placing objects into containers can often be stated in terms of Markov chains. For example, suppose we are sequentially placing objects into n containers and that we want to know how fast the containers are being "filled." The *state* of this Markov chain is the number of containers having at least one object. The *transitions* are either from a state k to the same state if the next object goes into an already occupied container or to the state $k + 1$ if the next object occupies a new container. Since each container is equally likely to contain the next object, the probabilities for these two cases are k/n and $1 - k/n$, respectively. The network of this Markov chain is

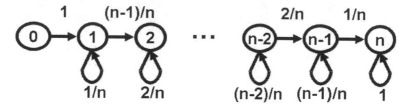

A Genetics Model

The laws of genetics in biology are intrinsically probabilistic. We will consider a very simplified model but one which exhibits the basic ideas. We imagine that a relatively small population of females is introduced to a large ambient population. Consider a single gene having two alleles: a dominant allele A and a recessive allele a. Suppose that the distribution of the three possible genotypes is $[p, q, r]$ in the ambient population, i.e., the fraction of the ambient population having genotype AA is p, having Aa is q and having aa is r. If the females mate the males randomly (with respect to this gene), then the distribution of the genotypes in successive generations of females in the subpopulation will form a Markov chain.

The *states* of the Markov chain are the three genotypes, and the *transitions* consist of the change of state from a mother to her daughter. The probabilities for parents having given genotypes are the well-known Mendelian laws:

parents		child		
		AA	Aa	aa
AA	AA	1	0	0
AA	Aa	1/2	1/2	0
AA	aa	0	1	0
Aa	AA	1/2	1/2	0
Aa	Aa	1/4	1/2	1/4
Aa	aa	0	1/2	1/2
aa	AA	0	1	0
aa	Aa	0	1/2	1/2
aa	aa	0	0	1

Since we know the distribution of the genotypes in the ambient population, and since we have assumed the females mate randomly with respect to this gene, we can compute the probabilities for each given female genotype to give rise to a given daughter genotype:

mother	daughter		
	AA	Aa	aa
AA	$p + \frac{q}{2}$	$\frac{q}{2} + r$	0
Aa	$\frac{p}{2} + \frac{q}{4}$	$\frac{1}{2}$	$\frac{q}{4} + \frac{r}{2}$
aa	0	$p + \frac{q}{2}$	$\frac{q}{2} + r$

This is the transition matrix of the Markov chain. The network of the Markov chain is:

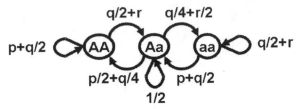

We leave it as an exercise to alter this model to include "preferences" of females of a given genotype for males of another (or the same) genotype as well as to include "survival probabilities" for each of the genotypes of the daughters. One can also construct a model that includes the variation of the distribution of the genotypes of both sexes. The resulting model is a pair of interacting Markov chains acting simultaneously.

Unlike our other examples of Markov chains, we have not considered these Markov chains as a "random walk." We could do so by considering only one "line" of females: a mother, her oldest daughter, *her* oldest daughter, etc. But as long as the number of children born by a female is independent of her genotype, it is more reasonable to regard this Markov chain as the sequence of distributions of the successive female generations.

More generally, we can view *any* Markov chain not as a random walk by a single particle but as a random walk by a very large population of particles all simultaneously "walking" on the network. It is this point of view that is best when we consider the long-term behavior of a Markov chain. The behavior of a single particle along its walk can be quite intricate. But the general behavior of the whole population of points can be very predictable and stable.

10.5 The Markov Sample Space

We have seen three definitions of a Markov chain so far. We first defined it to be a sequence $X_0, X_1, X_2, \ldots$ of random variables satisfying the Markov property. We then saw that it is equivalent to specify a stochastic matrix and a stochastic vector. Finally, we saw that we can visualize Markov chains as random walks on networks. But a Markov chain is a stochastic process, so it must have a sample space and a probability measure.

The sample space Ω of a Markov chain consists of all possible *infinite* paths along edges of its network. By "infinite" we mean that the path has a starting point but no ending point. In other words, the sample space is the set of all possible sequences $(i_0, i_1, i_2, \ldots)$ of states. This formulation is formally analogous to the definition of the Bernoulli process (coin tossing), which is a special case.

To define the probability on Ω, we need the transition matrix M and the stochastic vector $\vec{u_0}$ representing the initial distribution. It is standard notation to write $p_{ij}^{(n)}$ for the entries of the matrix M^n. The entry $p_{ij}^{(n)}$ is the probability for the Markov chain to be in state j given that it was in state i exactly n units of time previously. The components of $\vec{u_0}$ are usually denoted by a_i. The elementary events of the sample space Ω are the subsets $(X_n = i)$, i.e., the set of all paths whose n^{th} vertex corresponds to state i. The probability of an elementary event is given by:

$$P(X_n = i) = \sum_j a_j p_{ij}^{(n)}$$

So far, this process is defined analogously to the Bernoulli process. However, we do *not* postulate that elementary events are independent. Instead, we specify their probabilities to be:

$$P(X_{n_1} = i_1, X_{n_2} = i_2, \ldots, X_{n_k} = i_k) = \sum_j a_j p_{ji_1}^{(n_1)} p_{i_1 i_2}^{(n_2 - n_1)} \cdots p_{i_{k-1} i_k}^{(n_k - n_{k-1})}$$

In particular, we define

$$P(X_0 = i_0, X_1 = i_1, \ldots, X_n = i_n) = a_{i_0 j} p_{i_0 i_1} p_{i_1 i_2} \cdots p_{i_{n-1} i_n}$$

This last expression determines the preceding ones.

Definition. Given a stochastic matrix $M = (p_{ij})$ and a stochastic vector $\vec{u_0}$, the *Markov chain* whose transition probability matrix is M and whose initial distribution is $\vec{u_0}$ is defined by:

1. The sample space Ω is the set of all sequences of states $(i_0, i_1, i_2, \ldots)$,

2. The elementary events are the subsets $(X_n = i)$ of all sequences whose n^{th} entry is i, and

3. The probability is:

$$P(X_0 = i_0, X_1 = i_1, \ldots, X_n = i_n) = a_{i_0 j} p_{i_0 i_1} p_{i_1 i_2} \cdots p_{i_{n-1} i_n}.$$

Connectivity

We now classify Markov chains with respect to various properties relevant to their long-term behavior. The most obvious property is connectivity. If we draw the network of the Markov chain, it should be clear what we mean when we say the network is *disconnected*: it is made up of two or more parts having no edges between the parts. Here is an example:

If the network is not disconnected, we say the Markov chain is **connected**. Clearly, each connected part of a disconnected Markov chain acts like a Markov chain by itself, independent of the rest of the Markov chain. Because of this, we will always assume our Markov chains are connected.

Persistence and Transience

The next property we consider is recurrence. Having once occurred, a state must occur again or it may not. More precisely, let A_i be the event "state i eventually occurs at some time after 0." Then either $P(A_i \mid X_0 = i) = 1$ or $P(A_i \mid X_0 = i) < 1$. We call those two possibilities persistence and transience. More precisely,

Definition. A state i is **persistent** if the probability of returning to state i (after it has occurred at least once) is 1. A state i is **transient** if the probability is positive for the state i never to occur again.

Having once occurred, a persistent state must necessarily occur infinitely many times: each time it occurs, we repeat the argument that it must occur once more, so it can never "stop" occurring.

Consider the various random walks we have seen so far. In the random walk with absorbing barriers, the barriers are obviously persistent. Notice that in this case only one of the persistent states can occur during any one "walk." Although persistent states occur infinitely often, if they ever occur at all, it is quite possible in a connected Markov chain for a persistent state never to occur. The interior states in the random walk with absorbing barriers are all transient because we know with probability 1 that either one or the other barrier will be encountered eventually. In the random walk with one absorbing barrier and one reflecting (or elastic) barrier, there is just one persistent state. On the other hand, if both barriers are reflecting or elastic, all the states become persistent. We leave it as an exercise to prove these last two statements using the solution to the ruin problem.

Finiteness

With respect to persistence and transience, there is a striking difference between finite and infinite Markov chains. In a finite Markov chain, some state must be persistent. But for infinite Markov chains it is quite possible for every state to be transient. Consider the ordinary barrierless random walk. Suppose that the random walk is *not* symmetric, say $p>q$. Recall that in our solution to the ruin problem, we noted that there is a positive probability, $1 - (q/p)^j$, such that, starting in state j, the random walk forever drifts to the right and never encounters state 0. Now, if we start in state 0, the next state is state 1 with probability p, and from here the probability is $1 - q/p$ for 0 never to occur again. Therefore, with probability $p(1 - q/p) = p - q > 0$, state 0 never occurs again. By the same argument, *all* the states of a nonsymmetric barrierless random walk are transient. In the long run, a nonsymmetric random walk drifts forever either to the right if $p>q$ or to the left if $p<q$.

On the other hand, for the symmetric random walk, every state is persistent. In our solution to the ruin problem, we noted that starting in state $j>0$, state 0 eventually occurs with probability 1, and this generalizes to any two states

i and j. So not only is every state persistent, but also every state occurs infinitely often.

Periodicity

The last property we will consider is periodicity. Here is an example of a periodic Markov chain:

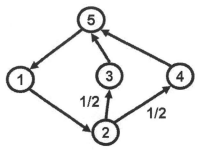

This Markov chain will "cycle" endlessly with period 4 in a merry-go-round fashion. These are not very interesting Markov chains with respect to long-term behavior, since they all essentially look more or less like this one. More precisely, one can prove that the states can be divided into classes $G_1, G_2, \ldots, G_t$ in such a way that the only edges in the network are from states in G_i to states in G_{i+1} or from states in G_t to states in G_1, as shown in this diagram:

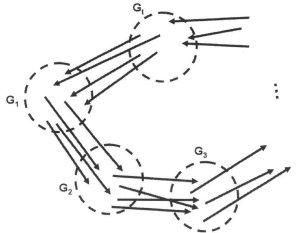

The long-term behavior of each piece G_i is just as if it were a Markov chain by itself. For this reason, we will assume that our Markov chains are not periodic.

Ergodicity

Having gone through all the above preliminaries we find that the most interesting Markov chains with respect to their long-term behaviors are finite, connected and nonperiodic. We will assume, in addition, that every state is persistent. We do this because no transient state can occur more than finitely many times in a finite Markov chain. Hence, all transient states eventually cease to be relevant. Such a Markov chain is said to be **ergodic**. We will give a more precise definition in Section 10.6.

10.6 Invariant Distributions

The most surprising fact about ergodic Markov chains is that the long-term behavior of such a Markov chain is independent of the initial distribution X_0. That is, no matter what the initial distribution is, the distributions of the X_n's as $n \to \infty$ will tend toward one particular distribution which we call the **invariant distribution**.

The best way to view this distribution is to take the point of view mentioned at the end of Section 10.3. Instead of thinking of the Markov chain as the motion of a single particle along the network, we think of an entire population of particles as simultaneously "walking" along the network. The invariant distribution has the property that although individual particles are in constant motion, the population as a whole has a fixed distribution. So if we choose not to distinguish one particle from another, we would perceive no change as the Markov chain proceeds in time.

Definition. For any Markov chain $X_0, X_1, X_2, \ldots$, a probability distribution for X_0 such that all the X_i's are equidistributed is called an **invariant distribution** of the Markov chain.

There are many other terms commonly in use for an invariant distribution, such as **stable state**, **stationary state** and **steady state**. These terms arise from the common practice of generalizing the term "state" to include probability distributions on the states. A state i, in the original sense of the term, is the special case in which the state i has probability 1 and states j for $j \neq i$ have probability 0. Such states are called "pure states," to distinguish them from states in the more general sense. To avoid the confusion of terminology that results from the ambiguity of the word "state," we will use the word "state" exclusively for pure states.

To find an invariant distribution we make use of the terminology of vectors and matrices that we introduced in Section 10.1. If we write $\vec{u_0}$ for the vector corresponding to the initial distribution and if M is the transition matrix, then

$$\vec{u_1} = \vec{u_0} M$$

is the vector corresponding to the distribution of X_1. Now if $\vec{u_1} = \vec{u_0}$, then clearly all subsequent distributions will be the same as the first two. Hence, an invariant distribution corresponds to an eigenvector of M whose eigenvalue is 1. Finding all such eigenvectors for a given matrix M is a exercise in linear algebra (simultaneous equations). Having found all such eigenvectors, the invariant distributions are those whose components are between 0 and 1 and add to 1.

Consider, for example, the machine operation model whose network is

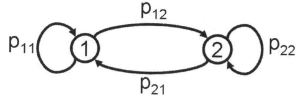

and whose transition matrix is

$$\begin{pmatrix} p_{11} & p_{12} = 1 - p_{11} \\ p_{21} = 1 - p_{22} & p_{22} \end{pmatrix}$$

To find an invariant distribution we must solve the matrix equation

$$(x_1 \quad x_2) = (x_1 \quad x_2) \begin{pmatrix} p_{11} & p_{12} = 1 - p_{11} \\ p_{21} = 1 - p_{22} & p_{22} \end{pmatrix}$$

for x_1 and x_2. This is the same as the system of equations,

$$x_1 = x_1 p_{11} + x_2 p_{21}$$

$$x_2 = x_1 p_{12} + x_2 p_{22}$$

This system of equations reduces to the single equation

$$x_2 = x_1 \frac{p_{12}}{p_{21}}$$

Therefore, the general eigenvector belonging to the eigenvalue 1 is

$$C\left(1 \quad \frac{p_{12}}{p_{21}}\right)$$

This will be a stochastic vector provided that:

$$C + C\frac{p_{12}}{p_{21}} = 1$$

Solving for C, we find that the invariant distribution for this Markov chain is:

$$\left(\frac{p_{21}}{p_{12}+p_{21}} \quad \frac{p_{12}}{p_{12}+p_{21}}\right)$$

or in terms of X_0,

$$P(X_0 = 1) = \frac{p_{21}}{p_{12} + p_{21}}$$

$$P(X_0 = 2) = \frac{p_{12}}{p_{12} + p_{21}}$$

For example, it the machine breaks down with probability 1/10 every hour that it is running, and if a broken machine will go back into service with probability 1/2 every hour that it is broken, then the machine will be running

$$\frac{p_{12}}{p_{12} + p_{21}} = \frac{0.5}{0.1 + 0.5} = \frac{0.5}{0.6} = \frac{5}{6}$$

of the time. Moreover, this will be true in the long run whether the machine is initially running or initially broken down.

Waiting Times and the Recurrence Theorem

In all the stochastic processes we have studied so far, the waiting times have played a crucial role. So it is also with Markov chains. For each state j, we define the random variable T_j to be the waiting time for return to state j, given that one starts in state j. More generally, we have the waiting times T_{ij} for the occurrence of state j starting from state i. In terms of Markov events,

$$(T_{ij} = n) = ((X_0 = i) \cap (X_1 \neq j) \cap \cdots \cap (X_{n-1} \neq j) \cap (X_n = j))$$

Of course, we have that $T_j = T_{jj}$. For Markov chains in general, these are not really random variables because one may have states for which $\sum_n P(T_{ij} = n) \neq 1$ (in fact, one can have $P(T_{ij} = n) = 0$ for all n). We call such an entity a **defective random variable**. However, we specifically choose to restrict attention to ergodic Markov chains, because in this case all the waiting times are ordinary random variables.

There is a slight technicality that we ought to mention briefly. The waiting times T_{ij} are not defined on the same sample space. In fact, T_{ij} is defined on the Markov chain for which X_0 takes the initial value i with probability 1.

The standard notation for $P(T_j = n)$ is $f_j^{(n)}$ and for $P(T_{ij} = n)$ is $f_{ij}^{(n)}$. The random variable T_j is also called the **recurrence time** of state j, and its expectation $E(T_j)$ is called the **mean recurrence time** of state j. We can now give a precise definition of what it means for a Markov chain to be ergodic.

Definition. A finite, connected, homogeneous Markov chain is **ergodic** if

Persistent   for every state j, $\sum_n P(T_j = n) = 1$, and

Nonperiodic   for every state j, $p_{jj}^{(n)} > 0$ except for at most a finite number of times n.

The most important facts about waiting times are contained in the following remarkable theorem:

Markov Chain Recurrence Theorem. For any finite, ergodic Markov chain,

1. The probabilities $p_{ij}^{(n)}$ converge to $\frac{1}{E(T_j)}$ as $n \to \infty$, for all i and j. This can be written as:

$$\lim_{n \to \infty} P(X_n = j) = \frac{1}{E(T_j)}.$$

2. The components of the invariant distribution are $\frac{1}{E(T_j)}$.

Consequently, no matter what the initial distribution is, the distributions of the random variables X_n of an ergodic Markov chain necessarily converge to the invariant distribution. In other words, ergodic Markov chains satisfy an analog of the Central Limit Theorem.

Furthermore, the invariant distribution may be regarded as specifying the average time the random walk exists in the various states. We then have that the average length of time between occurrences of state i is the inverse of the average time spent in state i. While this is an intuitively clear result, it is far from being easy to prove.

The Ehrenfest Diffusion Model

We illustrate the Recurrence Theorem for a nontrivial example. In the Ehrenfest Diffusion model, the transition probabilities are:

$$p_{ij} = \begin{cases} i/r & \text{if } j = i - 1 \\ 1 - i/r & \text{if } j = i + 1 \\ 0 & \text{otherwise} \end{cases}$$

We begin by computing the invariant distribution. We must solve the system of equations:

$$p_j = \sum_i p_i p_{ij} = \begin{cases} \frac{p_1}{r} & \text{if } j = 0 \\ \frac{p_{r-1}}{r} & \text{if } j = r \\ p_{j-1}\left(\frac{r-j+1}{r}\right) + p_{j+1}\left(\frac{j+1}{r}\right) & \text{otherwise} \end{cases}$$

Using the last equation above, we can solve for p_{j+1} in terms of p_j and p_{j-1}, obtaining:

$$p_{j+1} = p_j\left(\frac{r}{j+1}\right) - p_{j-1}\left(\frac{r-j+1}{j+1}\right)$$

Shifting indexes gives a recurrence formula for p_j in terms of earlier probabilities:

$$p_j = p_{j-1}\left(\frac{r}{j}\right) - p_{j-2}\left(\frac{r-j+2}{j}\right)$$

Since the solution will only be determined up to a constant multiple, it does not matter what we use for p_0. So let $p_0 = 1$, and then solve successively by applying the recurrence formula and the fact that $p_0 = \frac{p_1}{r}$,

$$p_0 = 1$$

$$p_1 = r$$

$$p_2 = r\left(\frac{r}{2}\right) - 1\left(\frac{r}{2}\right) = \frac{r(r-1)}{2}$$

$$p_3 = \frac{r(r-1)}{2}\left(\frac{r}{3}\right) - r\left(\frac{r-1}{3}\right) = \frac{r(r-1)(r-2)}{2\cdot 3}$$

A pattern is clearly developing. It seems that $p_j = \binom{r}{j}$. To check this, we substitute into the recurrence formula to get:

$$
\begin{aligned}
p_j &= \binom{r}{j-1}\frac{r}{j} - \binom{r}{j-2}\frac{r-j+2}{j} \\[2mm]
&= \frac{r\cdots(r-j+2)}{(j-1)!}\times\frac{r}{j} - \frac{r\cdots(r-j+3)}{(j-2)!}\times\frac{r-j+2}{j} \\[2mm]
&= \frac{r\cdots(r-j+3)}{(j-2)!}\times\frac{r-j+2}{j-1}\times\frac{r}{j} - \frac{r\cdots(r-j+3)}{(j-2)!}\times\frac{r-j+2}{j} \\[2mm]
&= \frac{r\cdots(r-j+3)}{(j-2)!}\left(\frac{r-j+2}{j-1}\times\frac{r}{j} - \frac{r-j+2}{j}\right) \\[2mm]
&= \frac{r\cdots(r-j+3)}{(j-2)!}\left(\frac{r-j+2}{j}\left(\frac{r}{j-1}-1\right)\right) \\[2mm]
&= \frac{r\cdots(r-j+3)}{(j-2)!}\times\frac{r-j+2}{j}\times\frac{r-j+1}{j-1} \\[2mm]
&= \frac{r\cdots(r-j+1)}{j!} = \binom{r}{j}
\end{aligned}
$$

So the formula we conjectured holds by induction.

Therefore, the eigenvectors belonging to eigenvalue 1 of the transition matrix for this Markov chain have j^{th} component $C\binom{r}{j}$ for any constant $C\neq 0$. For the invariant distribution, we must choose C so that $\sum_j C\binom{r}{j} = 1$. But

we know that $\sum_j \binom{r}{j} = 2^r$ by the Binomial Theorem. Therefore, $C = 2^{-r}$, and we conclude that the invariant distribution is:

$$
P(X_0 = j) = \binom{r}{j}2^{-r}
$$

This is none other than the binomial distribution for r tosses of a fair coin! In other words, it is as if we placed the particles of the Ehrenfest model into the two urns one at a time according to the toss of a fair coin.

We know that the binomial distribution is closely approximated by the normal distribution. Therefore, X_0 is very close to having the distribution $N(r/2, r/4)$, and $\frac{X_0 - r/2}{\sqrt{r/4}}$ is approximately $N(0,1)$. Suppose that we use a confidence level of 99.9%. Then

$$
P\left(-3.3\leq\frac{X_0 - r/2}{\sqrt{r/4}}\leq 3.3\right) = 0.999
$$

implies that

$$
P\left(|X_0 - r/2|\leq 1.65\sqrt{r}\right) = 0.999
$$

In an actual experiment, r will be of the order 10^{24}. In this case, when we replace the barrier between the urns, we will find with probability 0.999 that

$$\left| X_0 - 5 \times 10^{23} \right| \leq 1.65 \times 10^{12}$$

Although 1.65×10^{12} is a very large number, it is less than 10^{-11} of the number of particles in either urn, so that any departure of X_0 from being exactly $r/2$ would be very difficult to detect.

The value of $P(X_0 = j)$ when X_0 has the invariant distribution is $1/E(T_j)$ by the Recurrence Theorem. Again using the normal approximation, we have that:

$$E(T_j) \approx \frac{\sqrt{2\pi r}}{2} \exp\left(\frac{2}{r}\left(j - \frac{r}{2} \right)^2 \right)$$

If $j = r/2$, we get $E(T_j) \approx \frac{\sqrt{2\pi r}}{2}$. When $r \approx 10^{24}$, this is about 10^{12}. Again, this may seem to be very large, but this is only because the unit of time we are using is very small. On the other hand, if $j = 0$ and $r = 10^{24}$, then:

$$E(T_0) = \frac{1}{P(X_0 = 0)} = 2^r = 2^{10^{24}} \approx 10^{10^{23}}$$

Even if our time scale is as small as 10^{-50} sec, this waiting time dwarfs the waiting time mentioned in Section 5.6 (on the writing of *Hamlet* at random by a million monkeys). Clearly this state does not occur very often.

10.7 Monte Carlo Markov Chains

One of most common computations in probability theory is to find the expectation of a random variable. We have seen quite a large variety of such computations. Our goal in such cases is to express the expectation either as a number or as a formula depending on parameters. Unfortunately, as we began developing more complex stochastic models, it became much more difficult to achieve the ideal of reducing the expectation to a simple formula. This was especially true for randomized processes, but even card games can be very difficult to analyze in this way, because of their complex rules and exceptions.

In 1946, only one year after the first stored program electronic computer, ENIAC, was developed, Stanisław Ulam introduced the idea that one could use a computer to approximate an expectation by randomly generating sample points with a computer and then estimating the expectation by computing the sample mean. He called it the **Monte Carlo method** in a humorous reference to the casino, and it quickly became popular and continues to find new applications today.

One important example of the use of the Monte Carlo method is for approximating the invariant distribution of a Markov chain. We have seen a number of examples in which we explicitly computed the invariant distribution of a Markov chain, and there are more examples in the exercises. However, in general it can be difficult to compute the invariant distribution by solving the system of equations explicitly. To use the Monte Carlo method, start with a randomly chosen initial state, and "run" the Markov chain for a specified number of steps. This process is repeated. The multiset of states where the Markov chain was stopped is an approximation to the invariant distribution. This technique is the **Monte Carlo Markov Chain** (or **MCMC**) method.

While computing an invariant distribution is difficult in general, there is one kind of stochastic matrix for which this task is easy. A stochastic matrix M_{ij} is *reversible* for a distribution x when:

$$x_i M_{ij} = x_j M_{ji} \tag{10.1}$$

for every i and j. Note that in the definition of reversibility, Equation 10.1 is always true when $i = j$. This is an important feature of reversibility. If the transition matrix M is symmetric, then it is reversible for the equally likely distribution. So reversibility is a generalization of symmetry of a matrix. The most extreme example of a reversible transition matrix is the identity matrix. A Markov chain with such a transition matrix is totally disconnected, i.e., every state is disconnected from every other state. Such a totally disconnected Markov chain is reversible for *every* distribution on the states, and every distribution is invariant.

What makes reversibility useful is the following:

Invariant Distributions of Reversible Stochastic Matrices. If a stochastic matrix M is reversible for x, then x is an invariant distribution of M.

Proof. A distribution x is invariant if $xM = x$. In other words,

$$\sum_i x_i M_{ij} = x_j$$

for every j. Now,

$$
\begin{aligned}
\sum_i x_i M_{ij} &= \sum_i x_j M_{ji}, \quad \text{by reversibility,} \\
&= x_j \sum_i M_{ji} \\
&= x_j, \quad \text{because } M \text{ is a stochastic matrix}
\end{aligned}
$$

The result then follows.

It might seem that reversibility is rather rare and hard to achieve. However, as a result of a truly remarkable result due to Metropolis and his colleagues, later generalized by Hastings, one can construct a reversible stochastic matrix for *any* probability distribution. This reverses the way that we have been using Markov chains. Instead of modeling some system using a stochastic matrix, and then computing the invariant distribution, we can *start* with a probability distribution, and then construct a stochastic matrix that has the distribution as an invariant distribution. In other words, Markov chains can be "made to order." Even more remarkably, the Metropolis-Hastings algorithm allows one to start with any stochastic matrix whatsoever (as long as it has the same states as the target distribution), and to adjust the stochastic matrix so that it has the desired invariant distribution.

Let x be a probability distribution supported on $\{1, 2, \ldots, n\}$, and let M be any $n \times n$ stochastic matrix. The Metropolis-Hastings algorithm constructs a new stochastic matrix $MH(x, M)$ which is reversible for x. The idea behind the Metropolis-Hastings construction is to "fix" reversibility Equation 10.1, for every pair of states i and j. If equality holds, then we do not have to make any alterations. In particular, as we already noted, equality always holds when $i = j$. Now suppose that $i \neq j$ are a pair of states for which equality does not hold. We can assume that the left-hand side of Equation 10.1 is strictly greater than the right-hand side, for if they are not, then just reverse i and j. Now form the ratio of the two sides:

$$R_{ij} = \frac{x_j M_{ji}}{x_i M_{ij}}$$

Since both sides of Equation 10.1 are nonnegative and the left-hand side of the equation is strictly larger than the right-hand side, the denominator is nonzero and the ratio is strictly less than 1. If we now reduce the transition probability M_{ij} by this ratio, leaving the other transition probability M_{ji} alone, then the reversibility equation will hold because

$$x_i R_{ij} M_{ij} = x_j M_{ji}$$

Of course, by reducing a lot of transition probabilities, the resulting matrix is no longer going to be stochastic. However, as we have pointed out already, the reversibility equation will always hold when $i = j$, i.e., the reversibility equation is never affected by the diagonal entries of the matrix. So we increase all of the diagonal entries until the matrix is stochastic.

Metropolis-Hastings. Let x be a probability distribution supported on $\{1, 2, \ldots, n\}$, and let M be any $n \times n$ stochastic matrix. Construct a new stochastic matrix $Q = MH(x, M)$ in two steps:

$i \neq j$ If $x_i M_{ij} > x_j M_{ji}$, then set

$$Q_{ij} = \frac{x_j M_{ji}}{x_i}$$

otherwise set Q_{ij} to M_{ij}.

$i = j$ After defining all of the off-diagonal elements of Q, set

$$Q_{ii} = 1 - \sum_{\substack{k=1 \\ k \neq i}}^{n} Q_{ik}$$

The matrix Q constructed in this way is a stochastic matrix that is reversible for x.

The ratio R_{ij} is usually written $\alpha(i, j)$ and is extended to all i and j by setting it to 1 when the transition probability is not modified. Note that the x_i's only occur as ratios of two of them. Consequently, one can apply the Metropolis-Hastings construction to an unnormalized distribution. It can be very difficult or even infeasible to normalize a distribution when the sample space is very large, so this is a significant advantage of this construction.

The Metropolis-Hastings construction is often expressed algorithmically as a rejection sampling algorithm as follows:

Markov transition step From state i, randomly select the next state j according to the transition probability matrix.

Rejection step Compute $\alpha(i, j)$. If $\alpha(i, j) = 1$, then proceed to state j as usual. Otherwise, toss a coin with bias $\alpha(i, j)$ (i.e., perform a Bernoulli trial). If the toss is heads, then proceed to state j as usual; otherwise *reject* the transition and stay in state i.

The rejection step accomplishes the reduction in the transition probability required to ensure reversibility. At the same time, it increases the diagonal probability so that the matrix is stochastic. The advantage is that it is never necessary to explicitly compute the transition probability matrix.

While the Metropolis-Hastings construction is versatile, one must be careful not to use it carelessly. Although we know that the target distribution x is an invariant distribution of $MH(x, M)$, that does not mean that an initial state will converge to the target distribution. For example, if M is the identity matrix, then $MH(x, M)$ will also be the identity matrix, which will not converge to any nontrivial distribution. To ensure convergence, one must still show that $MH(x, M)$ is ergodic, and this can be difficult in general.

Nevertheless, Metropolis-Hastings has been an effective technique for evaluating very large and complex stochastic processes. This includes Bayesian networks and stochastic programs. For such processes, the random walk goes from a node to another node over the edges in the network. Metropolis-Hastings is very well suited to this case because the factors in the ratio $\alpha(i, j)$ all cancel, except for nodes that are nearby.

As we come to the end of this textbook, we find that the many threads that were developed can be woven together and joined. A stochastic program can link together discrete and continuous processes that are Poisson, normal, Uniform, Bernoulli or sampling processes. The stochastic program can then be evaluated by creating a Metropolis-Hastings Markov chain. Yet deep down, they all ultimately rely on a sequence of random coin tosses.

10.8 Exercises

Exercise 10.1 Compute the probability of the gambler's ruin for a gambler having initial fortune \$500 and upper limit on winnings \$1000, who is playing roulette and who is making bets of \$10 on red or black; do the same for bets of \$100. What advice would you give to the gambler? To the gambling casino?

Exercise 10.2 In doing homework problems, each success improves the chance of another successes, while each failure tends to increase the chance of subsequent failure. Build a Markov chain model for this.

Exercise 10.3 Consider the following model of the spread of disease. There are N persons in the population. Some are sick and the rest are not.

1. When a sick person meets a healthy one, the healthy one becomes sick with probability α.
2. All encounters are between pairs of persons.
3. All possible encounters of pairs are equally likely.
4. One such encounter occurs per unit time.
5. During each unit of time, each sick person recovers with probability β independently of the other cases above as well as of the previous time spent sick.

Let X_n be the number of sick persons at time n. Write the transition matrix for this Markov chain and draw its network.

Exercise 10.4 Alter the genetics model in Section 10.4 to include a genetic advantage for one of the genotypes (say, for the Aa genotype). How would you include a preference by females having certain genotypes for males having certain other genotypes? Assume that the genotypes AA and Aa are indistinguishable from one another.

Exercise 10.5 Prove that in a finite random walk without absorbing barriers, all states are persistent.

Exercise 10.6 A man has two girlfriends A and B, one living uptown and one living downtown, respectively. He either visits one of his girlfriends on a given evening, or he stays at home. The day after an evening at home he goes to the bus stop at a random time and takes whichever bus comes first, the bus uptown or the bus downtown, visiting A or B, respectively. The buses run in both directions during every 15 minute interval, on a fixed schedule. The man is not too compatible with A, for after a visit to her, he stays home the next evening with probability 9/10 and visits her again with probability 1/10. On the other hand, he is quite compatible with B; after a visit to her, he visits her again the next evening with probability 9/10 and stays home with probability 1/10. Set up the Markov chain for this process. Much to the man's surprise, he spends as many evenings with A as with B on the average. Compute how frequently he spends his evening at home, on the average. See Exercise 2.15.

Exercise 10.7 Shogi or Japanese chess is a popular game in Japan. In a shogi tournament, games that end in a draw are immediately replayed with the colors reversed, and the drawn games are ignored for the purposes of scoring. Although shogi tournaments actually consist of an odd number of games, suppose that it was decided that a particular tournament would consist of an even number of games. In this tournament, the current champion must win at least half the games to retain the championship title, while the challenger must win *more* than half the games. In other words, if they each win the same number of games, then the current champion is declared the winner. To compensate for this disadvantage, the challenger is allowed to choose the number of games N to be played, so long as $N \leq 30$. The challenger is somewhat weaker than the champion, being able to win only a dozen games out of every 25 games. What value for N should the challenger choose?

Exercise 10.8 Compute the invariant distribution of the genetics model in Section 10.3. Notice that it is not in general the same as the distribution of genotypes in the larger population.

Exercise 10.9 Compute the invariant distribution for the more general genetics models in Exercise 10.4.

Exercise 10.10 Compute the invariant of the symmetric finite Markov chain with reflecting barriers. How often does "Daddy" advance you a loan an the average?

Exercise 10.11 In Exercise 3.2, the San Francisco bar in question is 100 yards uphill from the Bay, but the drunk's home is only 10 yards uphill from the bar. How probable is it that the drunk falls into San Francisco Bay before finding his way home?

Exercise 10.12 In a chemical solution there are initially N molecules, each being one of types A, B, C or D. During every unit of time, exactly one collision occurs between a pair of these molecules, all possible collisions being equally likely. During such a collision nothing happens unless the colliding molecules are A and B or are C and D. If A and B collide, there is a probability α that they react and become a pair of C and D molecules. If C and D collide, there is a probability β that they become A and B. In chemical symbols:

$$A + B \underset{\beta}{\overset{\alpha}{\rightleftharpoons}} C + D$$

The state of this system is totally determined by the number of A molecules. Let the number of A molecules at time n be X_n. What is the transition matrix for this Markov chain? What is the invariant number of molecules of each kind?

Exercise 10.13 Same as Exercise 10.12 above, but for the autocatalytic reaction

$$A + A \underset{\beta}{\overset{\alpha}{\rightleftharpoons}} A + B$$

Exercise 10.14 Generalize Exercises 10.12 and 10.13 to an arbitrary number of initial molecules of each kind. Can the reaction rate constants be determined from the invariant distribution?

10.9 Answers to Selected Exercises

Answer to 10.1. This is an easy computation using the solution to the Ruin Problem. In the notation of Section 10.2, for bets of $10, we have $j = 50$,

$c = 100$ and $p = 18/38$, so the probability of eventual ruin is:

$$P(A \mid X_0 = 50) \quad = \quad \frac{(20/18)^{50} - (20/18)^{100}}{1 - (20/18)^{100}}$$

$$= \quad 0.995$$

For bets of \$100, we have $j = 5$, $c = 10$ and $p = 18/38$, so the probability is:

$$P(A \mid X_0 = 5) \quad = \quad \frac{(20/18)^5 - (20/18)^{10}}{1 - (20/18)^{10}}$$

$$= \quad 0.629$$

On the other hand, if the gambler would bet everything on a single turn of the wheel, the probability of ruin would be only $q = 20/38 \approx 0.526$.

I would advise the gambler either to read a good book on probability or to seek salvation by some other route. I would advise the casino to keep the "house limit" low.

Answer to 10.2. This is a two-state Markov chain, which can be depicted as:

where both p_s and p_f are larger than $1/2$. The invariant distribution of this Markov chain can be computed as in Section 10.6. We find that, on the average, out of n homework problems, $\dfrac{(1 - p_f)n}{(2 - p_s - p_f)}$ will be answered correctly. Here is the program for computing the number of correct answers.

```
> correct <- function(n,ps,pf) (1-pf)*n/(2-ps-pf)
```

Answer to 10.6. This random walk has three states:

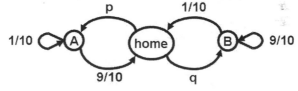

where $p = P(X_t = A \mid X_t = \text{home})$ and $q = P(X_{t+1} = B \mid X_t = \text{home})$. We only know that $p + q = 1$. However, we are given the information that the probabilities for states A and B in the invariant distribution are the same. This additional information is enough to compute p. To compute the invariant distribution we must solve this system of equations:

$$0.1x_1 + px_2 \quad = \quad x_1$$

$$0.9x_1 + 0.1x_3 \quad = \quad x_2$$

$$qx_2 + 0.9x_3 \quad = \quad x_3$$

As usual, the answer will be determined only up to a constant factor which is determined by the condition $x_1 + x_2 + x_3 = 1$. So we may set one variable equal to 1. Let us use x_2. Then the above system becomes:

$$p = 0.9x_1$$
$$1 = x_2$$
$$q = 0.1x_3$$

which is equivalent to:

$$x_1 = 10\frac{p}{9}$$
$$x_2 = 1$$
$$x_3 = 10q$$

To normalize so that $x_1 + x_2 + x_3 = 1$, divide by the sum $\frac{10p}{9} + 1 + 10q = (10p + 90q + 9)/9$. So the invariant distribution is:

$$p_A = \frac{10p}{10p + 90q + 9}$$
$$p_{home} = \frac{9}{10p + 90q + 9}$$
$$p_B = \frac{90q}{10p + 90q + 9}$$

Therefore, since we were given that $p_A = p_B$, we conclude that $10p = 90q = 90 - 90p$, or $p = 0.9$. We can now answer the real question: how often he spends his evening at home, on the average. This is

$$p_{home} = \frac{9}{10p + 90q + 9} = \frac{9}{9 + 9 + 9} = \frac{1}{3}$$

Indeed, we find that

$$p_A = p_B = p_{home} = 1/3$$

Of course, this model is not entirely realistic. One would naturally expect that one or the other states A and B will eventually "absorb," i.e., one should really consider a Markov chain like this:

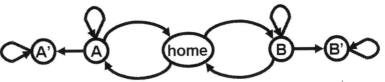

The question of the long-term behavior of this Markov chain is a "ruin problem" like that treated in Section 10.2 (although "ruin" is not the best term in this case).

Answer to 10.7. Now that your intuitions have been "adjusted" by the solution to the ruin problem, it should be intuitively obvious that fewer games would be better, because the challenger is weaker than the current champion. However, the constraints in this problem have a significant impact, so you will need to adjust your intuition yet again. To see why, suppose that the challenger and current champion were equally strong players. In this case, the only advantage to the current champion is the probability of a tie. This is the middle probability of the binomial distribution, and it decreases as N increases. The challenger should therefore choose to play as many games as possible. In the actual situation, the challenger is slightly weaker than the current champion, so there should be a point where the advantage gained by decreasing the probability of a tie will exactly counterbalance the disadvantage due to the ruin problem. This phenomenon was first discovered by Fox in 1959 and analyzed (independently) by Mosteller in 1961. The value of N that maximizes the probability of a win for the challenger is easily computed as follows:

```
> range <- (1:20)*2
> win <- sapply(range, function(N) 1-pbinom(N/2, N, 12/25))
> max(win)
[1] 0.3440288
> range[win==max(win)]
[1] 26
```

The challenger should choose to play a 26-game tournament and will win with probability around 34.4%. While this probability is much smaller than the challenger's probability, 48%, of a win in a single game, it is much better than the probability, 23%, of a win for a pair of games. As one might expect, neighboring choices of N will give nearly the same probability as the maximum. In fact, the probability at $N = 24$ is the same as the probability at $N = 26$. The reason for R selecting 26 is due to a tiny round-off error in the computation.

Answer to 10.8. The genetics Markov chain of this problem is given in Section 10.3. Its transition probability matrix is:

$$\begin{pmatrix} p+\frac{q}{2} & \frac{q}{2}+r & 0 \\ \frac{p}{q}+\frac{q}{4} & \frac{1}{2} & \frac{q}{4}+\frac{r}{2} \\ 0 & p+\frac{q}{2} & \frac{q}{2}+r \end{pmatrix}$$

To find the invariant distribution, we must solve the system of equations

$$\left(p+\frac{q}{2}\right)x_1 + \left(\frac{p}{q}+\frac{q}{4}\right)x_2 = x_1$$

$$\left(\frac{q}{2}+r\right)x_1 + \left(\frac{1}{2}\right)x_2 + \left(p+\frac{q}{2}\right)x_3 = x_2$$

$$\left(\frac{q}{4}+\frac{r}{2}\right)x_2 + \left(\frac{q}{2}+r\right)x_3 = x_3$$

As usual, we may set $x_2 = 1$. The system then reduces to

$$\frac{p}{2} + \frac{q}{4} = \left(1 - p - \frac{q}{2}\right)x_1$$

$$\frac{q}{4} + \frac{r}{2} = \left(1 - \frac{q}{2} - r\right)x_3$$

Hence,

$$x_1 = \frac{2p + q}{4 - 4p - 2q}$$

$$x_2 = 1$$

$$x_3 = \frac{q + r}{4 - 2q - 4r}$$

Using the fact that $p + q + r = 1$, these simplify somewhat to

$$x_1 = \frac{2p + q}{2q + 4r}$$

$$x_2 = 1$$

$$x_3 = \frac{q + r}{4p + 2q}$$

To normalize, divide by:

$$\frac{2p + q}{2q + 4r} + 1 + \frac{q + r}{4p + 2q} = \frac{2(2p + q)^2 + 4(2p + q)(q + 2r) + 2(q + 2r)^2}{4(2p + q)(q + 2r)}$$

$$= \frac{2(2p + q + q + 2r)^2}{4(2p + q)(q + 2r)}$$

$$= \frac{2}{4(2p + q)(q + 2r)}$$

Therefore, the invariant distribution is

$$p_{AA} = \frac{(2p + q)^2}{4}$$

$$p_{Aa} = \frac{(2p + q)(q + 2r)}{2}$$

$$p_{aa} = \frac{(q + 2r)^2}{4}$$

We may write this more succinctly as

$$p_{AA} = \left(p_A\right)^2 \qquad p_{Aa} = 2p_A p_a \qquad p_{aa} = \left(p_a\right)^2$$

where $p_A = p + \frac{q}{2}$ and $p_a = \frac{q}{2} + r$. One can interpret p_A and p_a as the probabilities that a randomly chosen gene in the ambient population is the dominant allele A or the recessive allele a, respectively.

The ambient population will, of course, follow the same genetic rules as the smaller population introduced to it. As a result, one would expect that the invariant distribution of the Markov chain which includes both males and females will be the same. This is, in fact, true. What is remarkable is that even more is true: the invariant distribution is achieved after a *single generation!* This is the content of the Hardy-Weinberg Principle of genetics. You are encouraged to work it out yourself. The full Markov chain has nine states (one for each possible couple). In any case, these considerations lead one to the following important generalization:

Law of Conservation of Genetic Material. In the absence of selective pressures, in a large population, genetic material is conserved, on the average.

One should be careful not to overestimate the power of this law, as it is only a probabilistic one. Furthermore, if a small amount of *drift* occurs, where this is defined as any change in p_A or p_a not due to selective pressure, then there is no pressure to return to the former state. One may think of p_A as performing a (continuous) random walk whose rate is larger when the population is smaller. As a result, the chance of absorption into one of the states $p_A = 0$ or $p_a = 0$ becomes much larger in smaller populations, and instead of an invariant distribution we have a ruin problem. This phenomenon is known as the **random genetic drift**. It is quite possible that genetic evolution owes as much to random genetic drift as it does to mutation and natural selection.

Answer to 10.10. The transition matrix of this Markov chain is:

$$\begin{pmatrix} 0 & 1 & 0 & 0 & & & \\ 0.5 & 0 & 0.5 & 0 & & & \\ 0 & 0.5 & 0 & 0.5 & & & \\ & & & & \ddots & & \\ & & & & 0.5 & 0 & 0.5 \\ & & & & & 1 & 0 \end{pmatrix}$$

Assume that there are n states in all. To compute the invariant distribution we must solve this system of equations:

$$0.5x_2 = x_1$$
$$x_1 + 0.5x_3 = x_2$$
$$0.5x_2 + 0.5x_4 = x_3$$
$$\cdots \qquad \cdots$$
$$0.5x_{n-3} + 0.5x_{n-1} = x_{n-2}$$
$$0.5x_{n-2} + x_n = x_{n-1}$$
$$0.5x_{n-1} = x_n$$

Rewrite the system of equations so that in any equation, the highest numbered variable occurs alone on the left-hand side:

$$x_2 = 2x_1$$

$$x_3 = 2x_2 - 2x_1$$

$$x_4 = 2x_3 - x_2$$

$$\cdots \qquad \cdots$$

$$x_j = 2x_{j-1} - x_{j-2}$$

$$\cdots \qquad \cdots$$

$$x_{n-1} = 2x_{n-2} - x_{n-3}$$

$$x_n = x_{n-1} - 0.5x_{n-2}$$

$$x_n = 0.5x_{n-1}$$

Now set $x_1 = 1$ and solve for the remaining variables inductively:

$$x_2 = 2$$

$$x_3 = 4 - 2 = 2$$

$$x_4 = 4 - 2 = 2$$

$$\cdots \qquad \cdots$$

$$x_j = 4 - 2 = 2$$

$$\cdots \qquad \cdots$$

$$x_{n-1} = 4 - 2 = 2$$

$$x_n = 2 - (0.5)(2) = 1$$

Except for x_1 and x_n, all of the variables have value 2, so the sum of all variables is $2n - 2$. Thus, the invariant distribution has $p_1 = p_n = \dfrac{1}{2n - 2}$, and $p_j = \dfrac{1}{n - 1}$, for $j = 2, \ldots, n - 1$.

To find out how often "Daddy" advances you a loan, we must use the Recurrence Theorem. The expected waiting time between advances, $E(T_1)$, is equal to $\dfrac{1}{p_1} = 2n - 2$. In other words, Daddy will advance you a loan every $2n-2$ units of time, on average. Note that the whole subtlety of the Recurrence Theorem lies in carefully distinguishing between the "average" amount of time spent in a state and the average time between occurrences of the state. The Recurrence Theorem states that these two differently defined averages have the obvious relationship.

Answer to 10.11. This is a ruin problem for a nonsymmetric random walk. In the notation of our solution to the ruin problem, we have $p = 1/4$, $q = 3/4$,

$j = 100$, $c = 110$. Thus, the probability of ruin is

$$\frac{3^{100} - 3^{110}}{1 - 3^{110}} = 1 - 3^{-10} \approx 0.999983,$$

i.e., the drunk has about 1 chance in 60,000 of negotiating the 10 yards to his home. We wish we could give the same advice to the drunk as we did to the gambler in Exercise 10.1, but it seems a bit too much to ask that the drunk take 10-yard steps.

Answer to 10.12. The number of molecules is always exactly N, so there are always $\binom{N}{2}$ possible collisions during each unit of time. By assumption, each possible collision is as likely as any other. If there are M molecules of type A, there are also M of type B, and there are $\frac{N-2M}{2}$ molecules of types C and D, respectively. The number of A-B collisions is M^2, and the number of C-D collisions is $\left(\frac{N}{2} - M\right)^2$. If the system is in state M at time n, there are three states possible at time $n + 1$: $M - 1$, M or $M + 1$. The transition probabilities are given by

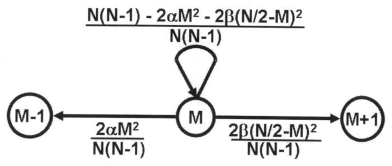

To compute the invariant distribution we must solve a rather formidable system of equations in $n = N/2$ variables:

$$x_0 = \left(1 - \frac{\beta N}{2(N-1)}\right)x_0 + \left(\frac{2\alpha}{N(N-1)}\right)x_1$$

$$x_1 = \left(\frac{\beta N}{2(N-1)}\right)x_0 + \left(1 - \frac{4\alpha + \beta(N-2)^2}{2N(N-1)}\right)x_1 + \left(\frac{8\alpha}{N(N-1)}\right)x_2$$

$$\cdots$$

$$x_j = \left(\frac{\beta(N-2j+2)^2}{2(N-1)}\right)x_{j-1} + \left(1 - \frac{4\alpha j^2 + \beta(N-2j)^2}{2N(N-1)}\right)x_j$$
$$+ \left(\frac{2\alpha(j+1)^2}{N(N-1)}\right)x_{j+1}$$

$$\cdots$$

$$x_{n-1} = \left(\frac{8\beta}{N(N-1)}\right)x_{n-2} + \left(1 - \frac{\alpha(N-2)^2 + 4\beta}{2N(N-1)}\right)x_{n-2}$$

$$+ \left(\frac{\alpha N}{2(N-1)}\right)x_n$$

$$x_n = \left(\frac{\beta}{N(N-1)}\right)x_{n-1} + \left(1 - \frac{\alpha N}{2(N-1)}\right)x_n$$

As usual, we solve for the highest numbered variable in each equation.

$$x_1 = \frac{\beta N^2}{4\alpha}x_0$$

$$x_2 = \frac{\beta(N-2)^2 + 4\alpha}{16\alpha}x_1 - \frac{\beta N^2}{16\alpha}x_0$$

$$\ldots \qquad \ldots$$

$$x_{j+1} = \frac{\beta(N-2j)^2 + 4\alpha j^2}{4\alpha(j+1)^2}x_j - \frac{\beta(N-2j+2)^2}{4\alpha(j+1)^2}x_{j-1}$$

$$\ldots \qquad \ldots$$

$$x_n = \frac{4\beta + \alpha(N-2)^2}{4\alpha N^2}x_{n-2}$$

$$x_n = \frac{4\beta}{\alpha N^2}x_{n-1}$$

Again, as usual, we set $x_0 = 1$ and compute the other x_j's successively. What we find for the first few x_j's is

$$x_0 = 1$$

$$x_1 = \frac{\beta}{\alpha}\cdot\frac{N^2}{4}$$

$$x_2 = \frac{\beta^2}{\alpha^2}\cdot\frac{N^2(N-2)^2}{4\cdot16}$$

This suggests that

$$x_j = \left(\frac{\beta}{\alpha}\right)^j\left(\frac{N(N-2)\cdots(N-2j+2)}{2\cdot4\cdots(2j)}\right)^2$$

$$= \left(\frac{\beta}{\alpha}\right)^j\binom{n}{j}^2$$

Substituting this into the general equation above, we find that it works. Note that α and β cannot be determined from the invariant distribution; only their ratio appears. Their values can only be determined by observing the rate of approach to equilibrium of the solution. Because of this, and to simplify the subsequent computations, we write $\rho = \beta/\alpha$ for this ratio.

It remains to normalize the x_j's. To do this we must resort to an approximation. We know that S_n in the Bernoulli process is close to being $N(np, npq)$. So we can substitute into the density function for the normal distribution to get:

$$\binom{n}{j}p^j q^{n-j} = q^n \binom{n}{j}\left(\frac{p}{q}\right)^j \approx \frac{1}{\sqrt{2\pi npq}}\exp\left(-\left(\frac{j-np}{2npq}\right)\right)$$

Now squaring this gives:

$$q^{2n}\binom{n}{j}^2\left(\frac{p}{q}\right)^{2j} \approx \frac{1}{2\pi npq}\exp\left(-\left(\frac{j-np}{npq}\right)\right)$$

or

$$\binom{n}{j}^2\left(\frac{p^2}{q^2}\right)^j \approx \frac{1}{2\pi npq^{2n+1}}\exp\left(-\left(\frac{j-np}{npq}\right)\right)$$

Thus, we would like to choose p and q so that $\dfrac{p^2}{q^2} = \rho$. Since $q = 1-p$, we can solve for p and q. This gives

$$p = \frac{\sqrt{\rho}}{1+\sqrt{\rho}} \qquad \text{and} \qquad q = \frac{1}{1+\sqrt{\rho}}$$

With these choices,

$$\binom{n}{j}^2 \rho^j = C\exp\left(-\frac{j-np}{npq}\right)$$

where C is a constant. This tells us that when the x_j's are normalized, we will get an invariant distribution that is approximately $N\left(np, \frac{npq}{2}\right)$, where p and q are chosen as above. Substituting these choices, we have

- The average number of molecules of type A is $\dfrac{N\sqrt{\rho}}{2+2\sqrt{\rho}}$

- The standard deviation of the number of molecules of type A is $\dfrac{\sqrt{N}\sqrt[4]{\rho}}{2+2\sqrt{\rho}}$

Answer to 10.13. This is very similar to Exercise 10.12, so we do not write out all the details. Let N be the total number of molecules, and let the state be determined by the number of B molecules. If there are M molecules of type B, there are $N - M$ molecules of type A. The transition probabilities are:

$$P(X_{N+1} = M - 1 \mid X_N = M) = \frac{2\beta M(N-M)}{N(N-1)}$$

$$P(X_{N+1} = M \mid X_N = M) = 1 - \frac{\alpha(N-M)(N-M-1)+2\beta M(N-M)}{N(N-1)}$$

$$P(X_{N+1} = M + 1 \mid X_N = M) = \frac{\alpha(N-M)(N-M-1)}{N(N-1)}$$

The system of equations has typical equation

$$\left(\frac{\alpha(N-j+1)(N-j)}{N(N-1)}\right)x_{j-1} + \left(1 - \frac{\alpha(N-j)(N-j-1)+2\beta j(N-j)}{N(N-1)}\right)x_j$$
$$+ \left(\frac{2\beta(j+1)(N-j-1)}{N(N-1)}\right)x_{j+1} = x_j$$

When we solve for x_{j+1}, we find that:

$$x_{j+1} = \left(\frac{\alpha(N-j)(N-j-1) + 2\beta j(N-j)}{2\beta(j+1)(N-j-1)} \right) x_j$$

$$- \left(\frac{\alpha(N-j+1)(N-j)}{2\beta(j+1)(N-j-1)} \right) x_{j-1}$$

If we set $x_0 = 1$, then:

$$x_1 = \frac{\alpha}{\beta} \cdot \frac{N}{2}$$

$$x_2 = \frac{\alpha^2}{\beta^2} \cdot \frac{N(N-1)}{2 \cdot 4}$$

One conjectures that $x_j = \left(\dfrac{\alpha}{2\beta} \right)^j \dbinom{N}{j}$, and substituting this into the above equation verifies the conjecture. This is actually easier than Exercise 10.12 because this formula is almost exactly the same as the binomial distribution, with parameter p chosen so that $\dfrac{p}{q} = \dfrac{\alpha}{2\beta}$, i.e., $p = \dfrac{\alpha}{\alpha + 2\beta}$. The invariant distribution is then given by:

$$p_j = \binom{N}{j} \left(\frac{\alpha}{\alpha + 2\beta} \right)^j \left(\frac{2\beta}{\alpha + 2\beta} \right)^{N-j}$$

The expected number of molecules of type B is $\dfrac{\alpha N}{\alpha + 2\beta}$, and the expected number of molecules of type A is $\dfrac{2\beta N}{\alpha + 2\beta}$.

Random Walks

We present several results about random walks that are highly counterintuitive. They have the effect of changing one's intuition about what it means to be random.

A.1 Fluctuations of Random Walks

Recall that the basic random variables of the random walk sample space are X'_n for $n = 1, 2, \ldots$ These are independent random variables taking values ± 1 with probability p and q, respectively. They represent the direction taken during the nth step of the random walk. The position of the random walk after the nth step is then the random variable $S'_n = X'_1 + X'_2 + \cdots + X'_n$. We computed the distribution of S'_n in general in Section 3.2. For the special case in which $p = q = 1/2$, the distribution is

$$P(S'_n = x) = \binom{n}{\frac{n+x}{2}} \frac{1}{2^n}$$

We write $p(n, x)$ for this probability. For the rest of this section, we will consider only the case of a *symmetric* random walk, i.e., one for which $p = q = 1/2$. Note that S'_n takes only even values for even n and only odd values for odd n.

First Passage Time and the Reflection Principle. The event $(S'_n = 0)$ means that the random walk has returned to the origin after n steps. However, it could have returned many times before. When was the first time it returned to the origin (or more generally, any point $a > 0$)? We answer this by computing the probability distribution of the random variable T_a, the time when the random walk first encounters the point a, i.e., the first time n such that $(S'_n = a)$.

To compute the distribution of T_a, we use an important principle called the *reflection principle*. Consider the event $C_{n,a,x}$ = "The random walk is at position x at time n *and* at some previous time was at position a." The following is the graph of a random walk in $C_{n,a,x}$:

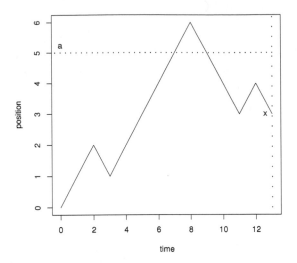

Now observe that for every random walk in $C_{n,a,x}$, there is a first time when it is at position a. We take each random walk in $C_{n,a,x}$ and "pivot" or "reflect" it up to the first time that it reaches position a:

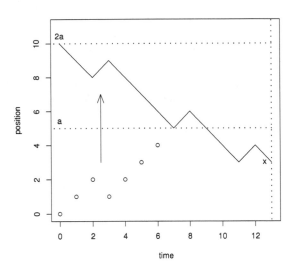

In this way, we get a random walk from $2a$ to x. Conversely, *any* random walk from $2a$ to x necessarily crosses a at some time, so *every* random walk from $2a$ to x is *uniquely* determined in this way! Now shift the axis so that $2a$ becomes the origin and x becomes the point $x - 2a$. Then we conclude that $P(C_{n,a,x})$ is the same as $P(S'_n = 2a - x)$. By symmetry, this is the same as

$P(S'_n = x - 2a)$. Thus,

$$P(C_{n,a,x}) = p(n, 2a - x)$$

We are now ready to compute $P(T_a = n)$. We first note that $(T_a = n)$ necessarily implies that the random walk moved from $a-1$ to a at step n. Prior to step n, the random walk *never* achieved position a, but ends at position $a - 1$ at step $n - 1$. This is just the "complement" of the event $C_{n-1,a,a-1}$.

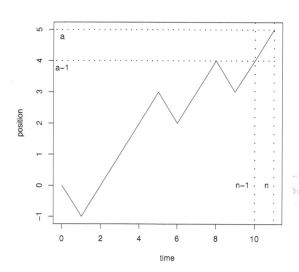

Note that this random walk is at $a-1$ at time $n-1$ so it is in $(S'_{n-1} = a-1)$. It never reaches position a before time $n - 1$ so it is *not* in $C_{n-1,a,a-1}$. More precisely, it is the difference of events:

$$(S'_{n-1} = a - 1) - C_{n-1,a,a-1}$$

Putting this all together:

$$(T_a = n) = ((S'_{n-1} = a - 1) - C_{n-1,a,a-1}) \cap (X'_n = 1)$$

This is the intersection of independent events. Therefore,

$$
\begin{aligned}
(T_a = n) &= P\big((S'_{n-1} = a - 1) - C_{n-1,a,a-1}\big)P(X'_n = 1) \\
&= \frac{1}{2}P\big((S'_{n-1} = a - 1) - C_{n-1,a,a-1}\big)
\end{aligned}
$$

Now $C_{n-1,a,a-1}$ is a subevent of $(S'_{n-1} = a - 1)$ so

$$
\begin{aligned}
P(T_a = n) &= \frac{1}{2}\big(P(S'_{n-1} = a - 1) - P(C_{n-1,a,a-1})\big) \\
&= \frac{1}{2}[p(n - 1, a - 1) - p(n - 1, 2a - (a - 1))]
\end{aligned}
$$

$$= \frac{1}{2}[p(n-1,a-1) - p(n-1,a+1)]$$

Maximum Position. We next ask how high the random walk travels (i.e., the maximum position achieved). Let M_n be this maximum for an n-step random walk. Conveniently, the events $C_{n,a,x}$ are just what we need to compute the distribution of M_n. Namely, we use the same "trick" of subtracting one of the $C_{n,a,x}$; but this time from another event of this kind.

First note that the event $C_{n,a+1,x}$ is a subevent of $C_{n,a,x}$; for if a random walk achieves position $a+1$, then it must have some time previously been at position a. Thus,

$$P(C_{n,a,x} - C_{n,a+1,x}) = P(C_{n,a,x}) - P(C_{n,a+1,x}) = p(n, 2a-x) - p(n, 2a+2-x)$$

But the event $C_{n,a,x} - C_{n,a+1,x}$ means that the random walk achieved position a but *never* achieved position $a+1$, i.e., the maximum achieved by the random walk is *precisely* a. The only distinction between this event and the event $(M_n = a)$ is that the latter does not specify the ending point x specified by the former. So to get $P(M_n = a)$ we add up the $P(C_{n,a,x} - C_{n,a+1,x})$ for all possible values of x. The permissible values of x range from a down to any reachable negative point on the x-axis.

Thus, $P(M_n = a)$ is this sum:

$$p(n, 2a-a) - p(n, 2a+2-a) \qquad (x=a)$$
$$+ \quad p(n, 2a-(a-1)) - p(n, 2a+2-(a-1)) \quad (x=a-1)$$
$$+ \qquad\qquad \cdots$$

which equals

$$p(n,a) - p(n, a+2)$$
$$+ \quad p(n, a+1) - p(n, a+3)$$
$$+ \quad p(n, a+2) - p(n, a+4)$$
$$+ \quad p(n, a+3) - p(n, a+5)$$
$$+ \qquad\qquad \cdots$$

Canceling in the obvious way, we get:

$$P(M_n = a) = p(n,a) + p(n, a+1)$$

Note that for each a only one of the summands on the right is nonzero. Here is a summary of the computations:

$$T_a \quad = \quad \text{time of first passage to or through position } a$$
$$\tfrac{1}{2}[p(n-1, a-1) - p(n-1, a+1)]$$
$$M_n \quad = \quad \text{maximum position achieved up to time } n$$
$$p(n, a) + p(n, a+1)$$

Random Walk Distributions

A.2 The Arcsine Law of Random Walks

The arcsine law is the distribution of the time of the last visit of a random walk to the origin. More precisely, consider a random walk up to time $2n$, and ask when the last time was that the random walk visited the origin. Let L_{2n} be the time of the last visit. Clearly, the random walk can only visit the origin during even-numbered times. We want, therefore, to compute $P(L_{2n} = 2k)$ for all k between 0 and n.

Now examine the event $(L_{2n} = 2k)$. We can rephrase this event as saying that the random walk was at the origin at time $2k$, and that from then on the random walk never visited the origin:

$$(L_{2n} = 2k) = (S'_{2k} = 0, S'_{2k} \neq 0, S'_{2k} \neq 0, \ldots, S'_{2k} \neq 0)$$

The Law of Successive Conditioning tells us that

$$P(L_{2k} = 2k) = P(L_{2k} = 2k \mid S'_{2k} = 0)\, P(S'_{2k} = 0)$$

Now $P(L_{2k} = 2k \mid S'_{2k} = 0)$ is the same as $P(L_{2n-2k} = 0)$. This follows from the independence of the steps of the random walk. We know how to compute $P(S'_{2k} = 0)$ so we must find a way to compute

$$P(L_{2n-2k} = 0) = P(S'_1 \neq 0, S'_1 \neq 0, \ldots, S'_1 \neq 0)$$

For this, we use the Law of Alternatives, conditioning on which way the walk went during the first step:

$$
\begin{aligned}
P(L_{2n-2k} = 0) \quad = \quad & P(L_{2n-2k} = 0 \mid X'_1 = +1)\, P(X'_1 = +1) \\
& + P(L_{2n-2k} = 0 \mid X'_1 = -1)\, P(X'_1 = -1) \\
= \quad & \frac{1}{2} P(L_{2n-2k} = 0 \mid X'_1 = +1) \\
& + \frac{1}{2} P(L_{2n-2k} = 0 \mid X'_1 = -1)
\end{aligned}
$$

By symmetry, both of the above conditional probabilities are the same. Thus,

$$P(L_{2n-2k} = 0) = P(L_{2n-2k} = 0 \mid X'_1 = -1)$$

If we now change coordinates, we may consider the "walk" as starting at $(1, -1)$ as in the diagram,

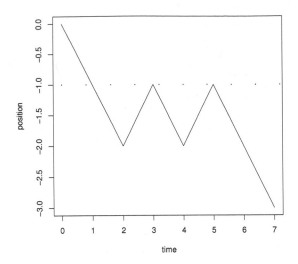

We now see that we have a familiar situation. $P(L_{2n-2k} = 0 \mid X_1' = -1)$ is the probability that the random walk travels no farther to the right than the origin in the first $2n - 2k - 1$ steps, i.e., $P(M_{2n-2k-1} = 0)$, where M_n is the maximum position of the random walk in the first n steps (see Section A.1). Thus

$$P(L_{2n-2k} = 0 \mid X_1' = -1) = P(M_{2n-2k-1} = 0) =$$
$$p(2n - 2k - 1, 0) + p(2n - 2k - 1, 1)$$

It is easy to see that $p(2n - 2k - 1, 0) = 0$. Thus

$$
\begin{aligned}
P(L_{2n-2k} = 0) &= p(2n - 2k - 1, 1) \\[2mm]
&= \binom{2n - 2k - 1}{n - k} \frac{1}{s^{2n-2k-1}} \\[2mm]
&= \frac{(2n - 2k - 1)!}{(n - k)!(n - k - 1)!} \times \frac{1}{s^{2n-2k-1}} \\[2mm]
&= \frac{(2n - 2k)!}{(n - k)!(n - k)!} \times \frac{n - k}{2n - 2k} \times \frac{1}{s^{2n-2k-1}} \\[2mm]
&= \binom{2n - 2k}{n - k} \frac{1}{s^{2n-2k-1}} \\[2mm]
&= p(2n - 2k, 0)
\end{aligned}
$$

Returning to our original problem, we find that

$$P(L_{2n} = 2k) = P(P_{2n-2k} = 0)\, P(S_{2k}' = 0)$$

$$= p(2n - 2k, 0)p(2k, 0)$$

$$= \binom{2n - 2k}{n - k} \frac{1}{2^{2n-2k}} \binom{2k}{k} \frac{1}{2^k}$$

The two factors in the formula above are immediately recognizable as the middle terms of two binomial distributions. By the Central Limit Theorem, we can approximate these terms using the normal density function. More precisely, for any large integer m (i.e., more than about 5),

$$\binom{2m}{m} 2^{-2m} = P(S_{2m} = m) \approx \operatorname{dens}(X = m)$$

where X has distribution $N(m, m/2)$. The density is:

$$\operatorname{dens}(X = m) = \frac{1}{\sqrt{2\pi}\sqrt{m/2}} \exp\left(-\left(\frac{m - m}{2(m/2)}\right)\right) = \frac{1}{\sqrt{\pi m}}$$

Applying this to the earlier computation, we have

$$
\begin{aligned}
P(L_{2n} = 2k) &= p(2n - 2k, 0)p(2k, 0) \\
&= \binom{2n - 2k}{n - k} \frac{1}{2^{2n-2k}} \binom{2k}{k} \frac{1}{2^{2k}} \\
&\approx \frac{1}{\pi\sqrt{k(n - k)}}
\end{aligned}
$$

Now set $x = \frac{k}{n}$. Then

$$P(L_{2n} = 2k) \approx \frac{1}{\pi n \sqrt{x(1 - x)}}$$

Thus, when n is large, the distribution function of L_{2n}; namely, $P(L_{2n} \leq 2k)$, is approximately equal to the area from 0 to k/n of the function $f(x) = \frac{1}{\pi\sqrt{x(1 - x)}}$. In other words,

$$P(L_{2n} \leq 2k) \approx \int_0^{k/n} \frac{dx}{\pi\sqrt{x(1 - x)}}$$

Now from calculus, we know that the derivative of arcsin (t) is $\frac{1}{\sqrt{1-t^2}}$, so the derivative of arcsin $(\sqrt{x})$ is $\frac{1}{2\sqrt{x(1 - x)}}$. Using the substitution $y = \sqrt{x}$, we find that:

$$
\begin{aligned}
\int_0^{k/n} \frac{dx}{\pi\sqrt{x(1 - x)}} &= \frac{2}{\pi} \int_0^{k/n} \frac{dx}{2\sqrt{x(1 - x)}} \\
&= \frac{2}{\pi} \left[\arcsin\left(\sqrt{x}\right) \right]_0^{k/n}
\end{aligned}
$$

$$= \frac{2}{\pi} \arcsin\left(\sqrt{k/n}\right)$$

Summarizing,

Arcsine Law. Let L_{2n} be the time of the last visit of a $2n$-step random walk to the origin. Then

$$P(L_{2n} = 2k) \approx \frac{1}{\pi\sqrt{k(n-k)}} \quad \text{and} \quad P(L_{2n}\leq 2k) \approx \frac{2}{\pi} \arcsin\left(\sqrt{k/n}\right)$$

Here is an example of this law. A gambler plays a fair game, betting 1 dollar every 10 seconds on the toss of a fair coin. If the gambler plays for a whole year, what is the probability that the last time the gambler "broke even" occurred after 1 day of play (i.e., the gambler had either a "winning streak" or a "losing streak" for 364 days). The arcsine law provides an excellent approximation of this probability:

$$P(L_{3153600}\leq 8640) \approx 0.0333,$$

i.e., about 1 chance in 30. This is amazingly large. One can analyze the fluctuations of coin tossing in even more detail, and the surprising conclusion is that it is very unlikely for a random walk to spend close to the same amount of time on both sides of the origin. Yet the expectation of S'_n is zero for all n. Thus a naive observer would regard the behavior of a typical observation of a random walk as being decidedly *unrandom*.

Memorylessness and Scale-Invariance

We prove two theorems that characterize families of distributions. The first one characterizes the distributions that are memoryless; the second characterizes the distributions that are scale-invariant, also called self-similar. The two families of distributions are not related to one another. The only connection is that the proofs of their characterizations are almost the same.

B.1 Memorylessness

Definition. A positive continuous random variable W is said to have the **exponential distribution** when

$$P(W>t + s \mid W>s) = P(W>t)$$

for all positive t, s.

We wish to characterize the probability distribution of random variables that satisfy this condition. We first restate the condition in terms of the complementary distribution function, which will turn out to be more convenient. Accordingly, let $G(t) = P(W>t)$. The condition $P(W>t + s \mid W>s) = P(W>t)$ may also be written

$$P(W>t + s, W>s) = P(W>t)\, P(W>s)$$

but the event $(W>t + s)$ is a subevent of the event $(W>s)$ Therefore, $(W>t + s) \cap (W>s) = (W>t + s)$. So we may equally well characterize a continuous memoryless waiting time by the condition:

$$P(W>t + s) = P(W>t + s)\, P(W>s)$$

or in terms of G:

$$G(t + s) = G(t)G(s)$$

From this equation alone, we will show that $G(t) = Ke^{Ct}$ for suitable constants K and C. If we think of $G(t + s)$ as a function of two variables t and s, we may compute the partial derivatives by the chain rule of calculus:

$$\frac{\partial}{\partial t}G(t + s) = G'(t + s)\frac{\partial(t + s)}{\partial t}$$
$$= G'(t + s)$$

Similarly, $\frac{\partial}{\partial s}G(t+s) = G'(t+s)$. Next, we differentiate $G(t)G(s)$ with respect

351

to both t and s:

$$\frac{\partial}{\partial t}G(t)G(s) \;=\; G'(t)G(s)$$

$$\frac{\partial}{\partial s}G(t)G(s) \;=\; G(t)G'(s)$$

Therefore,

$$G'(t)G(s) = G'(t+s) = G(t)G'(s)$$

Dividing both sides of this equation by $G(t)G(s)$, we get:

$$\frac{G'(t)}{G(t)} = \frac{G'(s)}{G(s)}$$

Now this must hold no matter what t and s are. Therefore,

$$\frac{G'(t)}{G(t)} = C$$

for some constant C. Finally, if we integrate both sides, we get:

$$\log|G(t)| = Ct + D$$

for some constant D, and this is the same as

$$G(t) = Ke^{Ct}$$

for some constant K. Therefore, the distribution of W is:

$$F(t) = P(W{\le}t) = 1 - G(t) = 1 - Ke^{Ct} \qquad \text{for } t{>}0$$

Since we must have $\lim_{t\to\infty} F(t) = 1$, the constant C must be negative. Write $\alpha = -C$. Since we must also have $\lim_{t\to 0} F(t) = 0$, the constant K must be 1. We conclude that the probability distribution of a continuous waiting time W is

$$F(t) = P(W{\le}t) = 1 - e^{-\alpha t}$$

for some positive constant α, and its density is

$$f(t) = \alpha e^{-\alpha t}$$

B.2 Self-Similarity

Definition. A positive continuous random variable S is said to be **self-similar** or **scale-invariant** when it is a continous on an interval $[0, a]$ and satisfies

$$P(S{\le}st/a \mid S{\le}s) = P(S{\le}t)$$

for all t, s in $(0, a]$.

The term **scale-invariance** arises from the fact that conditioning on any initial subinterval produces a random variable which is stochastically identical to the entire probability distribution. The distribution models phenomena for which the behavior is determined by the same laws at all scales. The condition for memorylessness is very similar. It also involves a readjustment, but instead of magnifying the interval, one shifts the interval to a new starting time. Both operations are special cases of linear affine transformations.

Now if S is scale-invariant on $[0, a]$, then it is easy to see that S/a is scale-invariant on $[0, a]$, and conversely. Accordingly, we may assume without loss of generality that $a = 1$. The scale-invariance property is then

$$P(S \leq st \mid S \leq s) = P(S \leq t)$$

Since the event $(S \leq st)$ is a subevent of $(S \leq s)$, the property is equivalent to

$$P(S \leq st) = P(S \leq s)\, P(S \leq t)$$

or in terms of the distribution function: $F(ts) = F(s)F(t)$, for all s, t in $(0, 1]$. This is exactly the same equation as the one characterizing memoryless distributions, except that s and t are multiplied rather than added, and it uses the distribution function rather than the complementary distribution function. We would like to apply the characterization of memoryless distributions, so we "change variables" to get a function that is memoryless. Accordingly, we define a function G by

$$G(x) = F(e^{-x}) \quad \text{for } x \text{ in } [0, \infty)$$

We now check that G satisfies the memoryless condition:

$$G(x)G(y) = F(e^{-x})F(e^{-y}) = F(e^{-x}e^{-y}) = F(e^{-x-y}) = G(x+y)$$

Applying the characterization theorem in Section B.1, we conclude that $G(x) = Ke^{Cx}$, for some constants K and C. By definition of G, we know that $G(0) = F(e^{-0}) = F(1) = 1$, but we also have that $G(0) = Ke^{C \cdot 0} = K$. So $K = 1$. We now change variables to $t = e^{-x}$. This maps the interval $[0, \infty)$ to the interval $(0, 1]$. The distribution function of S can then be computed as follows:

$$F(t) = F(e^{-x}) = G(x) = e^{Cx} = (e^{-x})^{-C} = t^{-C}$$

Since $F(t) \rightarrow 0$ as $t \rightarrow 0$, C must be negative, so we introduce a new parameter $\alpha = -C$ in the formula for $F(t)$.

Characterization of Scale-Invariant Distributions. A continuous random variable S on $[0, a]$ is scale-invariant if and only if for some $\alpha > 0$

$$P(S \leq t) = \left(\frac{t}{a}\right)^{\alpha} \quad \text{for all } t \text{ in } [0, 1]$$

or equivalently,

$$\text{dens}(S = t) = \alpha \left(\frac{t}{a}\right)^{\alpha - 1}$$

This family of distributions is the special case of the beta distribution for which the first shape parameter is α, and the second shape parameter is $\beta = 1$.

References

R. Eckhardt. Stan Ulam, John von Neumann, and the Monte Carlo method. *Los Alamos Science*, pages 131–137, 1987. Special Issue.

P. Fox. A primer for chumps. *Saturday Evening Post*, November 21 1959.

W. Hastings. Monte Carlo sampling methods using Markov chains and their applications. *Biometrika*, 57(1):97–109, 1970.

N. Metropolis, A. Rosenbluth, M. Rosenbluth, A. Teller, and E. Teller. Equations of state calculations by fast computing machines. *J. Chem. Phys.*, 21(6):1087–1092, 1953.

N. Metropolis and S. Ulam. The Monte Carlo method. *J. Amer. Statis. Assoc.*, 44(247):335–341, 1949.

F. Mosteller. Optimal length of play for a binomial game. *The Mathematics Teacher*, 54:411–412, 1961.

J. Pearl. *Probabilistic Reasoning in Intelligent Systems: Networks of Plausible Inference*. Morgan Kaufmann, San Francisco, 1988.

G. Zipf. *Human Behavior and the Principle of Least Effort*. Addison-Wesley, Reading, MA, 1949.

Index